BIOLOGY
OF
FLOWERING PLANTS

PLATE I

[*Frontispiece.*

Flower Movements of Mesembryanthemum.

Below, flowers open in sun at noon; above, flowers closed in shade at 5 p.m.

BIOLOGY
OF
FLOWERING PLANTS

By
M. Skene

DISCOVERY PUBLISHING HOUSE
NEW DELHI-110002

Published by:
Namit Wasan

DISCOVERY PUBLISHING HOUSE PVT. LTD.
4383/4B, Ansari Road, Darya Ganj
New Delhi-110 002 (India)
Phone : +91-11-23279245; 23253475; 43596065
E-mail : discoverybooksindia@gmail.com
discoverypublishinghouse@gmail.com
namitwasan9@gmail.com
web : www.discoverypublishinggroup.com

Edition: **2020**

ISBN: 978-81-7141-205-1

Biology of Flowering Plants

Printed at:
Infinity Imaging Systems
Delhi

PREFACE

THIS book is an attempt to give an account of the way in which the flowering plant lives, especially in relation to its environment. This, it might be said, is the aim of *ecology*; but ecology approaches the plant as a member of a community, while *biology*, as it is understood here, is interested rather in the plant as an individual. In its methods biology has, during the last generation, become more and more experimental; it builds on a foundation of physiology. A certain amount of pure physiology must therefore be introduced. The difficulty of giving enough to make the foundation sound, and yet not so much as to obscure the picture, has been fully realised, though perhaps not overcome.

The great majority of the drawings in the text are the work of Miss A. M. Davidson; about one-third are original, and these bear her initials, the remainder are copied, sometimes with modifications, from various sources which are acknowledged in the underlines. I am indebted to Professor J. E. Weaver for permission to reproduce Figs. 1, 2, 4, and 5 from his *Ecological Relations of Roots*. Fig. 50 is taken from Miss E. Kirkwood's *Plant and Flower Forms*. The photographs are original, and for help in their preparation I am indebted to Mr. J. G. Taylor.

I wish to express my thanks to Professor J. Arthur Thomson, the general editor of this Series, for his kind and stimulating criticism; to Professor W. G. Craib for much information I have received from him in the course of many discussions. My thanks are also due to the publishers for their friendly advice and for drawing my attention to various obscurities; and to my wife for her constant aid throughout the preparation of the book.

ABERDEEN,
May, 1924.

CONTENTS

LIST OF PLATES

DRAWINGS IN THE TEXT

THE BIOLOGY OF FLOWERING PLANTS

CHAPTER I

THE ABSORPTION OF WATER AND SALTS

§ 1. The Necessity of Water for the Plant. § 2. The Origin and Nature of Soil. § 3. The Root System. § 4. The Absorption of Water. § 5. The Absorption of Salts from the Soil. § 6. Exceptional Means of absorbing Water and Salts.

§ 1. The Necessity of Water for the Plant

The necessity of water for the plant is fourfold. (1) Abundant water is essential to the active life of the protoplasm. The immense and varied chemical activity, the metabolism, of the living substance proceeds only when the colloids of the protoplasm are saturated with water. The resting seed, with metabolism reduced to a minimum, is characteristically dry; germination and renewed activity set in after the absorption of an amount of water frequently greater than the total weight of the dry seed. (2) In particular a sufficient supply of water is necessary to the most obvious part of the growth process—extension. The embryonic plant cell is filled with protoplasm. The great increase in size which takes place as the embryonic cells pass into one or other of the types of adult tissue, is linked to the formation of vacuoles containing a watery solution—the cell sap. The increase in volume of the vacuole is due to absorption of water through osmotic pressure; to the development of the turgor pressure of the sap is due the extension of the elastic cell wall. The cell sap of the maturing cell occupies much the largest fraction of its volume. So we find that the greater portion of the fresh weight of the plant consists of water; an average figure is 80 per cent., but in succulent organs such as the leaf of a lettuce or the fruit of the strawberry, water may account for more than nine-tenths of the total weight. (3) The subaerial organs of the plant, in particular the leaves, constantly lose water vapour to the atmosphere; they *transpire*. This loss must be made good, and, normally, a very large supply is necessary. It has been reckoned that in the course of an 18-weeks' growing season a sunflower loses 6 gallons of water in transpiration. The water transpired during the growing season is several times the total weight of the mature plant in herbaceous species, so that the supply required to cover this loss accounts for much the larger fraction of the water absorbed by the plant. (4) With the water supply from the soil the plant obtains the elements potassium, magnesium, calcium, iron, phosphorus, sulphur, and nitrogen, which are essential to its development.

On germination the first part of the seedling to leave the seed coats is the radicle; before the young shoot appears above the soil the root may be several inches long and have already started to branch. The immediate demand for water is emphasised by this early making of contact with the source of supply, the soil.

§ 2. The Origin and Nature of Soil

The soil covers the fertile land surface of the globe in a layer which in general extends from a few inches to a few feet in depth. Not only is it the normal medium of plant growth, it is also a product of the vegetation it supports. The commonest obvious distinction between the soil proper and the sub-soil which underlies it is that the soil is darker brown or almost black in colour because of the presence of humus, the disintegrating and altered remains of dead vegetation.

In the formation of soil from an exposed rock mass, the first stage, in a temperate climate such as ours, is the work of frost. Water accumulating in minute crevices of the rock surface or soaking into joints and fissures expands with enormous force on freezing, with the result that greater or smaller fragments are loosened and the disintegration of the rock begins. Even this early stage is assisted by plants. Lichens and simple algæ can live on bare rock surfaces; on chalk and limestone especially lichens live actually *in* the rock substance. In the former case thin fixing hyphæ, the colourless filaments of the fungus constituent, in the latter a considerable portion of the vegetative body, penetrate the rock, chiefly by the solvent action of water rich in carbon dioxide. The rock surface is thus eroded by the plant and at the same time a further entrance for water which will freeze in winter is gained.

With this initial breaking down of the rock the solvent action of water containing carbon dioxide in solution becomes more marked. The extent of this action depends, of course, on the minerals of which the rock is composed; quartz sand grains are unaffected and mica is very slowly attacked; the felspars give up their alkali metal constituents and are reduced to the aluminium silicates which form the basis of clay; the calcium carbonate in limestone and chalk is rapidly dissolved.

As the rock mass begins to disintegrate it offers a foothold to an exiguous vegetation of modest requirements; the lichens and algæ persist, mosses come in, and with them hardy higher plants. The action of such a vegetation is manifold. Roots and rhizoids penetrate rock crevices, enlarging them by solvent action and even helping to split considerable masses by the force of their growth expansion. The aerial parts of the vegetation arrest blown dust, which gets washed down and adds fine material to the substratum. As roots die off and leaves wither and fall they are partly decomposed by the action of bacteria, and the products become incorporated—the beginning of humus formation.

The resultant mass now begins to take on the characters of soil. Its basal constituent, the refractory portion of the original rock, has been broken down and is mixed with finer material of the nature of clay, derived from other more soluble rock constituents, and with humus, derived from the plant covering; the whole is permeated by water held especially by the finer material. Perhaps about this stage a further important factor enters in the shape of the earth-worm, the nature and magnitude of whose operations should be followed in Darwin's monograph on vegetable mould (1881). It brings down large fragments of dead plants from the surface into the soil. It subsists on vegetable debris which passes through the digestive tract where it is triturated with mineral particles and whence it is ejected more finely divided and more intimately mixed with these. The worm-casts of fine earth—thrown on the surface in such numbers that the face of a green lawn may be almost blackened in the course of a single night—continually turn over the soil as with a slow, invisible, but efficient plough. Burrowing in every direction, 50,000 individuals in the acre keep the soil light and prevent its rapid compacting to a solid medium unfit for plant growth. Only in suitable soil can the earth-worm thrive, but, as in the case of the plant itself, the worm is largely the moulder of its proper medium.

Bacteria, too, are important soil organisms; along with fungi they are responsible for the breaking down of dead organic matter to the humus stage, and also for the disappearance of the humus, for in good soil the humus supply is constantly renewed and yet does not increase. Certain species have an effect of the first importance in controlling the supply of nitrogenous compounds in the soil, by assimilating atmospheric nitrogen or by various conversions of nitrogenous compounds.

Not all soils are formed in this fashion *in situ*. The greater part of the British Islands is covered with soil transported from a distance. The particles of sand and silt and clay and, it may be, humus too, have been carried by water or by ice to regions often far removed from the place of their original formation. Thus river sands and gravels and the mud of estuaries and salt marshes have been transported by rivulet, stream, and river from higher levels to be deposited where the slower flow of the water has let them settle. Great tracts of country are covered with glacial drift of gravel or clay, planed off higher levels and carried to where the melting face of the ice-sheet stood for the moment. In drier climates soils may be transported by wind, a case known in this country only in the relatively small movements of the white sand dunes.

In such cases the soil may bear no relation to the underlying rock. A prolonged geological history may intervene between the initial degradation of the rock mass and the final colonisation by an advanced plant community, yet the various intermediate stages and the final result are essentially the same as when the soil is found *in situ*.

Soil Structure.—The chief solid constituents of a fertile soil are : (*a*) *quartz and mica*, in the form of sand and the finer silt ; (*b*) *clay*, the very finely divided colloidal silicates of aluminium derived from felspar and chalk ; (*c*) *humus*, organic matter, also colloidal, derived from the remains of dead plants ; (*d*) sometimes a proportion of *calcium carbonate* from chalk or limestone. The mass is moist and when in good condition crumbles in the hand.

This crumby property is an expression of the fact that the soil is not a simple mixture of its various solid constituents moistened with water, but that it has a definite *structure*. That this is so is demonstrated very clearly either by shaking the soil with water or by drying it out completely. In the former case drying the sediment of mud does not restore it to its proper condition ; in the latter it is difficult to wet the dusty mass thoroughly, and when this has been done only a sticky paste is produced. The soil crumbs have been destroyed, and simple wetting or drying, though it may restore the original degree of moisture, does not reconstruct the crumbs. In agricultural practice this fact is of great importance, for a soil if worked when too wet becomes puddled, quite unfit for plant growth, and may require prolonged treatment before it is again fit to bear crops.

The details of this structure are far from being fully elucidated but its general nature may be sketched. The coarser grains of sand, ranging up to particles of 3 mm. diameter, form a skeleton round which are built up the crumbs by the glueing action of colloidal particles of clay and humus fully imbibed with water, and by the surface tension of fine water films ; the crumbs are aggregates of the larger particles with intermixture of the finer grades of sand and silt and of clay and humus. The precise state of the colloidal constituent is not known ; it may consist of the clay and humus particles as a whole, or of a gelatinous and indefinite layer surrounding these particles.

We do not know whether there is a fundamental unit size of soil crumb, nor, if there is, how it varies in different soils. The crumby structure is evidently the effect of natural causes of soil formation, the earthworm being probably the most important agent ; but the maintenance of this structure under the special conditions of agriculture—the keeping of the soil in good *tilth*—is the chief aim of cultivation.

The moisture of the soil is not, of course, pure water. It is a solution of salts, organic compounds, and gases. The inorganic salts are of special importance in forming the source of supply of the essential mineral elements of the plant. The soil water is conveniently referred to as the *soil solution*.

Soil Solution.—The composition of the soil solution is not known in any single case, the reason being that there at present exists no method of separating it from the soil in its original state. In fact a little consideration shows that even for a single soil the composition of the soil solution must vary with changing conditions in the soil and probably very rapidly. The solution tends to come to an equilibrium concentration of solutes ; but this equilibrium depends not only on the nature of the soil minerals, and on the solvent power of the water—into which will enter the factor of carbon dioxide content—but also, and perhaps most largely, on the nature and amount of the soil colloids. The colloids retain salts by adsorption, and in contact with free water the partition of salts between the colloids and the liquid depends on the amount of liquid present, that is on the dilution of the solution. The composition of the soil solution therefore varies with the amount of moisture in the soil, and of course this is constantly changing.

The methods which have been employed to separate the soil solution are discussed by Stiles and Jorgenson (1914), and may be summarised here. (*a*) Collection of drainage water from field drains. (*b*) Collection of water run slowly through a column of soil. Only the first portion coming through is taken as representing the water originally present in the soil. This method is in effect a collection of artificial drainage. Some investigators have used liquid paraffin to displace the soil water. (*c*) Extraction of a definite weight of soil with a definite amount of solvent—either distilled water or an acid. (*d*) Removal of the soil moisture by high centrifugal force. (*e*) Squeezing out the solution by high pressure. (*f*) Drawing the solution into a porcelain filter candle by suction. (*g*) Recently attempts to estimate the concentration of the soil solution have been made by determining the lowering of the freezing-point of the moisture in the soil. It will be seen that the first three methods do not give a true sample of the solution present in the soil, though they have a certain use for comparative purposes. The second three do extract the soil solution but only a fraction, the size of which will depend on the precise condition of soil and of experiment.

The kind of result obtained by three of these methods is shown in Table I. The soil in each case is a loam, though of different origin. The figures are taken from Russell's book, " Soil Conditions and Plant Growth."

TABLE I

COMPOSITION OF SOIL SOLUTION

Soil.	Method of extraction.	Parts per million of solution of				Total concentration.
		K	PO_4	Ca	N	
Michigan Loam ..	Oil displacement	71·1	12·2	68·2	3·2	154·7
New Ferry Loam	Centrifugal ..	33·6	7·2	44·4	1·6	86·8
Rothamstead Loam	Drainage ..	4·5	—	122·0	14·0	140·5

The differences exhibited by similar soils of different fertility are illustrated by Table II :

TABLE II

COMPOSITION OF SOIL SOLUTION

Sassafras Loam, New Jersey : centrifulgal extraction.

Fertility of sample.	Parts per million of solution of				Total concentration.
	K	PO_4	Ca	N	
Good	33·6	7·2	44·4	1·6	86·8
Poor	24·4	7·0	26·9	0·1	58·4

It will be noted that the total concentration of the solutes estimated ranges from 58 to 155 parts per million, or from 0·006 to 0·015 per cent. The whole of the solutes are not, however, included in these analyses. A complete analysis of the drainage from unmanured agricultural land at Rothamstead quoted by Russell shows about 250 parts per million or 0·025 per cent. of dissolved solids ; a manured soil gave 0·04 per cent.

Bouyoucos and McCool (1915), who devised the freezing-point method, find a depression in dry soils of round about 1° C. ; this is equivalent to that given by a 1·7 per cent. solution of salt. In moist soils the depression is much less—round about 0·05° C., equivalent to that given by a 0·1 per cent. salt solution. This is a concentration of the same order as that given by the extraction methods, which of course deal with soils in a moist or wet condition. Bouyoucos found great differences in different soils, as might be expected. The freezing-point method is the most expeditious yet invented, and the best, in that it permits of the examination of the soil solution *in situ* and under different conditions of moisture.

In normal moist soils, then, the solution is certainly very dilute. The solutes are constantly removed and constantly renewed, though not necessarily at equal rates. The removal takes place partly by the plant, and partly by washing down of the solution to the subsoil in drainage after rain. This action is complex, for certain constituents —notably potassium and the phosphates—tend to be retained in the soil by the adsorptive action of the colloids, and by interaction with calcium compounds. This fact is illustrated by the relatively small amount of potassium and phosphorus found in the drainage water. The high figures for calcium and nitrates show that these are easily washed out (cp. Table II).

Under natural conditions the material removed by the plant is ultimately returned to the soil in plant remains or in animal droppings. Furthermore, the material washed into the subsoil tends to be brought up again by the action of deeply penetrating roots. The soil minerals are subject to the slow solvent action of the soil moisture. The most serious wastage occurs in the loss of nitrates, and the balance here is largely restored by bacterial action.

Reaction of Soil Solution.—The effect of soil reaction has had much attention drawn to it in recent years, consequent on the recognition of the profound influence exerted on many physiological processes by the reaction of the medium. This influence depends on the *strength* of the acid or base present. It is well known that although equimolecular quantities of hydrochloric and of acetic acids neutralise the same quantity of a base, the former is a *strong* acid, while the latter is *weak*. The difference lies in the greater degree of dissociation of the former, which leads to a high concentration of hydrogen ions in a solution. It is the concentration of hydrogen ions present which is the important thing, irrespective of the actual amount of acid present. This concentration depends not only on the amount and nature of the acids, but also on the presence of other substances, such as salts, which may keep the hydrogen ion concentration from changing much over a wide range of concentration of the acid—a so-called "buffer action."

The expression of hydrogen ion concentration as a fraction of normality is attended with a certain inconvenience, for the amounts are very small. The concentration in a nearly neutral soil might be for example, 0·000000109 or $\frac{1 \cdot 09}{10,000,000}$. This may be written $1 \cdot 09 \times 10^{-7}$, or still more simply as $10^{-6 \cdot 96}$.

It is possible, therefore, to express all hydrogen ion concentrations as negative powers of 10. In practice the exponent alone is used and is written simply as a positive integer. This is called the p^H value of the medium. The acidity, as hydrogen ion concentration, in the example taken, would be written $p^H = 6 \cdot 96$. In using p^H values two points must be remembered; (1) that the acidity or hydrogen ion concentration increases as the p^H value decreases; (2) that for every decrease by unity the concentration of the hydrogen ions has increased tenfold, *i.e.* $p^H = 5 \cdot 0$ means ten times as high a concentration as $p^H = 6 \cdot 0$. The neutral point lies at 7·07, and values above this represent *increasing alkalinity*, the reason being that the hydroxyl ions (of the constitution OH) which determine alkalinity are now increasingly preponderant.

Hydrogen ion concentration may be determined by electrometric methods, which yield absolute values of great accuracy. The apparatus required is expensive and the determination somewhat troublesome. It is also possible to use the colour changes of certain indicators added to the solution. The shade of colour produced in the solution to be tested, by an appropriate indicator, is compared with a series of standard solutions of known p^H value made up with the same indicator. The values of these standard solutions have been accurately determined by electrometric methods. The indicator method is not so accurate, but, with the necessary precautions, it gives satisfactory results for soil work and is both cheap and convenient. Determinations are made in water extracts of the soil; the p^H values of such extracts differ little from those of the solution obtained by pressure from the soil, although of course the extracts are much more dilute (Olsen, 1923). This is due to strong buffer action of the soil solution. For details of methods the student should consult Clark (1920), and Atkins (1922). A general account of the biological relations is given by Bayliss (1920).

Only soils which contain much carbonate—calcium carbonate is much the most important—tend to have an alkaline reaction, and a p^H value of over 7·0. The natural processes in the soil always produce acids, and in absence of abundant neutralising base the soil has an acid reaction.

Soil acidity may be due to the presence of mineral acids. If ammonium salts are present, the base is absorbed more readily by plant roots and a balance of mineral acid is left in the soil. Atkins (1922) has shown that in certain soils which contain iron pyrites, sulphuric acid is present. Of more general importance is the acidity produced by carbon dioxide in solution, which acts as a weak acid and is always present. Olsen (1923) has shown that the dissolved carbon dioxide increases the acidity of nearly neutral soils by $p^H = 0{\cdot}5$, the effect being less marked in more acid soils. Much the most effective cause of acidity is, however, the presence of the organic constituent, the humus. Humus contains organic acids which may act directly. It also interacts with salts—silicates, nitrates, etc.—in solution, removing the base and liberating the acid, either as free acid or in combination with aluminium as acid salts of this metal. The mechanism of the reaction is obscure; it may be a chemical reaction with the humus acids, or it may be the result of colloidal adsorption (Skene (1915), Russell). In any case, the action of humus compounds is the chief cause of soil acidity. Soils that contain large amounts of undecomposed humus—the most extreme is peat—are remarkable for their very acid properties. The various factors, therefore, which lead to an accumulation of humus are indirectly responsible for increase in soil acidity.

Many investigations have been made on the hydrogen ion concentration of various natural and agricultural soils. The relation to carbonate content is well shown in an investigation by Salisbury (1921). On a calcareous heath, on a steep gradient, in Hertfordshire the exposed top of the slope, subject to extreme leaching of soluble constituents, showed a p^H value of 5·1–5·4, with a carbonate content of 0·02 per cent.; halfway down the slope the carbonate content was 0·68–1·0 per cent. and the p^H value 7·3; near the base the carbonates reached 30 per cent., and the p^H value 7·6.

The effect of humus may be illustrated by the example of the different layers of soil in a birch wood in Epping Forest summarised in Table III.

TABLE III

HYDROGEN ION CONCENTRATION AND HUMUS CONTENT

Depth of sample.	Humus content.	p^H value.
0–1½ in. ..	5·5 per cent.	4·9
4½ „ ..	3·3 „ „	5·2
7 „ ..	2·5 „ „	5·3
10 „ ..	3·0 „ „	5·2

In the different types of soil in the same locality, bearing different types of vegetation, the acidity may be markedly different, though in each soil a moderately wide range may exist. Thus the surface soil in a beech wood growing on chalk showed p^H values ranging from 6·1 to 7·4. In an extensive investigation of Danish soils, Olsen (1923) found an extreme range from p^H 3·5 on moorland peat to p^H 7·9 on calcareous soils.

As has been said, apart from abundance of chalk or other carbonates, soils are naturally acid. It is clear that even soils of an extremely acid character can support vegetation, though of a specialised character. Such soils are agriculturally barren ; of the direct effect of a high hydrogen ion concentration on the plant, however, we know very little. It may raise the hydrogen ion concentration of the root cells slightly (Truog, 1919), and this no doubt affects metabolism and growth. It may also affect the entry of salts and of water into the root. But the poverty of acid soils is not due to the acidity alone. Such soils, if derived from granite or schist rocks, are also very poor in nutrient salts. Acid clay soils are physically very sticky and heavy. The " liming " of a soil, so important in agricultural practice, not only reduces hydrogen ion concentration, but adds nutrient bases, alters the adsorptive properties, particularly of clay soils, and makes the soil lighter. Much remains to be done before the actual effects of acidity as such can be properly understood.

State of Soil Moisture.—Moisture is retained in the soil in a variety of ways. After rain more water is present than the soil can hold ; the excess sinks under the action of gravity and drains away, and this may be referred to as *gravitational water*. There then remains a fraction held in the minute crevices between the fine particles and as films round them to which the term *capillary moisture* is applied. Moisture is also retained absorbed by the soil colloids, and as water of hydration in the silicates of the clay. This fraction has been called the *hygroscopic moisture*, being held to be equivalent to the moisture taken up by a dry soil from a saturated atmosphere. The hygroscopic moisture is held by very high imbibition forces, amounting to as much as 1000 atmospheres. The surface tension retaining the capillary moisture is relatively small, equal to 2 or 3 atmospheres. The gravitational water is not held at all.

The determination of the proportions of these fractions is a matter of difficulty, and recently Bouyoucos (1921) has attempted a new classification based on his freezing-point method. The amount of moisture in the soil freezing out at any given temperature, may be estimated by the expansion of the soil (the pores of which are previously filled with a non-freezing organic fluid, ligroin) determined in a special instrument, the dilatometer. Bouyoucos finds that a fraction of the moisture freezes out from 0° C. to −1·5° C., a further fraction between −4° C. and −78° C., and a third fraction not even at the latter low temperature. The fraction which freezes at −1·5° C. is termed *free water*, the remainder *unfree water*. The free water is to be identified with what we have termed capillary moisture; of the unfree water, that fraction which freezes between −4° C. and −78° C. is termed *capillary absorbed*, and corresponds to the water absorbed by the colloids. The water which does not freeze is *water of hydration* and *solid solution*. Bouyoucos holds that the amount of unfree water is a constant for a given soil independent of the total moisture present; but Keen (1919, 1922) has obtained evidence from the depression of the freezing-point at different moisture contents that this is not the case. Bouyoucos's method has the advantage of enabling us to separate the water exactly into the different fractions; the divisions do not correspond exactly to those of the older classification.

We know, however, principally from the work of Keen (1914, 1922) on the mode of evaporation of water from soil, that such fractions are not in any case really sharply separated; the one merges into the other. If we think of a colloid like glue in a dry condition absorbing water, we see that after it has become saturated the surface is no longer sharply delimited; the imbibition water of the soil colloids which coat the sand grains, therefore. passes without a break into the thicker film of moisture which may be retained by capillarity. On drying out there can be no sudden change from film water to imbibed water. There is, however, a range of water content in which the forces retaining water in the soil increase very rapidly though continuously. The amount of capillary water is relatively greater in thin layers of soils ; it is not sharply separated from the gravitational water.

The importance from the botanist's point of view of any classification of soil moisture into different fractions lies not in the objective reality of these, but in the practical use to which it may be put in studying the relation of the water supply to the plant in different soils, and at different degrees of moisture. This relation we shall consider later in the present chapter.

Relation of Soil Constitution to Retention of Water.—The amount of water which can be retained is very different in different soils ; it is a complex function of the soil constitution. It is clear that the greater the proportion of fine particles in the soil, the greater will be the total surface and the greater the amount of water retained as surface film. Again, the greater the proportion of colloidal constituents, clay and humus, the more imbibition water will there be. This relation between soil constitution and water content has been long known. Pfeffer quotes experiments by Meister in 1859 which showed that a sandy soil could absorb 30·4 parts of water, and a peaty soil 105·2 parts per 100 parts dry weight. Later investigators have extended such observations to other soils. In a series of determinations on American soils, Briggs and Shantz (1912*a*) found the moisture retained by fine sandy soils to range from 4·7 to 6·7 per cent., by sandy loams from 9·7 to 11·9 per cent., by loams from 18·9 to 27 per cent., and by clays from 27·4 to 30·2 per cent.

Attempts have been made in recent years to correlate the amount of water retained with one or all of the constituents of the soil. Kraus (1911), in an intensive investigation of the soils near Carlstadt, found that the water content was inversely proportional to the amount of " soil skeleton," by which is understood the coarser particles which cannot pass through a sieve with half-millimetre meshes. Thus two basalt soils with 41·08 and 80·1 per cent. of coarse particles contained respectively 13·09 and 6·39 per cent. of water. This means, of course, that the amount of water is proportional to the amount of fine particles with a diameter of less than 0·5 mm. Crump (1913*a*) established a relation between the amount of humus and the water content of various English soils. He calculates a *coefficient of humidity* by dividing the weight of water by the weight of humus. Table IV gives some of his results.

TABLE IV

Nature of soil.	Percentage of dry weight.		Coefficient of humidity.
	Water.	Humus.	
Loose peat..	174	79	2·23
Underlying compact peat ..	61	17	3·52
Underlying sandy peat	21	8	2·59

The approximate equality of the coefficients indicates that the humus content largely controls the amount of water. The coefficient for other types of soil, *e.g.* for those containing clay, is different.

Briggs and Shantz (1912*a*) have attempted to work out a more exact relation between soil moisture and all the constituents of the soil. They express the water-retaining power as the *moisture equivalent* which is the percentage of water retained against a centrifugal force of 1000 g. This quantity is more easily determined and is less liable to experimental error than the amount of water retained against gravity. The constitution of the soil is expressed as the percentages of sand, silt, and clay, defined as particles between 2 and 0·05 mm., 0·05 and 0·005 mm., and less than 0·005 mm. in diameter, respectively. They give the equation

$$\text{Moisture equivalent} = 0{\cdot}02 \times \text{sand} + 0{\cdot}22 \times \text{silt} + 1{\cdot}05 \times \text{clay}$$

as relating the water-retaining power to the soil constitution. The equation is empirical, and the factors would require modification to suit different series of soils. It should be noted that the large factor for the finest particles expresses the controlling influence of these.

These various results are attempts to amplify and put into mathematical form the familiar fact that a sandy soil is dry and a clay or humus soil wet, and to enable us to make finer distinctions.

The Atmosphere of the Soil.—If the presence of fine particles in the soil is chiefly of importance in retaining water, the crumby structure provides for a free circulation of air and for ready penetration of roots. The amount of air in a given volume will tend to be greater in light dry soils and, in a particular soil, an increase of the water is accompanied by a decrease in the amount of air. Not only is the amount of air in a given volume of soil limited and subject to variation, its composition is different from that of the open atmosphere and is influenced by various factors. Russell and Appleyard (1915) found that in the top 6 in. the percentage of oxygen is about 20·6 and of carbon dioxide 0·25, while in the lower layers the former gas tends to decrease and the latter to increase. The cause of this is, of course, the respiratory activity of roots and soil organisms, and the decreased mobility of the gases. In grassland the carbon dioxide content tends to be higher; in one water-logged soil it rose to 9·1 per cent., while the oxygen fell to 2·6 per cent. Now, as the roots of higher plants and many soil organisms require an adequate supply of oxygen for respiration, a condition which lowers the oxygen content is potentially deleterious. Moreover carbon dioxide has a narcotic action and may slow down root growth, or inhibit the germination of seeds. Certain soil bacteria are anaërobes (living only in absence of oxygen); these probably find a normally suitable medium in the film water, which, in a garden soil examined by Russell and Appleyard, contained only 0·2 per cent. oxygen and as much as 99 per cent. carbon dioxide. Water-logging and an abundance of organic matter, which favours an excess of micro-organisms and so a great consumption of oxygen, are the two factors most concerned in reducing the supply of oxygen available to the roots of the higher plants.

Soil Organisms.—The organisms of the soil are numerous

both in species and individuals, and possess an importance the magnitude of which is being more and more realised. The earthworm has already been mentioned as one of the chief agents in soil formation. Numerous other animals occur. Buckle (1923) has found representatives of forty-nine genera of insects or insect larvæ on or in the soil of agricultural land. These include carnivorous, phytophagous, and scavenging species. Some are pests; others play a part in breaking down organic matter and incorporating it with the soil. There is an abundance of other arthropods, and there are many worms (in the wide sense) and Protozoa. Some are agricultural pests; probably all are active in breaking down organic matter into particles and simpler compounds.

The study of soil Protozoa is in its earlier stages and offers many difficulties. According to Kopeloff and Coleman (1917), about twenty-seven species, flagellates, ciliates, and rhizopods, have been identified. Some of these may exist only encysted, while others are certainly active. A gramme of soil may include anything from 10 to 100,000 individuals. Interest centres in the relation to the bacterial flora which forms the chief food of many species. Too vigorous a protozoan fauna may seriously reduce the numbers of the bacteria which deal with organic matter, especially nitrogenous compounds, and a decrease of fertility may result. Excessive numbers of Protozoa are also associated with the " sickness " to which the rich soil of hot-houses is subject. This condition can be controlled by partial sterilisation of the soil by heat or volatile disinfectants. Cutler, Crump, and Sandon (1922) found about thirty species of Protozoa in an English field soil; of these, six (two amœbæ and four flagellates) were constantly present in considerable numbers. There were great fluctuations in the numbers from day to day and a distinct seasonal change, the highest number being recorded in November and the lowest in February. On the whole bacteria were least numerous when amœbæ were most abundant. Russell's book and the papers quoted in it should be consulted.

The soil flora includes algæ, fungi, and bacteria : to these might be added the comparatively rare saprophytic species and developmental stages of higher plants, and of course plant roots. Algæ are very common on the surface of damp soils, diatoms and blue-green species often forming a slimy covering. Green algæ are also abundant—*e.g.* Vaucheria and, especially on damp peaty soils, Chlorococcum. The unicellular algæ seem to be washed down to considerable depths and to retain their vitality for prolonged periods. Miss Bristol (1920) has described sixty-four species of algæ from different soils. Some of them have withstood desiccation for over sixty years. The presence of algæ in the surface regions has doubtless an important influence in reducing the carbon dioxide and in increasing the percentage of oxygen in the soil atmosphere.

Particularly in soils rich in humus, fungi are abundant : the mushrooms of an old pasture and the toadstools of a wood are the fruiting bodies of subterranean mycelia of great extent. As well as Basidiomycetes there occur many other forms, such as species of Mucor, and Aspergillus, yeasts, and Actinomycetes. Along with the bacteria these are active in breaking down higher organic compounds, and especially woody tissues.

Some conception of the numbers of bacteria present may be gained from Table V, compiled from data given by P. E. Brown (1913).

TABLE V

BACTERIA IN AN IOWAN LOAM SOIL

Depth.	Number per 1 gramme dry soil.	Water per cent.	Humus per cent.	Nitrogen per cent.
4 in.	1,752,000	12·5	3·55	0·2465
12 „	546,000	12·5	3·21	0·2305
24 „	93,000	9·5	2·38	0·0883
36 „	31,000	8·5	1·93	0·0337

The relations to certain properties of the soil are also shown. Many factors affect the numbers and nature of the bacterial flora, *e.g.* moisture, temperature, aeration, reaction, amount of organic compounds.

Most soil bacteria are destructive, in the sense that they carry on various stages of the disintegration of complex organic bodies, such as proteins and carbohydrates. The end product of one bacterial species serves as the raw material for another, and in the end a carbohydrate may be entirely resolved into carbon dioxide and water, or a protein into carbon dioxide, nitrogen, water, and inorganic compounds of sulphur and phosphorus. Such reactions, at all stages, are continually in process, and may represent the activities of the majority of the soil bacteria. One group deserves special mention as being most intimately related to fertility—the bacteria which deal with nitrogen compounds. Proteins are broken down to amino-acids, and these, and urea, to ammonia (*ammonification*) by various putrifying forms. The ammonia is oxidised to nitrite, and the nitrite to nitrate, by two autotrophic (assimilating carbon dioxide) bacteria, Nitrosomonas and Nitrobacter (*nitrification*), the most favourable source of nitrogen for higher plants being thus formed. Acting in the opposite and, to higher plants, unfavourable, direction are bacteria which reduce nitrates to gaseous nitrogen (*denitrification*). The denitrifying bacteria are most active in badly aerated soils. Finally, several soil bacteria (*Azotobacter chroococcum* and *Clostridium Pasteurianum*) assimilate atmospheric nitrogen, converting it into organic compounds which ultimately add to the combined nitrogen of the soil (*nitrogen fixation*). The extent of nitrogen fixation or of nitrification in a soil is a good measure of the bacterial activity favourable to higher plants and, indeed, to the fertility of the soil.

Soil Temperature.—The temperature of the soil has marked characteristics which change with constitution, degree of moisture, and the nature of the vegetable covering. Russell sums up as follows : " The temperature curve of the soil at a depth of 6 inches below the surface somewhat resembles that of the air in summer, but it lacks the sharp peaks and depressions. The soil minimum is always greater than that of the air, especially in summer; the maximum is also usually greater in winter, although it is sometimes below in summer. In winter time, however, the curve is often flat all the twenty-four hours, and sometimes shows no variation for two or three days together." At that depth the soil tends to have a higher and more uniform temperature than the air; at greater depths—which may be more important for root growth—the temperature is still more uniform but with a lower mean. Cannon (1915) has recorded very high temperatures in desert soils at depths to 15 to 30 cms.: an average maximum of over 30° C. may be maintained for four months of the year.

Kraus (1911) has studied the temperature of the surface regions of the soil on the Wellenkalk, at Gambach on the Main. He shows that in bare soil, with scanty vegetation, the temperature of the surface layers (2–5 cms.) habitually rises very high in summer in sunny weather; it is frequently 10° C. higher than the air temperature. In dull weather and at night, air and soil temperatures are about the same. The amount of soil moisture has a very marked effect on the temperature; as the specific heat of water is very much higher than that of dry soil, dry soil heats up much more. Kraus compared the temperature of moist soil near a spring with that of the dry ground a short distance away; on sunny days the dry soil was frequently 5° C. higher than the wet. Here the cooling effect of evaporation also comes into play. Similarly bare soil shows a temperature exceeding that of soil with a vegetable covering (Grass, Thymus, *Hieracium Pilosella*, etc.) by 2°–8° C. The soil temperature below grass is, through the day, often as much below the air temperature as the temperature of bare soil is above it. Calcareous and sandy soils which drain rapidly reach a higher temperature than clay and humus soils which retain more water.

Summary.—The soil, then, is a shallow layer of loose material covering the fertile surface of the land. It is derived from the weathered debris of the rocks, which is often transported to great distances by wind, water, or ice. Through the action of earthworms humus is incorporated with this basic material, and produces a medium of a peculiar crumby structure largely dependent on the properties of the colloid constituents. The importance of this structure is threefold. It provides a medium which is well aerated, well watered—retaining moisture, yet allowing it to move freely—and easily pervious to plant roots. The soil water is a dilute solution which contains the salts necessary to plant growth. It supports an extensive fauna and flora, largely microscopic, playing an essential part in the maintenance of fertility. The soil is penetrated by the roots of the higher land plants of which it is the characteristic medium of growth.

§ 3. The Root System

Penetration and Direction.—It is a familiar fact that the primary root of the seedling grows into the soil ; if the seed happens to be inverted, a curvature of the radicle brings the point vertically downwards. The most important directive influence acting on the radicle is the force of gravity ; if the radicle lies vertically it is in a position of equilibrium ; if not, then the unilateral stimulus of gravity produces an excitation which results in a growth curvature and brings the tip into the vertical position. This reaction is called *geotropism*, a tropism being a movement of a plant organ induced and directed by an external stimulus. It has been shown that the stimulus is perceived solely, or at least preponderatingly, in the apical 1½ mm. of the root tip. The excitation is conducted to the growing zone just behind, where a differentiation of the growth rate on the upper and lower sides produces the reaction. Many roots are also phototropic, curving away from a source of light, but the geotropic reaction is the one chiefly responsible for the vertical direction of growth. The side roots do not grow vertically ; they extend at a definite angle from the main root. This position is a resultant of gravitational excitation and of the influence of the main root. If the tip of the main root is removed, one or more of the side roots curve downwards and continue in a vertical direction. The roots of higher orders are ageotropic and grow outwards from their respective parent axes.

In the soil other stimuli are effective in modifying the direction of growth. A slight wounding of the tip leads to a *traumatotropic* curvature away from the exciting influence. More gentle contact with a solid body leads to a *haptotropic* curvature towards the stimulus. A gradation in the degree of moisture induces a *hydrotropic* curvature towards the greater saturation. An *aerotropic* curvature may bring the root towards a higher concentration of oxygen. Under such influences, the root tip grows downwards through the soil, moving slightly from side to side, and thus aided in its penetration.

Penetration is further aided by the mode in which growth takes place. The region of active elongation lies 3–5 mm. behind the tip. The pushing force is thus applied close behind the penetrating point. This avoids the danger of buckling which would be present if the growing zone were long or situated further back, as is the case in many shoots and in such roots as do not require to overcome the resistance of the soil, *e.g.* the aerial roots of the epiphytic orchids. The structure of the root tip also aids penetration. It is covered by the root cap, an organ about 0·5–1·0 mm. long, the outer cells of which are constantly sloughed as they rub against soil particles and constantly renewed by a special meristem at the root tip. The slimy nature of these cells lubricates the tip. The force developed by growing roots is considerable; it is a familiar fact that they are capable of splitting drains and stone walls if they gain entrance into a crack. The exact extent of the force has been investigated by Pfeffer; a pressure of 100 lbs. to the square inch and more is commonly developed.

Branching.—A few days after germination the growth rate of the radicle slows down and it produces side roots. These arise, as do all except the primary root, endogenously from the pericycle. They burst through the cortical tissue and appear in a definite number of rows corresponding to the groups of protoxylem, the wood vessels first formed, opposite which they grow out. In the pea, for example, there are four rows, in the beet two. This position favours the rapid transference of water to the conducting tissues of the main root. In the further development of the root system of many plants, such as the lupins, the mallows, the dandelion, the primary root remains predominant and is the centre of a system of branches of the second and higher orders. This, however, is not always the case.

The most notable departure from this type is shown by the monocotyledons. The radicle does not retain its supremacy beyond the first stages of germination. In the maize it may be identified for several days ; in most cases it is scarcely recognisable. Its place is taken by a number of roots of equal value produced adventitiously, first from the cotyledonary node of the hypocotyl, the seedling stem between radicle and cotyledons, and later from the higher nodes of the stem. The distinction between these two types of root system may be seen very clearly by comparing the common dandelion and the annual meadow grass.

The later development of the root system is very diverse. In many dicotyledons the main root ceases to be of primary importance very early, side roots of the second and higher orders equalling or exceeding its growth ; an example is the common groundsel. In this, and in many other, dicotyledons adventitious roots are formed in addition. In Rhizophora and other mangroves the main root fails to develop, and its place is at once taken by rapidly developing side roots. Where vegetative multiplication by runners, etc., occurs, the root systems of the new plants are entirely adventitious. In many root systems there is a differentiation into long anchoring and conducting roots, and short and short-lived absorbing roots. The two types may also show structural differences (Tschirch, 1905.) Transitions occur between the two. Goebel regards the differences as merely quantitative.

The root systems of different plants thus come to differ both in plan and in extent, and these differences are largely specific, though the type shown by any species, and more particularly its extent, may be influenced by external conditions. As the success of a plant may be determined by the suitability of its root system to exploit the particular soil conditions of the habitat, certain habitats are associated with plants of particular root types.

Types of Root System.—The roots of herbaceous plants have been studied by Freidenfeldt (1902) in relation to the plan of their systems. He distinguishes and classifies thirteen types with various sub-types. These do not all seem to be very well marked, nor is the classification satisfactory, but a description of the most distinctive gives an idea of the variety to be met with.

1. *Main root types :*

(*a*) The primary root does not persist and is soon replaced by a strongly developed system of laterals, typically shallow rooting ; characteristic of weeds of cultivated soil or fertile waste ground. Examples are *Galeopsis Tetrahit*, *Veronica agrestis*, *Chenopodium album*, *Stellaria media*. (*b*) Deep systems with a permanent and persistent primary root and few laterals ; shown by annuals such as *Reseda Luteola*, *Plantago Coronopus*, and perennials such as *Plantago media*, *Taraxacum officinale*. In most biennials and some perennials the tap root forms a fleshy store, *e.g.* carrot, parsnip. (*c*) Intermediate between (*a*) and (*b*) with a persistent and well-developed primary and also well-developed laterals, *e.g.* pea, *Atriplex sps.*, *Polygonum aviculare*.

2. *Adventitious root types :*

(*d*) A number of strong descending adventitious roots with sharply differentiated, smaller absorption roots, each resembling a system of type (*c*). Examples are the nettle, primrose, *Plantago lanceolata*, and *Plantago major*. (*e*) Adventitious roots produced on long rhizomes, fine, very numerous but little branched ; whole system shallow. Examples are *Anemone nemorosa*, *Maianthemum bifolium*, *Paris quadrifolia*. This type is characteristic of rich humus soils. The rhizome replaces the roots as an anchoring organ. (*f*) Adventitious roots of bulbous or corm plants are numerous and more or less equal, with few or no branches and few root hairs, *e.g.* Crocus, Hyacinthus, Scilla, Tulipa, etc. (*g*) The adventitious roots are very numerous and richly branched. This type approaches (*d*), but there is less distinction between the main roots and the laterals, which are much more numerous. The majority of adventitious root systems probably belong to this type, which is subject to much variation in the number and degree of branching of the laterals ; examples are the sedges, grasses, and cereals. Very long unbranched roots are produced, among the others, by some cereals. In species growing in sand, *e.g.* *Carex arenaria*, the branches are often extremely fine. (*h*) The roots are strong and penetrate deeply without branches (Orchis, Epipactis), with few branches (Hemerocallis, *Asparagus sps.*), or with fairly numerous branches (Helleborus). The roots are frequently fleshy and act as storage organs. (*i*) The root system of aquatics may be vigorous ; the laterals may be numerous but are generally unbranched as in *Ranunculus Flammula* and Nymphæa ; frequently there are no side roots, and in the case of submerged plants the roots are often few and quite simple. Lemna has a single root.

As regards trees, a comparison between the root systems of *Picea excelsa*, *Abies pectinata*, and *Pinus sylvestris* was made by Nobbe (1875). In the first year the pine has the longest primary and much the largest number of secondaries and tertiaries ; the fir is least branched. Later the side roots of the pine dominate and form a very widespread root system, while the primary continues in its growth and forms a strong tap. The depth of the root system and the relation of the tap to the laterals depend very largely on the type of soil. The extent and plasticity of the pine root system partly account for the success of the

PLATE II

Root System of Scots Pine.

A tree growing in deep peat. The tap root is not exposed, but the extent of the system is seen; the cutting is about 3 feet deep.

tree in diverse habitats. The fir is characterised by a dominant tap and deep root system. The primary of the spruce ceases growth after five years, and the root system is entirely composed of shallow laterals. A spruce uprooted by the wind is a common sight, and the shallowness of its root system is then very evident.

Broad-leaved trees have been investigated by Büsgen (1905), who divides their root systems into two classes on a different principle from Nobbe's. The ash has a system in which the laterals are very long, moderately branched, and with long terminal branches. The beech has a system in which the laterals are not so long, but have very numerous short and extremely fine terminals. The maples occupy an intermediate position, and other trees show various gradations. Of tropical trees investigated by Büsgen most—such as the coffee and cinchona—belong to the ash type; the cocoa approaches the beech. Büsgen points out that the ash type exploits a large volume of soil *extensively* and the beech type a smaller volume *intensively*. He reaches the conclusion that the two are equally efficient as water absorbers in different ways.

From yet another point of view, Cannon (1911) has divided root systems into *specialised* and *generalised*. In the former there is predominance of either a tap root or of the laterals; in the latter both tap and laterals are well developed. The generalised systems are the more plastic, and plants with such can occupy more varied habitats.

Further Types of Root Systems and Their Extent.—Such investigations yield information as to the plan of development of root systems. We get indications, too, of the depth to which a root penetrates or of the effectiveness with which it exploits the soil. Many attempts have been made to give a more exact account of the extent of various root systems. The first of these occurs in one of the earliest works on plant physiology, Stephen Hales' "Vegetable Staticks" (1727): "I dug up another Sun-flower, nearly of the same size, which had eight main roots reaching fifteen inches deep and sideways from the stem; it had besides a very thick bunch of lateral roots, from the eight main roots, which extended every way in a hemisphere about nine inches from the stem and main roots.

" In order to get an estimate of the length of all the roots I took one of the main roots with its laterals and measured and weighed them ; and then weighed the other seven roots with their laterals, by which means I found the sum of the length of all the roots to be no less than 1448 feet.

" And supposing the periphery of thin roots at a medium to be 0·131 of an inch, then their surface will be 2276 square inches, or 15·88 square feet ; that is equal to 0·4 of the surface of the plant above ground."—This estimate has the added interest that it is one of the first examples of the application of quantitative methods to physiological problems.

Nobbe, in the paper quoted above, measured the numbers and the total lengths of the root systems of the one-year seedlings of the three common conifers, and compares the effectiveness of the root system of the pine, with over 3000 roots having a combined length of 1198 cm., with that of the fir, which has only 134 roots in all 99 cm. long. Such differences throw light on the ability of the pine to thrive in dry barren soils. The root systems of our broad-leaved trees are usually rather shallow, but may extend horizontally for great distances ; that of the elm has been recorded with a spread of more than 50 feet. Sachs reckoned the volume of soil exploited by the roots of a good-sized sunflower to be a cubic metre.

Interest has always attached to the depth to which roots penetrate. Our native plants, living usually in well-watered and rather shallow soils, are seldom very deep rooted. Hannig (1912) found the roots of *Convolvulus arvensis* penetrating to the unusual depth of 2–2·30 m. On deep peat the roots of sedges and other plants may attain depths of 6 feet and more. There is at present little precise information about the root systems of our native plants.

The most exact and extensive investigation yet made

is that of Weaver (1919, 1920), on the plants of the prairies and chaparral of Nebraska, the prairies of Washington and Idaho, the plains, sandhills, gravel-slides, and forests of Colorado. Beside the plant to be studied a trench was dug and the whole root system was carefully excavated, charted, and photographed in its finest details. Weaver has charted the systems of about two hundred plants. Besides giving an accurate picture of the architecture and extent of the root system, his work is concerned with the relation of definite types to their special habitats, and also with variations in the root system of a single species under different conditions.

The root systems of the prairie plants of Nebraska are characterised by their very great depth. Of forty-three perennials investigated only six, all grasses, have root systems less than 3·5 ft. deep; such are *Elymus canadensis* (1·8 ft., 1·3 ft.), *Kœleria cristata* (1·8 ft., 1·3 ft.), and *Stipa spartea* (2·2 ft., 1·5 ft.). Nine have roots reaching depths of 3·5 to 6 ft.; such are *Andropogon scoparius* (5·4 ft., 3·2 ft.), *Solidago rigida* (5·2 ft., 3·3 ft.), *Verbena stricta* (4·3 ft., 3·7 ft.). The roots of the remaining twenty-eight species reach greater depths than 6 ft., most about 10 ft., and several from 17 to 20 ft.; examples of these are *Agropyrum repens* (8 ft., 6 ft.), *Aster multiflorus* (8 ft., 5 ft.), *Ceanothus ovatus* (14·5 ft., 11 ft.), *Rosa arkansana* (21·2 ft., 16 ft.), *Solidago canadensis* (11 ft., 8 ft.). Of greater importance than the maximum depth is the *working depth*, by which Weaver means "the average depth reached by a large number of roots or branches of the root system, and to which depth considerable absorption must take place. It has no absolute value, like maximum depth of penetration, but can usually be determined for most root systems with a considerable degree of accuracy." The working depth of the first group ranges from 1·2 to 1·5 ft., of the second from 1·7 to 3·8 ft., and of the third from 3·3 to 16 ft. The second figures quoted for the examples given above are the working depths. An example of a plant of the third group, *Liatris punctata*, is shown in Fig. 1.

The lateral spread in the surface regions is always small. Lower down the roots may spread through a radius of 4 ft. from the centre of the system, though generally the spread is less than this, and frequently less than 1 ft.

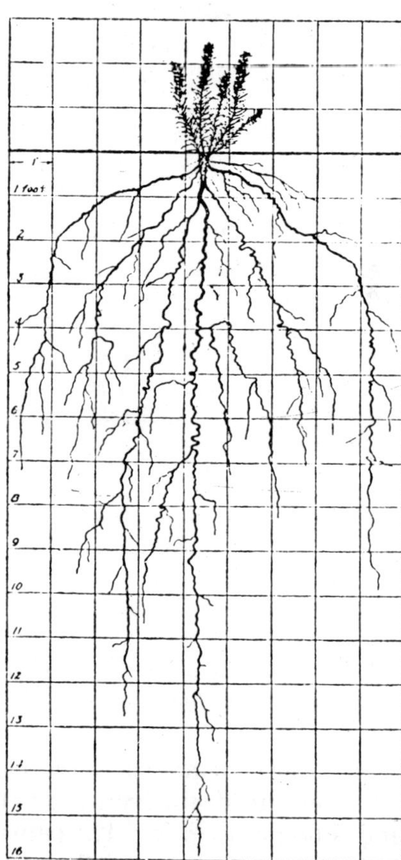

FIG. 1.—Root system of *Liatris punctata*, from the prairies of Nebraska. (From Weaver.)

The great depth usually attained is only possible in the very deep soils of the prairies: "The fertile dark-coloured prairie soil of the region is the type commonly called loess, much of which, however, is confounded with glacial drift. The loess covers the hills and valleys alike to a depth of from 20 to 100 ft., being much thicker than this in places and much thinner in others." The region has an annual rainfall of nearly 30 in., most of it in the growing (summer) months; but there is much run-off in the heavy storms. This, coupled with high evaporation, prevents the surface regions from retaining much moisture. Frequently there is little water available for plant growth in the first 5 ft. of soil. This accounts for the deep penetration of most root systems. The shallow-rooted grasses complete their active growth in the early summer months, and lie dormant through the droughts of July and August.

Contrasted with this community is the prairie of the Pacific Northwest (Washington) with a different floristic composition, characteristic plants being *Sieversia ciliata*, *Wiethia amplexicaulis*, *Lupinus leucophyllus*, *Lupinus ornatus*, *Poa sandbergii*, *Leptotænia multifida*, *Agropyrum spicatum* (Fig. 2). There are three grasses with a root system confined to the first 18 in. of soil; these, again, lie dormant

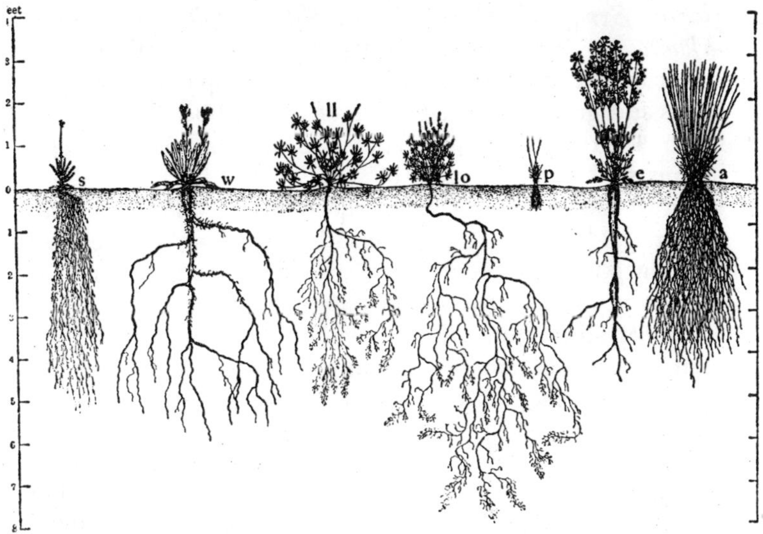

FIG. 2.—Root systems of typical plants of the Washington prairies; s, *Sieversia ciliata;* w, *Wiethia amplexicaulis;* ll, *Lupinus leucophyllus;* lo, *Lupinus ornatus;* p, *Poa sandbergii;* e, *Leptotænia multifida;* a, *Agropyrum spicatum.* (From Weaver.)

through the summer months. The remainder of the species examined had root systems of medium depth from 4 to 6 ft., only one or two penetrating as deeply as 10 ft. The rainfall in this region is about 20 in., and occurs chiefly in the winter months. The soil is a friable, dark brown, silt loam originating from the decomposed underlying basalt, and is many feet deep. It absorbs water during the winter and acts as a reservoir; during the growing season it is gradually depleted. By June, on exposed slopes, there is no water available for plant growth in the top 6 in., by July there is none in the top 2 ft., but even in August there is still a reserve at 4 to 5 ft. Comparing the water relation of this soil with that of the eastern prairies, it will be seen how the shallower root systems are related to the altered conditions.

On the eastern borders of the eastern prairies there is a chaparral or scrub community with such shrubs as *Rhus glabra*, *Symphoricarpus vulgaris*, *Vitis vulpina*, and *Rosa arkansana*. The community also exists in the prairie itself, where moister patches of soil allow it to supplant the prairie herbs and grasses. The growth of the shrubs increases the soil moisture by lowering evaporation, and the formation of a humus mulch. The roots of some of these shrubs extend to great depths, to 21 ft. in the case of *Rosa arkansana*. Generally, however, a large absorbing system is developed near the surface and the lateral spread is sometimes great—over 20 ft. in the case of *Rhus glabra*.

Contrasting more strongly with the root systems of the prairie plants are those of plants inhabiting the Colorado sandhills. Depths of 8 or 9 ft. are attained by some plants (*Psoralea lanceolata*, *Artemisia filifolia*, *Eriogonum microthecum*), but of nineteen species examined eight had roots entirely confined to the surface 2 ft. Of the deep-rooted species *Psoralea lanceolata* alone has a working depth of 7 ft.; the others all show a predominant development of absorbing roots within 3 ft. of the surface. The rainfall is about 23 in., but it is all absorbed rapidly, and the first few inches of the soil act as an efficient mulch, so thoroughly preventing evaporation that below 6 in. the soil remains more or less uniformly moist (cf. Fig. 3).

Three communities on the east slope of the Rockies at an elevation of 8000 ft. may be considered. On the gravel slides, with a moving surface of small stones overlying a shallow (4 to 6 in.) soil, and coarse subsoil that passes into rock at 2 to 4 ft., a sparse vegetation of dwarf shrubs (*Apocynum androsæmifolium*, *Eriogonum flavum*, etc.) is developed.

The roots show a marked lateral development a few inches below the surface, though there may be, in addition, a well-branched system penetrating 2 or 3 ft. to the maximum depth of the subsoil. The gravel surface layer preserves an even moisture in the underlying soil. The " half-gravel slide " community exists on stabilised slopes with a deeper and richer soil. Here again the surface root system is well developed, but is more markedly supplemented by a deep and well-branched system draining water from depths of 3 ft. A forest community with *Pinus ponderosa* and *Pseudotsuga mucronata* finally supplants the half-gravel slide community. The soil here consists of 1 to $1\frac{1}{2}$ ft. of rich humus covered by litter, and the root systems of the undergrowth plants are almost entirely confined to this shallow region.

Roots of Desert Plants.—The root systems of desert plants, exposed as these are to the most extreme arid conditions, have always attracted attention. The most extensive investigation is that carried out by Cannon (1911) in the neighbourhood of the Desert Laboratory at Tucson, Arizona. In this region, with a rainfall of about 11 in. occurring chiefly in midwinter and midsummer, extreme desert conditions are not exhibited, but, except in the flood plain of the Santa Cruz River, there is no water available for plant growth in the upper layers of the soil within a few weeks after rains. The summer rainfall is torrential, and there is a large run-off, while in winter the rainfall is better distributed and absorbed. There are over two hundred species of annuals, of which one-fifth have their growth period in summer, and the remainder in winter. None seem to possess a deep root system, the depths measured being always under 1 ft. The winter annuals have in general a well-developed tap root with sparing laterals. The summer annuals, on the other hand, belong to Cannon's "generalised" type with vigorous and much-branched laterals. This difference is related by Cannon to the more favourable conditions of absorption in the warmer summer soil which, along with temporarily high atmospheric humidity following the torrential rains, promotes shoot development and necessitates a better absorbing system. Both classes must be regarded as short-lived plants, which draw their water from the upper layer of soil during the brief periods when this is relatively moist.

Of the desert perennials the most distinctive root systems are shown by the Cactuses. Thus *Opuntia versicolor* (Fig. 3) sends down a stout anchoring tap root with few

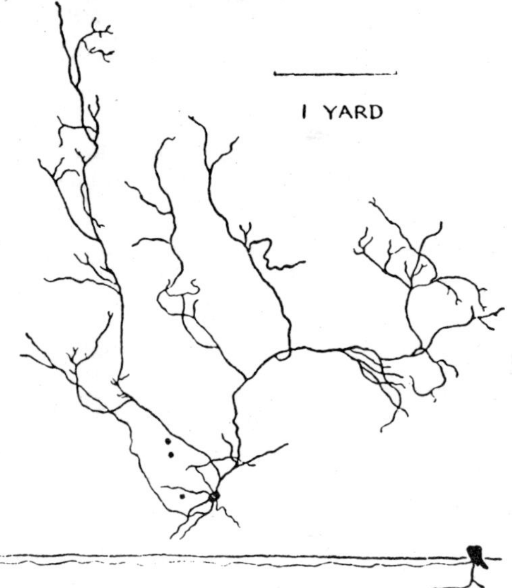

Fig. 3.—Root system of *Opuntia versicolor*; above, plan of about half the root system, the large black dot representing the position of the shoot; below, section showing the short anchoring root and the depth of the shallow system on one side. (After Cannon, modified.)

branches to a maximum depth of 1 ft. A small number of laterals arise from the tap just below the surface of the soil. These do not branch much, but they give rise to bunches of fine rootlets which are renewed yearly. The laterals and their branches spread out in every direction, keeping within a few centimetres of the surface, for a very great distance; in one instance they reached a distance of 14 ft. from the tap root. The absorbing system of this cactus, which is typical of the family, is thus very superficial; in *Opuntia arbuscula* the roots lie within 2 cm. of the surface. Obviously such a system can serve to absorb water only when the surface layers of the soil are saturated after rains. The immense water storage capacity of these plants makes this method possible. The cactus root system is very strongly specialised in Cannon's sense.

A specialised root system with a strongly developed tap is exhibited by *Koeberlinia spinosa*, a leafless spiny member of the Capparidaceæ. Most of the desert shrubs have, however, generalised root systems, or systems more or less like that of the cactus. The mesquite or locust, *Prosopis velutina*, grows as a small tree on the flood plain of the rivers, where its tap root may reach to ground water at a depth of 25 to 40 ft. On the shallow soils of the desert it remains a small bush; the tap may penetrate deeply if the soil permits, but it is often kept short by the underlying hard pan. A mass of laterals is then developed close to the surface and may stretch to the enormous distance of 50 ft. from the parent root. The creosote bush, *Larrea tridentata*, the most abundant shrub of the region, grows both on the flood plain and on the true desert. In the shallow soils of the latter it sends a strong tap to the depth of 2 ft.; where this meets the pan it forks and runs horizontally for a considerable distance. A number of sparingly branched laterals arise within 6 in. of the surface and spread to about 12 ft. from the plant. Most of the shrubs resemble the cactus in producing bunches of fine absorbing rootlets which last through the periods when water is available and then die off.

It will be seen that the water supply of all these plants is typically drawn from the surface 2 to 3 ft. of the soil. Deep-going roots are usually impossible. The same conclusion is reached by Fitting (1911) for the extreme desert near Biskra in Algeria. He did not examine the root systems in detail, but he points out that rock underlies the soil at depths of 3 to 9 ft., and that there is no water reserve at these depths. He found many of the root systems to be poorly developed. Cannon (1913), in a more detailed survey of conditions in the Algerian Sahara, describes many extensive root systems ; in plants growing on the deeper soils of the desert, deeply penetrating tap roots are met with. Volkens (1887), too, speaks of the shrubs of the Egyptian desert as characterised by very deep-going roots which tap deep-lying moist soil and water veins. "Some are probably to be found only in places where well-fed veins of water are present, so the widespread colocinth." Roots of Acacias penetrating to ground water at a depth of 40 ft. were observed during the excavation of the Suez Canal. Later investigators regard this arrangement as exceptional for desert plants.

Other Root Systems.—The root systems of the native and crop plants of Britain have been little studied ; this is practically a virgin field for investigation. Some observations have been made on the regions of the soil exploited by the systems of different plants living together.

A certain amount of information exists on the root systems of aquatic and marsh plants. In Kirchner and in Sherff (1912), the types of system for a number of monocotyledons and for the plants of an American marsh may be inferred from the figures. A short summary is given by Arber, whose account deals more with mechanical features.

As might be expected, the plant growing in water or mud has in general a reduced root system ; the absorbing surface need not be so great, and moreover the presence of water may reduce very considerably the necessity for resisting strains. In Potamogeton, both in species with floating leaves and in those entirely submerged, bunches of stout unbranched adventitious roots, a few inches long, arise from the nodes. Castalia produces abundant adventitious roots which may reach a foot in length and are sparingly beset with fine rootlets. Sagittaria and Alisma, sending leaves and shoots above the surface of the water, have many short stout roots sometimes with fine rootlets and sometimes without. Plants like *Ranunculus Flammula*, which grow in very wet soil, have systems of matted fibrous roots with numerous fine rootlets attaining the length of a foot ; while

Triglochin palustre, growing in similar situations, has sparse adventitious roots coming in a bunch from the rather bulbous base of the stem, without branches and only 2 or 3 in. long. *Ranunculus aquatilis* has fine, short and sparingly branched adventitious roots. *Elodea canadensis* has few, slender adventitious roots which are quite unbranched and usually not more than 4 in. long. Lemna produces a single unbranched root about an inch long.

Modification of the Root System.—From these descriptions it will be seen (*a*) that the type and extent of the root system are related to the conditions in which the plant lives, although, even in peculiar and extreme environments, very different types of root systems are to be found together; and (*b*) that the type of root system is to a high degree specific. We must now consider the modifications which the system of a species may undergo in response to environmental conditions. The factors which principally affect root development are: (1) soil texture, (2) soil moisture, (3) the supply of salts, (4) hydrogen ion concentration, (5) soil aeration, (6) soil temperature. In addition to these, it must be remembered that the conditions to which the shoot system is subjected, and the degree and rate of its growth and development, must affect the growth of the root system. Such effects can at present scarcely be disentangled from the effects of the immediate environment of the root. Further, it is often difficult or impossible to distinguish in nature between the direct effects of soil moisture, aeration, and texture on the root.

1. The *influence of soil texture* is most obvious when a layer of rock or of hard pan underlies the soil and prevents penetration. Not only is the depth of the system thereby limited, but changes in form may be caused, especially in systems with tap roots, by the breaking up of the tap into irregular branches. The presence of large stones and rocks is a potent means of distorting the symmetry of root systems. The great depth of the systems of prairie plants already noted is, of course, only possible in an easily penetrable deep soil. The stiffer clays may present considerable

resistance to penetration. But in the case of clays the exact cause of the effects is at once obscured by the different conditions of moisture and aeration.

2. *Soil Moisture.*—The necessities of the plant would, it might seem, require the development of a root system proportionately more extended the drier the soil. On the other hand, increasing drought means a decrease in the materials and conditions favourable to vigorous growth. As a result it seems that plants growing in extremely dry and extremely wet soil both produce relatively poor root systems, though in the latter case the total growth of the plant is much better. In intermediate moistures more extensive root systems occur. From his own observations, Freidenfeldt (1902) draws the conclusions: (1) that a plant shows a maximum root development at a particular water content for a given type of soil; (2) that on each side of this root development is diminished; (3) that dry soils tend to promote growth in length of the tap or chief laterals and so to produce a deep root system; (4) but that many native plants which grow both on very wet and very dry soils, such as *Nardus stricta*, *Festuca rubra*, species of Saxifraga, and of Alchemilla, show relatively small differences in the two kinds of habitat. These results, however, must be taken with a certain amount of caution, for other conditions than soil moisture are certainly operative.

Weaver (1919) examined the root systems of examples from two or more different stations of each of eleven species; in seven there were striking modifications of root habit. Various factors are concerned, but Weaver (1919, 1920) lays chief stress on the differences of water supply. In the plains, *Euphorbia montana* (Fig. 4) had a root system penetrating to 7·5 ft. with a working depth of 2 to 4 ft., while in the half-gravel slide the maximum depth was 4·5 ft., and the whole of this layer of soil was filled with rootlets. In general the most marked development of fine roots occurs where moisture supply is most favourable.

Certain species, however, such as *Koeleria cristata* with shallow branch roots and *Allionia linearis* with a deep

4

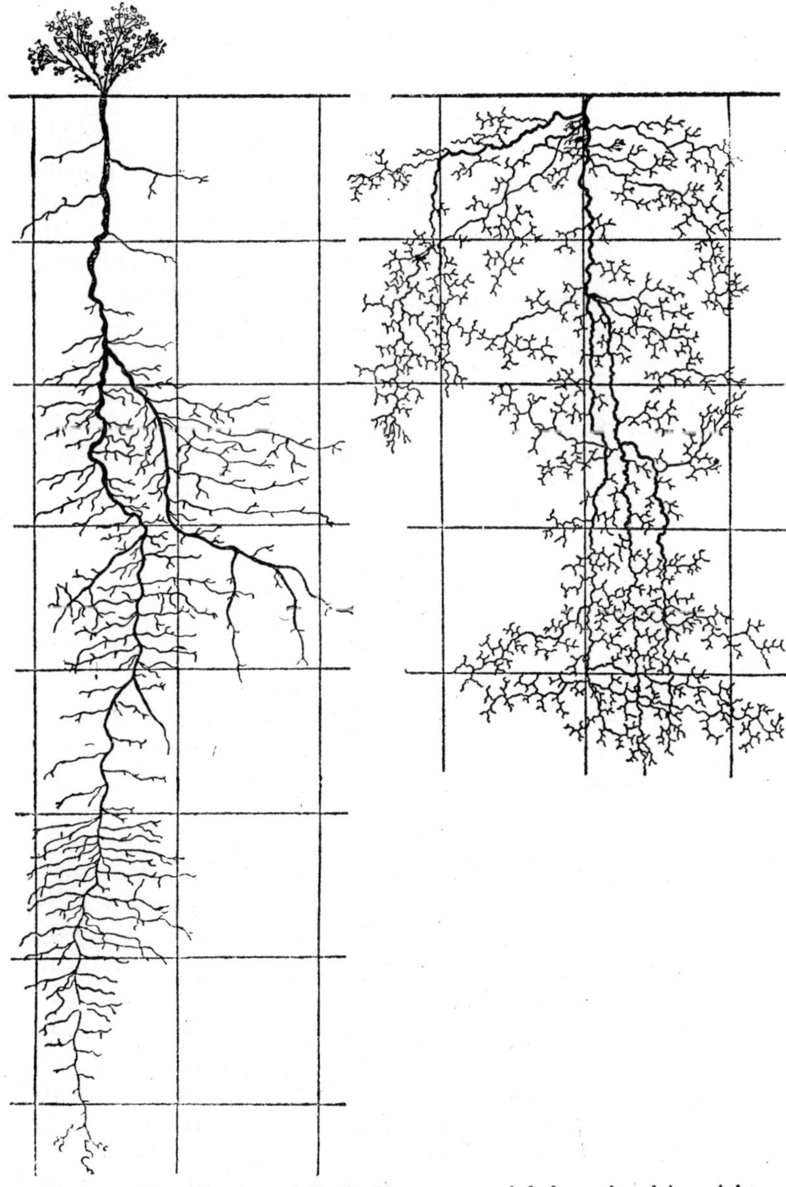

Fig. 4.—Root systems of *Euphorbia montana*; left from the plains, right from the half-gravel slide; the side of the squares measures 1 ft. (From Weaver.)

tap, show the same type and extent of root system under widely different conditions of moisture supply; they are profoundly conservative.

The relation of water supply to root development may be illustrated by the behaviour of three of the cereals examined in the prairies and the short-grass plains as shown in Table VI.

TABLE VI

ROOT SYSTEMS OF CEREALS IN DIFFERENT ENVIRONMENTS

Cereal.		Plains.	Prairies.
Rye	maximum depth	3·4 ft.	5·3 ft.
	working depth	2·8 „	4·0 „
Oats	maximum depth	3·4 „	4·4 „
	working depth	2·7 „	3·4 „
Wheat	maximum depth	2·7 „	5·4 „
	working depth	2·3 „	3·8 „

The rainfall on the plains, and the water supply in the soil, are much less favourable than on the prairies; the unfavourable conditions are reflected in a poorer shoot growth as well as in the less well-developed root systems.

3. *Mineral Nutrients.*—The action of the supply of mineral salts is peculiarly difficult to disentangle from other effects, not only because it is closely related to other soil conditions, and in particular to moisture content and reaction, but also because of secondary reactions through the effect on the shoot system. Nobbe (1862), who grew plants in soil, alternate layers of which had been watered with nutrient solution, found branching of the root system more vigorous in these. Weaver, Jean, and Crist (1922) used a similar method, but isolated the layers of soil fertilised with nitrate by thin layers of wax. They found that the roots of the potato and of various cereals always branched more vigorously in the fertilised layer. Frank (1893) found that the half of a pea root system grown in a soil rich in nitrates developed better than the other half in poor soil. Here the effect of the shoot is eliminated,

and apart from salt content the soils had probably very similar properties; the results therefore clearly indicate a direct beneficial action of the salts. In most investigations, however, comparison can be made only between the systems of different individuals. The effects of the addition of certain salts to an exhausted Rothamstead soil studied by Brenchley and Jackson (1921) are shown in Table VII.

TABLE VII

DEVELOPMENT OF ROOTS AND SHOOTS OF WHEAT (16 WEEKS OLD). DRY WEIGHT IN GRAMMES

	Exhausted soil.	+ Sodium nitrate.	+ Superphosphates.	+ Nitrates + Superphosphates.
Shoots	7·15	9·5	11·1	13·9
Roots	2·5	2·6	4·1	3·9
Shoot/Root ..	2·9	3·5	2·7	3·7

The increase in root development is marked with phosphate as compared with nitrate, and the diminution in the shoot/root ratio shows that it is not wholly due to a secondary effect through the shoot. Gericke (1921) found that wheat seedlings grown in culture solutions lacking nitrogen showed a remarkable root development though the shoots were stunted.

The effect of phosphates in promoting root growth has long been recognised in agricultural practice. It is of importance not only in promoting the swelling of such "roots" as the turnip or mangold, but in causing increased development of the branches of the absorbing system. Russell sums up: "Dressings of phosphates are particularly effective wherever greater root development is required than the soil conditions normally bring about. They are invaluable in clay soils, where the roots do not naturally form well. . . . They are used for all root crops like swedes, turnips, potatoes, mangolds. . . . Phosphates are needed for shallow-rooted crops with a short period of growth like barley. Further, they are beneficial wherever drought is likely to set in, because they induce the young roots to grow rapidly into the moister layers of the soil below the surface."

4. The *hydrogen ion concentration* of the soil has, undoubtedly, an important effect on root development. This is clearly shown by experiments carried out by Olsen (1923) on the growth of various plants in culture solutions identical except for their p^H values. Table VIII shows the dry weight attained by *Senecio sylvaticus*, which occurs naturally in acid soils of p^H 4·9–5·6, and by *Tussilago Farfara*, which grows in alkaline soils of p^H 7·0–7·9.

TABLE VIII

HYDROGEN ION CONCENTRATION AND ROOT DEVELOPMENT

p^H of nutrient solution		3·0	3·5	4·0	4·5	5·0	6·0	6·5	7·0	7·5
Dry weight in grammes of root system	Senecio	0·3	2·2	**3·4**	—	2·3	2·0	0·5	0·2	—
	Tussilago	—	0·1	0·8	1·2	1·4	**2·5**	—	2·5	—

This shows the effect of the p^H value on a particular species and also the different reaction of species naturally occurring in soils of different acidity. It does not, of course, enable us to say how far the effect is due to direct action on the roots.

Tottingham and Rankin (1922) find that growth in length of the roots of wheat is maximal in solutions of p^H 7·5. At that value the roots in their cultures attained a length of 30 cm. as compared with 24·5 cm. at p^H 6·4 and 10·6 cm. at p^H 5·3. The greatest dry weight, however, was made at p^H 6·4. Hoagland (1917) found the roots of barley to be injured at p^H 7·4, to show good growth from p^H 6·16 to p^H 7·07, with a maximum at the former, and injury again at p^H 3·54. These results are as yet not very extensive, but the immediate effect of the p^H value of the soil on that of the cell sap makes it certain that such reactions as growth rate must be affected. The hydrogen ion concentration of the soil is evidently one of the important factors affecting the development of the root system.

5. *Aeration.*—Many plants grow in stations where the supply of oxygen to the roots is very restricted. This is the case with the aquatics of still water and with plants of bogs and marshes. The root systems of such plants we have already seen to be relatively poorly developed, though in some cases they are extensive. Such roots show, in general, a well-developed aerating system of cortical spaces, intercellular or lysigenous (formed by the breaking down of cells). In secondary tissues an aerating system is developed by the action of the cambium or of a special phellogen, or cork-forming meristem (Arber). The system is in communication with the air spaces of the shoot, these too being very prominent in aquatic plants. In some plants there is considerable variation in regard to the extent to which the aerating system develops under different conditions. Batten (1918) has shown that the roots of *Epilobium hirsutum* when grown in clay have more extensive air spaces than when grown in sand. The total extent of the root system is much less in the former case although the shoot growth is rather better.

We may here refer to the *pneumatophores*, or special aerating roots, possessed notably by various mangroves. According to Schimper, these are simplest in *Carapa obovata*, where the upper edge of creeping roots projects above the surface of the mud or water. In Bruguiera, the horizontal roots bend out of the mud in knee-shaped structures. In Avicennia, special negatively geotropic lateral roots grow vertically above the surface for a few inches, while in Sonneratia they may be a few feet long. In the tropical shrubby *Jussieua repens*, there are normal laterals growing into the mud and others which rise to the surface, evidently floated by the air they contain in a remarkable cortical aerenchyma. All these roots possess very marked aerenchyma, or aerating tissue, developed in the primary cortex, and, while that cortex is retained, abundant lenticels (openings to the atmosphere).

The effect of placing an ordinary root system in water-logged soil depends on the type of plant used. Plants normally growing in ordinary soil usually suffer badly. Balls (1912), on excavating root systems of the cotton plant in Egypt, found that all roots which had been submerged by irrigation water for ten days were dead, though this region of the soil might be again exploited by new roots produced after the fall of the water-level. Bergman (1920) found that the roots of *Impatiens balsamina* and of Pelargonium when submerged died off, and that the leaves wilted, went yellow, and dropped off in a few days. After ten days, if the plant survived, new roots developed at the water-level and recovery took place. *Ranunculus sceleratus* and *Cyperus alternifolius* grew in saturated soil with no ill effects, and the roots grew throughout the submerged soil, these roots being provided with aerating tissue. The injurious effects were also produced in Impatiens and Phaseolus in a soil permeated with carbon dioxide instead of air. Noyes, Trost and Yoder (1918) found a retardation of growth of roots of radish and lupin grown in soil permeated with carbon dioxide. The fact that carbon dioxide has a narcotic effect must be taken into account here, but similar results have been obtained using the neutral gas nitrogen by Cannon (1920), and Cannon and Free (1920) using helium. The growth of most roots ceased when only 1–2 per cent. oxygen was present; in pure nitrogen the roots of the sunflower die and are replaced by new adventitious roots, short, thick, little branched and with no hairs. In the same conditions the roots of the maize do not die; their growth rate slows down and then increases again; a physiological adjustment has taken place.

The growth of land plants in water culture is a remarkable instance of accommodation. Such plants as the oat, barley and buckwheat may be brought to maturity with their root systems in the highly abnormal medium of a weak solution of salts. To ensure successful development, however, it is necessary to aerate the culture solution frequently by blowing air through it. The effect of aeration comes out very clearly in these conditions, as it is not complicated by changes in moisture or salt supply. Stiles and Jorgenssen (1917) found that well-aerated cultures showed a 70 per cent. increase of root growth over non-aerated.

In nature the effects of lack of aeration may occasionally be observed clearly. Thus, Emerson (1921) found that the tap root of seedlings of Picea and Pinus, growing in Sphagnum tussocks, ceased growth or bent sharply in a horizontal direction near the water-table.

Some indirect effects of bad aeration have already been referred to. They include modification of bacterial activity, inhibition of the destruction of humus compounds, and the production of organic toxins. Such effects have a very important influence on plant growth. For a detailed discussion the monograph on "Aeration" by Clements (1921) should be consulted.

6. The *influence of temperature* on root growth has been the object of frequent study. For the main root of Pisum, Leitch (1916) has shown that the growth rates at 29° C., 20° C., 10° C., and 0° C. are as 48, 24, 8, and 1. Above 30° C. (at which temperature irregularities occur), a progressive diminution takes place. As to the effect of different soil temperatures on the growth of root systems, there seems to be no definite information. One may be sure that in desert soils, in the surface layers of which temperatures of 50°–60° C. frequently occur, growth may be inhibited. Cannon (1915) found that root growth of such desert plants as Prosopis and Fouquieria is appreciable only above 15°–20° C. and is active at 35° C. In the soils of temperate regions inhibition must often be caused by low temperatures. The greater uniformity of soil temperature must mean a steadier though perhaps slower development of the root system. It has been generally believed that the growth of tree roots ceased in summer and winter in temperate climates, but McDougal (1916) has shown for certain American species that if water is abundant growth continues through the summer, and that the only regular cessation is that due to low winter temperatures.

Of alteration in the form or depth of root systems caused by temperature nothing is known. Mention may be made here of the view expressed by Diels (1918*a*), that the occurrence of a thick layer of light-coloured cork about the root neck and stem base in the desert plants of West Australia tends to protect the living tissues from too great overheating in the hot sand.

Interacting Conditions in Nature.—The system which actually develops in any given case in nature is the resultant of the interplay of these various factors, acting directly or indirectly on the growing regions of the roots, influenced of course by the development of the shoot, and working always on the material of the plant's inheritance. We have quoted instances where the root system was very plastic, external conditions changing it readily and profoundly, and others where it remained very constant under widely different circumstances. Various instances of modifications under natural conditions have been given, in the main such as seem to indicate the working of one particular factor. We may close this account with one or two cases where, as is usual, the active factor cannot be easily identified, or where more than one is involved.

Yapp (1908) describes two plants of *Lysimachia vulgaris* growing near each other in Wicken Fen. Numerous strong adventitious roots arise from the horizontal rhizomes. In the first plant the rhizome lay $1\frac{1}{4}$ in., in the second $3\frac{1}{2}$ in. deep. In the first the roots grew vertically down, in the second they spread out obliquely or almost horizontally. It is probable that we have here an effect of diminished aeration due to approach to the water-level.

In deposits of river sand thin layers of humus soil may often be found interbedded. The roots of trees and shrubs tend to be confined to such layers. In an exposed section the way in which the roots, *e.g.* of a birch, run along the dark-coloured humus is sometimes striking. Waterman (1919), in a study of the dune plants of Lake Michigan, found that the only cause of irregularity or asymmetry in the root systems was the presence of nests of humus, in which the roots gathered. The factor which is effective in these cases cannot be easily identified, for the humus soil differs from the sand in many respects ; it may modify and attract root systems either because of its greater water content, or its greater supply of nutrients ; its p^{H} value is also different from that of sand.

A particularly interesting case is figured by Weaver (1919), in a root system of *Kuhnia glutinosa* growing in a prairie soil of stratified sands and clays (Fig. 5). In two beds of clay lying 8 and 13 ft. deep the number of laterals is very much greater than in the sand. The clay is much wetter than the sand, but in this case, too, we may be inclined to suspect that the influence of salt supply is effective.

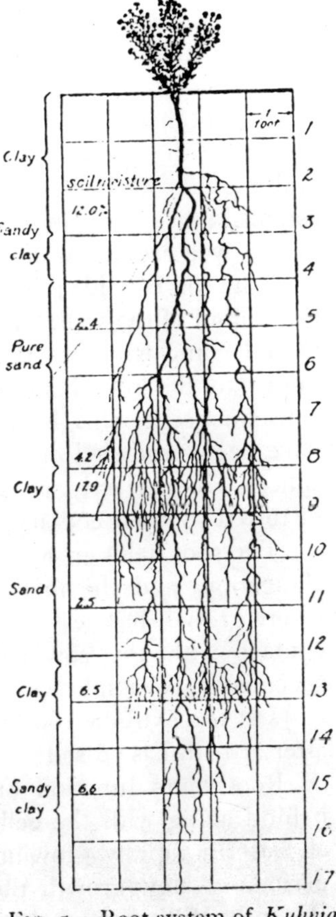

FIG. 5.—Root system of *Kuhnia glutinosa*, growing in alternate layers of clay and sand. (From Weaver.)

§ 4. THE ABSORPTION OF WATER

Root Hairs.—The absorption of water is only to a very limited extent carried on actually by the roots. These traverse the soil thoroughly and often extensively, but they provide neither the large absorbing area nor the intimate contact with the water films which are necessary in ordinary soils. Absorption is normally the work of the root hairs. These occupy a region a very few centimetres long, extending backwards from a point about a centimetre behind the tip. When they die away the external cortical cells become suberised or corky, and form an exodermis impermeable to water; the power of absorption is lost. As the result of secondary changes the cortex may die and be sloughed off, the protection of the living tissues within being taken on by a secondary corky skin.

A root hair is a protrusion of an epidermal cell, the lumina of the two being continuous. It has a delicate wall and moulds itself to the grains of soil which it touches, establishing a very intimate contact. The external layers of the wall are mucilaginous, and this favours close union with the soil colloids; indeed, the soil colloid may merge in the colloid of the wall. If a seedling is uprooted, the grains of soil come with it; they cannot be washed off without injury to the root hair; the two are glued together. A consequence of this is that whereas the root hairs which have been produced on a seedling grown in a moist atmosphere are perfectly regular, those produced in soil are distorted by their contact with the soil particles.

The root hairs grow out just behind the region of active elongation, and the importance of this is obvious. Only where growth has ceased is it possible for the hair to make contact with the soil particles without injury. Certain succulents (Crassulaceæ, Cactaceæ), however, bear root hairs at the extreme root tip; these are probably produced after growth has ceased.

Root hairs function only for a short time, dying off behind along with the cell layer which gives rise to them, at least in rapidly growing roots. It seems likely that at periods of slow growth the hairs may be more persistent. Recently Whitaker (1923) has shown that many herbaceous Compositæ have roots which bear root hairs over their entire length: these hairs may persist and function for two to three years. Roots with secondary thickening have hairs only at the tip. When seedlings are transplanted the hairs of the exposed roots die off, and are renewed as the plant establishes itself.

According to measurements of Schwarz (1883), root hairs produced in moist air reach lengths of 0·5 to 3 mm.: in earth the length is usually less; in water it is more. They may occur to the number of 200 to 400 per sq. mm. of root surface, and the increase of absorbing surface over that of a hairless root may be as much as eighteen times.

Many plants when grown in water produce no root hairs; this may well be due to reduction of the oxygen supply below the minimum necessary to the very active growth entailed in their production. The same lack of root hairs is seen in new roots produced by plants growing in soil saturated with nitrogen instead of air. Many aquatic plants, however, possess root hairs. In Hydrocharis (Arber) they are remarkably long, in Elodea they are produced only when the root enters the mud, in Lemna they are absent. In some marsh plants, such as *Myrica Gale*, no root hairs are formed. They are, as a rule, absent from mycorhizal roots.

Forces of Absorption.—The actual absorption of the water depends on the osmotic pressure of the cell sap of the root hair. The direction of the flow is determined by the osmotic gradient; water passes towards the solution with the higher osmotic pressure. Normally, of course, the vacuolar fluid of a plant cell, containing in solution small quantities of salts together with larger quantities of organic crystalloids such as sugar, has a very considerable osmotic pressure. We may take 10 atmospheres as a conventional figure, and this is roughly the pressure exercised by a 15 per cent. solution of cane sugar. As the concentration of the soil solution is a very small fraction of this, the mechanism of transference of water from soil to plant is provided for, at least when the soil contains abundant water. When the water supply runs low, however, complications arise.

In the first place it must be pointed out that the whole of the osmotic pressure need not be employed in drawing water into the cell. It is a familiar fact that, when a cell is plentifully supplied with water, hydrostatic pressure acts against the elastic pressure of the cell wall and distends it; this is the *turgor pressure* of the cell sap ; it is an expression by the solvent of the *osmotic pressure* which is due to the solute. As a result the cell becomes distended and stiff, or turgid. When the cell wall is extended to its maximum water of course ceases to enter, and the whole of the osmotic pressure is used up as turgor pressure acting against the cell wall. When the cell wall is not fully turgid, only a part of the osmotic pressure is used up as turgor pressure, and the balance is available to draw in water. The actual force—*suction force*—available for drawing water into the cell is therefore the difference between the osmotic pressure and the turgor pressure ; or rather, it is this difference *less* the opposing forces which tend to retain water in the soil.

A root hair freely supplied with water would soon become fully turgid, and no suction force would exist were it not that water is constantly passing from it into the root ; a small water deficit is therefore maintained, and this suffices to keep up a flow of water into the plant. This case is realised in water plants. In land plants the deficit is likely to be larger ; the more difficult the water supply, the greater, within the limits of the available osmotic pressure, will be the suction force, and if, as probably happens in very dry soils, the water deficit becomes so great that the cell borders on plasmolysis (shrinking of the plasma lining from the wall), the full osmotic pressure will be available as suction force.

In dry soils forces other than osmotic pressure are active, both in the cell and in the soil. It has already been noted that the mucilaginous cell wall unites with the colloidal constituents of the soil ; and thus the relatively great forces of imbibition of these colloids must come into play, if not continually, at all events when the soil moisture has been reduced to a certain point. The extent of these forces, so far as the cell is concerned, is quite unknown.

Osmotic Pressure of Root Cells.—As to the osmotic forces at the disposal of the cell we have a good deal of information. Measurements have usually been made of the osmotic pressure without reference to the suction force.

Hannig (1912) measured the osmotic pressures of the roots of a large number of plants; he used the cortical cells of the roots, as the root hairs of plants taken from the soil are too difficult to work with. But he remarks that occasional observations on the root hairs showed the osmotic pressure to differ little from that of the cortical cells. Table IX gives his results for a number of plants growing in different conditions in the Strasburg Botanic Gardens.

TABLE IX

OSMOTIC PRESSURES OF ROOT CELLS

Name of plant.	Situation.	Osmotic pressure in atmospheres.
Impatiens balsamina . .	Garden soil	5·3
Galanthus nivalis	Garden soil	8·7
Bellis perennis	Gravel path	10·6
Menyanthes trifoliata . .	Mud	5·3
Alisma plantago	Mud with water	7·1
Pistia stratiotes	Floating plant	5·3
Elodea canadensis	Submerged aquatic	5·3
Opuntia Raffinesquiana . .	Loose soil	8·7
Acacia lophantha	Pot shrub	7·1

For salt-marsh plants measurements have been made by T. G. Hill (1908), who found the osmotic pressure of the root hairs of Salicornia and of Suæda equivalent to 6·7 per cent. sodium chloride; a mesophyte (see pp. 199, 200) seedling from adjacent ploughed fields gave a value of 1·5 per cent. sodium chloride. These amounts indicate osmotic pressures of 41 and 9·1 atmospheres respectively. The values for the salt-marsh plants are thus very high, and this is of course related to the fact that they draw their supply from salt water with a high osmotic pressure. The water of the marshes may have a salt content of 3·67 per cent. The root hairs of these plants have great powers of accommodation. After soaking in fresh water for twelve hours, the osmotic pressure may fall as low as 11·3 atmospheres. This power of accommodation may be important in nature, as the salt content of the marsh water is subject to great fluctuations and may fall as low as 0·79 per cent. after heavy rains.

The osmotic pressures of the plants of the desert round Biskra have been investigated by Fitting (1911). He was unable to examine the roots, and made his determinations on the leaves. It has been shown by Hannig (1912) that the osmotic pressure of the leaves is in general 2 to 4 atmospheres higher than that of the root cells; but it is of the same order, and so may be taken as indicating the order of pressure probably obtaining in the root hairs.

An abstract of his results for plants growing in different situations is given in Table X. The numbers refer to the number of species found to have the osmotic pressures shown.

TABLE X

OSMOTIC PRESSURES OF DESERT PLANTS

Osmotic pressure in atmospheres.	106 and over.	71–106.	53–71.	35–53.	20–35.	10–20.
Extreme desert plants	13	5	8	17	19	9
Salt swamp plants ..	2	–	2	7	2	–
Dune plants ..	1	–	1	7	5	7

It will be seen that the great majority of these plants have very high osmotic pressures, and that a number have the enormous value of more than 100 atmospheres. These plants, too, have considerable powers of accommodation. Thus the date palm growing in the salt swamp had a pressure of about 50 atmospheres, while in the irrigated cultivated land round Biskra its pressure was between 28 and 42. *Mesembryanthemum nodiflorum* had a pressure of over 100 atmospheres in the desert, and of 35 to 40 in the cultivated land. Faber (1913) found the osmotic pressure of the leaf cells of the mangrove *Rhizophora mucronata* to range from 24 to 27 atmospheres. For steppe plants Iljin, Nazarova and Ostrovskaja (1916) found the osmotic pressure of the root cells to be about 15 atmospheres, compared with 10 atmospheres for meadow and 6 atmospheres for swamp plants. An individual growing on the steppe had a pressure of about 2 atmospheres more than that of an individual of the same species growing in the meadow.

Investigations in the coastal deserts of Jamaica by Harris and Lawrence (1917), using a freezing-point depression or cryoscopic method of determining the osmotic pressure of the leaves, gave values typically over 20 and usually between 30 and 50 atmospheres for plants with leathery leaves. Cactuses and other succulents had low osmotic pressures, less than 10 atmospheres.

The actual suction force has been measured by Ursprung and Blum (1921*a*). The measurement is made by determining the concentration of sugar solution at which no change occurs in the volume of the cell, measured in the first place under liquid paraffin; in other words, the concentration at which water passes neither into nor out of the cell gives a measure of the suction force of the cell sap. The suction force is not a constant for a given plant; it rapidly grows less as the cell is supplied with water, and it thus depends largely on the forces withholding water from the plant; indeed, as we shall see, it has been used to measure these. Table XI gives values obtained for the epiderm of roots of *Phaseolus vulgaris* and *Vicia Faba*, with the corresponding osmotic pressure.

TABLE XI

SUCTION FORCES OF ROOT CELLS

Plant and conditions.	Suction force.	Osmotic pressure.
Phaseolus in moist sawdust	0·9 atmos.	7·03 atmos.
,, in drier sawdust ..	1·6 ,,	—
Vicia in moist sawdust	0·7 ,,	8·03 ,,
,, in drier sawdust	3·2 ,,	11·20 ,,
,, in distilled water	0·0 ,,	7·56 ,,

Such measurements, then, show us that the root hair exercises a suction force on the soil water which is closely accommodated to the water supply available at the moment.

The osmotic pressure on which the suction force depends is also related to the requirements of the individual and varies with changing external conditions. It is, moreover, specific in the sense that plants habitually growing in extreme conditions have a characteristically high osmotic pressure.

Soil and the Water Supply.—We may now consider the way in which the forces which hold water in the soil are related to absorption by the plant. It has already been implied that the whole of the water in the soil is not equally available for plant growth. The water which the plant can readily absorb has been called *growth water*, *available water*, or *physiological water-content* ; the water not easily available has been called the *non-available water*, or expressed as the *wilting coefficient*. In the terminology of Clements (1905), the total water content of the soil is the *holard*, the water available for plant growth is the *chresard*, and the non-available water is the *echard*.

The point at which the plant seriously fails to cover its water requirements by absorption from the soil is indicated by the onset of *permanent wilting*, that is, wilting from which the plant does not recover if placed in a saturated atmosphere. In the case of ordinary mesophytes, the flagging of the leaves can be readily observed, though even with them the determination of the wilting point is, to a certain extent subjective. In the case of plants with stiff leathery leaves, or with leaves of the needle or heath types, the determination of wilting by inspection is much less certain, or may even be impossible. Such plants, if potted, may be balanced along a steelyard with the plant on one side of the fulcrum and the pot on the other. As long as the plant draws sufficient water from the soil, the pot end of the system rises ; when the plant end of the system rises it is indicated that the plant no longer draws sufficient water from the soil to cover its losses, and that a condition equivalent to wilting has set it. This method has been applied to the cactus among others.

Wilting depends not on complete failure of the water supply, but on failure to obtain a sufficiently rapid supply.

The plant draws water from the soil after it has wilted, and indeed after it has died as the result of desiccation. A series of determinations on wheat by Briggs and Shantz (1912*a*) showed that the amount of water in the soil on the death of the plant was 15 per cent. less than at wilting, and that four months later the dead plant had reduced the soil moisture by a further 20 per cent. A vivid illustration of the fact that wilting follows inadequacy rather than failure of supply is given by soft-leaved garden plants, which sometimes flag on a hot summer day and recover their turgor in the cooler and more humid night hours, although no addition has been made to the reserve of water in the soil. For these reasons terms like "available" or "non-available" water are not happy. The term "growth" water is better, for when wilting sets in the turgor essential for active growth is lost, and, while we know little of the relation of loss of turgor to other cell activities, we know that photosynthesis is stopped. Wilting is therefore an important physiological condition. It has been claimed, further, that permanent wilting is accompanied by plasmolysis of the root hairs, and by a break in the column of water in the conducting vessels (Bakke, 1918); but this is not quite certain. The term "wilting coefficient" introduced by Briggs, and now in general use to denote the percentage of moisture in the soil when wilting takes place, is the most expressive and accurate yet proposed.

Wilting is a result of a failure of water income to cover water outgo. If we analyse this statement we see that the processes involved are: (*a*) the rate of transpiration, depending on various atmospheric conditions, and also on the state of the plant; (*b*) the rate of conduction, depending on the structural features of the species; (*c*) the rate of absorption, depending on the suction force of the root hairs and the supply from the soil, which in its turn depends on (*d*) the amount of moisture present; (*e*) the forces with which it is held; and (*f*) the rate at which it moves. The last two conditions are at least partly functions of (*d*). Wilting is therefore a resultant of several factors.

Now wilting is an important if not quite definite physiological state, but we are here concerned with the attempt to use it to indicate a critical point in the relation of soil moisture to the plant. This can be done only with great caution, for it seems clear that wilting must occur at different moisture contents as the shoot of the plant is subject to conditions of drought or of humidity, and transpiration is rapid or slow. It is certain, too, that a limitation of water supply to the leaves is sometimes the result of the condition of the conducting vessels, and this may occur more frequently than is yet suspected. The necessity for using the wilting coefficient with caution will appear from the investigations we must now consider, on the effect of soil composition and of the forces retaining water in the soil.

We have already seen that the amount of water retained is a function of soil constitution, and the same is true of the proportion of water which the plant cannot readily utilise. The finer particles of the soil are active in retaining water against gravity; they also hold it against the plant. This was demonstrated as long ago as 1865 by Sachs (quoted from Jost) for the tobacco; his results are given in Table XII.

TABLE XII

Supply of Water to the Tobacco by Two Different Soils

Soil.	Water capacity per cent. of dry soil.	Water content at wilting.	
		Per cent. of dry soil.	Per cent. of original water.
Sand and humus	46·0	12·3	27
Coarse sand	20·8	1·5	7

The soil with humus withholds more water from the plant than the sand, though, as it has a greater water capacity, it also supplies much more. Recently much exact work has been done on this relation. Some of the results obtained by Crump (1913*b*), for moorland plants, are given in Table XIII.

TABLE XIII

WATER RELATIONS OF MOORLAND SOILS AND PLANTS

Soil.	*Calluna vulgaris.*		*Deschampsia flexuosa.*	
	Humus per cent.	Non-available water per cent.	Humus per cent.	Non-available water per cent.
Peat ..	54–78	27·0	74–80	54·0
Sandy peat	35	12·0	34·0	13·0
Loam ..	10–11	5·8	9·0	13·0
Sand ..	—	—	4·5	1·4

The most extensive work is that of Briggs and Shantz (1912, *a* and *b*), who have determined the wilting coefficients of many species in a variety of soils, and have studied their relation to the soil constitution. They found that for any given soil the wilting coefficient was practically a constant and did not vary for different plants. Calling the average coefficient of about 100 species studied unity, the extreme variants were Japanese rice with 0·9, and Colocasia with 1·13. Succulents like Echeveria, xerophytes like Artemisia, and mesophytes like the tomato and the clover, all reduce the soil moisture to practically the same point when wilting sets in. This result is, of course, contrary to experience, which recognises that some plants can exist in much more arid conditions than others, and does not agree with the results of Crump quoted in Table XIII. It is explained by the method employed by Briggs. His plants were allowed to exhaust the soil slowly under shelters or glass where evaporation was relatively low. These conditions throw the weight of influence on the set of factors acting in the soil, and are therefore calculated to make the best use of the wilting coefficient as an indicator of soil conditions. They do not, however, express the capabilities of the plant in natural conditions. This may be illustrated by further reference to Crump's work (1914). With experiments conducted in the open he found, for example, that in peaty soils one-quarter of the water was non-available for

Nardus stricta, and only a ninth to *Vaccinium Vitis-Idæa*; his plants were only protected from rain. Here the factors affecting the shoot also come into play; a better idea is obtained of the plant's reaction in a natural environment, but a less accurate indication of the action of the soil alone. The critical review by Blackman (1914) should be consulted.

Briggs found the wilting coefficient to be very different in different types of soil, and he related it to various physical soil constants. Table XIV gives some of his results, and shows also for comparison the coefficient as calculated from the equation which relates it to soil constitution.

$$\frac{\text{Sands} \times 0{\cdot}01 + \text{silt} \times 0{\cdot}12 + \text{clay} \times 0{\cdot}57}{1 \pm 0{\cdot}025} = \text{W. Coeff.}$$

Sands {coarse sand particles between 2 and 0·25 mm.
{fine ,, ,, ,, 0·25 and 0·05 mm.
Silt ,, ,, 0·05 and 0·005 mm.
Clay ,, less than 0·005 mm.

TABLE XIV

COMPOSITION AND WILTING COEFFICIENTS OF DIFFERENT SOILS

Soil.	Per cent. composition.				Wilting coefficient.			$\frac{\text{Moist. equiv.}}{\text{W. coeff.}}$
	Coarse sand.	Fine sand.	Silt.	Clay.	Observed.	Calculated.	Ratio.	
Fine sand (No. 8) ..	35·4	55·1	4·8	4·5	3·3	3·6	0·92	1·67
Sandy loam (No. 3) ..	33·1	50·0	8·6	7·5	4·8	4·9	0·98	2·02
Fine sandy loam (B)	15·8	42·4	28·7	12·9	10·8	10·7	1·01	1·84
Loam (No. 5)	2·0	48·8	37·7	12·3	13·9	13·5	1·03	1·80
Clay loam (No. 6) ..	4·4	20·5	52·6	22·0	16·3	16·6	0·98	1·85

The agreement between the observed results and those calculated is remarkably good. Formulas were also devised connecting the wilting coefficient with various soil constants. Thus, W. coeff. $= \frac{\text{moisture equivalent}}{1 \cdot 84(1 \pm 0 \cdot 007)}$. The factors actually found are also shown in Table XIV. The actual factors probably require modification for types of soil of origin widely different from those studied by Briggs, and the relation may not in general be very exactly expressed by such simple equations. As Keen (1922) has pointed out, it would be surprising if relations so intricate were to be susceptible of a simple linear expression. But Briggs' results do give us a very striking illustration of the effect of soil constitution on water supply, and they mark an important step forward in the endeavour to put these relations on a quantitative basis.

Soil Forces and Water Retention.—We may turn to the problem of what condition in the soil slows down the water supply and occasions wilting in such experiments as those of Briggs, where the soil factors are dominant. The most natural assumption is that at the wilting coefficient the soil water has been reduced to the point when it is present only as extremely fine films, and absorbed by the colloids, and is thus held by the high imbibition forces which amount to many hundreds of atmospheres.

It is a matter of much difficulty to determine the forces withholding water from the plant at any given degree of moisture. C.A.Shull (1916) determined the amount of water taken up from soil by the seeds of the cocklebur (Xanthium), which have semipermeable seed-coats. He also determined the amount of water taken up by these seeds from a series of salt solutions of graded osmotic strengths. He argues that the osmotic strength of the solution, from which the same amount of water is absorbed as from a soil of given moisture content, indicates the force with which that soil retains water. Using a heavy silt loam with a wilting coefficient of 19·1 per cent., he obtained the results shown in Table XV.

TABLE XV

RELATION OF SOIL MOISTURE TO WATER INTAKE BY XANTHIUM SEEDS

Soil moisture per cent.	Water intake by seeds per cent.	An equal per cent. of water is taken in from—	Osmotic pressure = retaining force in soil.
Saturated	51·44	Pure water	0·0 atmos.
20·04	50·00	—	—
19·31 = wilting coefficient	49·31	—	—
18·87	47·26	0·1 molar NaCl	3·8 ,,
18·07	45·51	0·2 ,, ,,	7·6 ,,
17·93	43·23	0·3 ,, ,,	11·4 ,,
17·10	37·70	0·5 ,, ,,	19·0 ,,
14·88	28·61	1·0 ,, ,,	38·0 ,,
13·16	21·36	2·0 ,, ,,	72·0 ,,
11·79	11·94	4·0 ,, ,,	130·0 ,,
9·36	6·47	Sat. NaCl	375·0 ,,
5·83 (air dry)	0·00	Sat. LiCl	965·0 ,,

The relation for some other soils at a water content of about the wilting coefficient is given in Table XVI.

TABLE XVI

INTAKE OF WATER BY XANTHIUM SEEDS FROM VARIOUS SOILS AT WILTING COEFFICIENT

Soil.	Wilting coefficient.	Water content.	Intake by seeds.
Coarse sand	0·83	0·80	40·98
Fine sand	3·21	3·19	49·77
Sandy loam	8·33	7·86	48·38
Loam	12·93	12·66	49·02
Clay loam	16·12	16·01	49·49

Ursprung and Blum (1921*b*) observed that the suction force of the root cells became very quickly accommodated to the osmotic pressure of a solution to which they were transferred. Thus bean roots growing in sawdust had a suction force of 1·4 atmospheres; transferred to a 0·65 per cent. solution of sugar with osmotic pressure of 5·3 atmospheres, the suction force rose to 5·7 atmospheres in less than four days. These investigators consider, therefore, that the suction force of the root cells, which can be determined as already described, is a measure of the soil forces resisting the withdrawal of water, being just greater than these. They have not made many determinations, but they found, for example, that the suction force of a bean rose from 1·1 to 2·1 atmospheres as the soil dried out, the latter figure being reached when wilting set in.

Such widely different methods as those of Shull and Ursprung, give concordant results. The plant wilts when the forces retaining water in the soil are quite low, amounting only to 1 or 2 atmospheres. With appreciably less water in the soil than the wilting coefficient, Xanthium seeds withdraw nearly as much water as from saturated soil or pure water, and this from very different types of soil, coarse sand only excepted. As the moisture content falls the retaining forces increase very rapidly, until in air-dry soil they reach about 1000 atmospheres. It is, therefore, not an increase in the retaining forces which is responsible for wilting; film water is still present, and the force necessary to withdraw it is well within that at the disposal of ordinary plants. The reserve, however, cannot be large, for a further reduction of 3 or 4 per cent. enormously increases the holding forces—imbibitional forces begin to dominate.

The water films at the wilting point must therefore be very thin, and this indicates that the important factor is the *rate* of water transfer in the soil. Sachs recognised that when a root hair removes water from a particular small area, the equilibrium of surface forces is disturbed and a flow of water towards the root hair results. The flow is ready when abundant water is present, but when the films become thin the resistance offered to flow by surface tension and friction increases, and with very fine films the flow is very much impeded. The point, probably not well defined, at which this takes place has been called the *lento-capillary point*, and it must lie just above the wilting coefficient. When it is reached a sufficiently rapid absorption of water is impossible and shortly after wilting occurs. Only at water-contents well below the wilting coefficient does the plant require to exercise its full potential suction force to obtain further supplies. At still lower water-contents, after the root hairs are plasmolysed, and especially after the death of the plant, water is no longer absorbed osmotically; the imbibition forces of the colloids of the cell walls and of the protoplasm come into play. The osmotic absorption of water by normally functioning root hairs must go on much longer in such desert plants as those studied by Fitting than in ordinary mesophytes. Taking the soil used by Shull, one might expect a mesophyte to absorb water osmotically down to a moisture content of about 17 per cent., while a desert plant might do so down to 12 per cent.

Bouyoucos (1921) identifies his "unfree water" with the wilting coefficient. It seems more probable that it lies lower in the scale of soil moisture, and that the plant wilts while there is still "free water" in the soil, though Bouyoucos refers to this fraction as "physiologically very available." The capillary-absorbed fraction of the unfree water ("physiologically slightly available") which freezes below 4° C. must be held with a force of over 50 atmospheres and can hardly be absorbed osmotically. It is difficult, however, to give Bouyoucos' fractions a physiological meaning at present.

§ 5. The Absorption of Salts from the Soil

Soil Solution and Salt Supply.—The root hair absorbs not only water but also the necessary mineral nutrients. Something has already been said of the constitution of the soil solution. It is a weak solution of mineral salts momentarily in equilibrium with the soil colloids; its concentration varies primarily with the soil constitution, and also, and very markedly, with the degree of moisture; its constitution depends on the nature of the soil and, in cultivated ground, on the manuring; its reaction depends on its salt content, on the nature of the soil colloids, on the acids of the humus, and on the lime-content of the soil.

Hall, Brenchley and Underwood (1914) state that the concentration of the soil solution has an important effect on plant growth: the more concentrated the solution the better the growth. They obtained the same result in water cultures. But Stiles (1915) has shown for water cultures of barley that concentrations of the nutrient solution of 1·8, 0·36, 0·18, and 0·09 parts per thousand produced approximately equal crops if the solution were changed sufficiently frequently to maintain the supply; the lower concentrations are of the same order as that of the soil solution. The conditions in water culture and in soil are of course very different, but it is significant that in the simpler conditions the plant can thrive in an extremely dilute solution, if the amounts of the essential salts are sufficient (cf. also Brenchley, 1916, and Stiles, 1916). An exceptionally weak soil solution, however, generally indicates a soil poor in important minerals, and has, therefore, a starving effect due to actual lack of one or more of the necessary salts. The effects of such starvation may be seen especially in crop plants and weeds growing in light sandy soils; the shoots are stunted, the leaves small, and premature flowering and fruiting occur. In such conditions water supply may also play a part. Highly concentrated soil solutions are typical of saline soils of coastal regions and deserts, and of alkaline desert soils. The injurious effects of these on ordinary plants may be in part due to high osmotic pressure, but other factors—excess of toxic salts, extreme alkalinity, and disturbed water relations—are important. Gola has proposed a classification of plant communities based on the osmotic pressure of the soil solution; the account by Cavers (1914) should be consulted.

Balance of Essential Salts.—To ensure normal growth there must be an adequate supply of compounds yielding the seven essential elements: nitrogen, phosphorus, sulphur, magnesium, potassium, calcium, and iron. Silicon, chlorine, and sodium are always present both in soil and plant but are not essential, though silicon may play an important part in the formation of mechanical tissue. Some of the essential elements produce injurious effects when supplied to the plant singly, even in small concentrations. This is notably the case with the salts of magnesium, which are highly toxic. A pea seedling grown in 0·04 per cent. magnesium nitrate is injured in a few days; the leaves turn brown and wither, and the roots become slimy and cease growing. Sodium salts and ammonium salts are also highly toxic, potassium salts to a less extent. Loew (1892) showed that the toxic action of magnesium salts was entirely removed when calcium salts were added to the nutrient solution, and this antitoxic effect of calcium has since been demonstrated against salts of sodium, potassium, and ammonium. Calcium salts produce no ill effects unless supplied in relatively high concentrations. It should be noted that both the toxic and antitoxic effects are due to the metallic cation and not to the anion of the salt.

Much work has been done on this relation in recent years, especially in America. Osterhout (1906, 1907) showed the toxic effect of single salts applied to salt-water plants, such as Zostera, fresh-water plants, such as Vaucheria, and land plants in general. Solutions of a single salt have an injurious effect, and growth in these is worse than in distilled water; with pairs of salts growth is better; it is normal only when all the necessary salts are supplied in the appropriate concentrations. Osterhout describes such a solution as *physiologically balanced*; sea water is a balanced solution for marine plants.

In nature and in agricultural practice the toxic effect may be due, in some soils, to excess of sodium, magnesium, or ammonium. It may be prevented by application of calcium carbonate. The proportion of calcium to the toxic salt required to obtain antitoxic action is a matter of interest; it is in general very small. It is different for different plants, and also varies with the concentration of the nutrient (soil) solution and the stage of development of the plant (Shive, 1915).

The mechanism of the antitoxic effect is not fully understood. Loew sought an explanation in a definite combination of calcium with proteins which is destroyed by excess of magnesium. Osterhout (1922) considers that an undue increase of the permeability of the plasma membranes is caused by single salts. True (1922) supposes that insoluble calcium-pectin compounds in the cell walls of the root and root-hairs are changed into soluble potassium and magnesium compounds by excess of these elements. The relation is complex, and more than one effect may be involved.

Mineral salts may also neutralise the toxic effect of organic compounds formed as the result of decomposition processes in the soil. Schreiner and Skinner (1910, 1912) have shown that dihydroxystearic acid, which they isolated from the soil, has a toxic action which is prevented by ammonium salts, while the harmful action of cumarin is corrected by phosphates.

Lime and the Plant.—The importance of calcium, not only as an essential nutrient, but as an antitoxic agent, must be specially noted. We may here refer to one of the most discussed problems of plant biology, the question of the so-called *calcifuge* and *calcicole* species. Some plants, like the heather or ling, the sweet chestnut, the broom, the foxglove, grow only on soils poor in calcium carbonate or lime—they are calcifuge. As they therefore grow on markedly siliceous soils they have been termed *silicole.* Others, like the box, the rock-rose, the kidney vetch, and the bee orchis, occur only or chiefly on soils rich in calcium carbonate—they are calcicole. The classical case is that of two Swiss alpine species of Achillea: *Achillea moschata* is characteristic of siliceous soils, and *Achillea atrata* of calcareous soils. Each is, however, capable of growth on either type of soil in absence of the other; in competition, *Achillea moschata* completely suppresses *Achillea atrata* on siliceous soils, and *vice versa.* A similar case has recently been described by Tansley (1917) for two English bedstraws. *Galium sylvestre* is a calcicole, *Galium saxatile* a calcifuge. Each can germinate and produce mature plants on both types of soil, although *Galium saxatile* is markedly weaker on calcareous soil, and *Galium sylvestre* somewhat weaker on sandy loam. In competition, *Galium sylvestre* completely suppresses *Galium saxatile* on calcareous soil ; on sandy loam, *Galium saxatile* becomes dominant.

The effect of the presence or absence of lime in such cases may be due to a variety of causes. In the first place, calcareous soils tend to be well drained and warm. It is in all probability because of this that they are preferred by such plants as the box and the rock-rose, for these grow perfectly well on well-drained siliceous gravels in sunny situations. In the second place, the hydrogen ion concentration of calcareous soils is low, that of soils poor in lime tends to be high. Olsen (1923) has found a remarkable correlation between the occurrence of Swedish plants and the p^H value of the soil. Salisbury (1921) has found a similar correlation for some English plants. Olsen has also shown experimentally that different species react differently to the same p^H value in identical nutrient solutions. When we find plants like *Calluna vulgaris* or *Galium saxatile* showing marked signs of weakening in calcareous soils, it is very likely that we are witnessing the effects of an unsuitable hydrogen ion concentration. In the third place, excess or deficiency of calcium may produce an " unbalanced " and toxic soil solution for a particular plant. It has been shown (Schimper, Jost) that excess of calcium depresses the absorption of potassium and iron, with the result that the broom, the sweet chestnut, and the maritime pine become stunted and chlorotic, forming little chlorophyll. Watering these plants with potassium and iron salts enables them to grow on calcareous soils. Mevius (1921) states, however, that for the broom and pine the hydrogen ion relation is important. For further details the critical survey by Salisbury (1920) should be consulted.

Absorption of Salts.—The regulation of the entry of the salts into the cell is carried out by the protoplasm or its limiting plasma membranes. On the classical de Vries-Pfeffer theory of the semi-permeability of the protoplasm, the proof of non-penetration by various substances is given by the production of plasmolysis in solutions of sufficient concentration. Among the substances which produce plasmolysis are included the common salts which are important plant nutrients, such as potassium nitrate and potassium sulphate; it has been assumed that the penetration of such salts, which necessarily takes place, is possible in the very dilute solutions in which they normally occur in the soil. As potassium nitrate, for example, plasmolyses a root hair having an osmotic pressure of 10 atmospheres in a 2·9 per cent. solution, and as the concentration of this salt in the soil must be less than 0·01 per cent., it is clear that no conclusions as to permeability in the latter case can be drawn from the behaviour in the former. Fitting (1915) has recently shown by careful measurements that penetration does, in fact, take place to an appreciable extent even in high concentrations, and that, moreover, the exposure of a cell to a highly concentrated salt solution makes it less permeable. Stiles and Kidd (1919 *a* and *b*) have studied the course of absorption of various salts, using the delicate electrical conductivity method, and, besides demonstrating absorption of various salts in all the concentrations used, obtained the interesting result that the final concentration of the salt inside the cell is many times that outside if weak solutions (N/5000) are used, while it is less than that outside in strong solutions (N/10). N/5000 calcium chloride (= 0·001 per cent.) gives equilibrium with the internal concentration 15·3 times the external, while for N/10 calcium chloride (= 0·55 per cent.) the ratio is 0·24. This is important, for it shows that from the weak soil solution the plant is able to absorb relatively large amounts of salts.

The absorption of salts *as such* does not, however, take place. In the soil, as in any other solution, salt molecules are ionised, and, as the soil solution is very dilute, the degree of ionisation is great. Taking any one salt, there are present two kinds of ions and undissociated molecules. That the ions pass into the cell independently follows from the fact that the two ions of a salt are absorbed to unequal extents. This is shown by the extensive investigations of Pantanelli

(1915) on seaweeds, yeasts, aquatic plants (Azolla) and land plants (bean, lupin, chick-pea). Analysis of the solutions in which these plants were grown showed great differences between the two ions of a salt, both in the rate of absorption and in the equilibrium finally reached. Stiles and Kidd (1919*b*) have shown that chlorides and nitrates are more rapidly absorbed than sulphates, and that, of cations, potassium is absorbed most rapidly, then sodium, and then calcium and magnesium. Into the mechanism of this relation we cannot go here, though we may note that these investigators agree that adsorption by colloids is involved. It may be noted that these results are a step in the analysis of the phenomena grouped under the head of the "selective permeability" of the protoplasm. A familiar example of this was given by the known facts that the constitution of the ash of a plant does not agree closely with that of the minerals in the water or soil in which it grows, and that many plants are able to accumulate large quantities of certain elements, the richness of the kelps in iodine and of the tobacco in nitrates being instances.

Solvent Action of Roots.—We may here refer to the action of the plant in making minerals available for absorption. In a classical experiment Sachs showed that if seedlings are grown with the roots touching a sloping plate of marble, the course of the root branches was revealed as an etching on the polished surface. The inference was that an excretion of the roots was responsible for dissolving the marble. This action might be due to a solution of carbon dioxide in water, and the carbon dioxide respired by the roots would be sufficient to explain it. Using a mixture of calcined gypsum and various other insoluble minerals, poured on a glass plate to obtain a polished surface, Czapek found that only those minerals which were soluble in water containing carbon dioxide in solution showed etching, *i.e.* carbonates, and phosphates of calcium, magnesium, and iron. Aluminium phosphate, which is insoluble in carbon dioxide water, was not etched.

Many investigators have endeavoured to show that other

acids are secreted by roots, but the proof is in most cases insufficient. Goebel (1893) showed that roots of cress and barley secrete formic acid, but this is stated by Czapek to be in the form of an alkaline salt; Schulow (1913), who has demonstrated the secretion of malic acid by peas and maize, is doubtful whether it is in the form of a free acid or of a salt. Recently Haas (1916) has shown by measurements of hydrogen ion concentration that carbon dioxide accounts for the increase of acidity produced by roots of wheat and maize. A great difficulty in the way of exact demonstration is that of raising plants in strictly sterile conditions. There is no doubt that in many cases the acids found have been bacterial products; of course, in nature, such acids may have an important action which is not, however, to the credit of the higher plant. This may help to explain such facts as the utilisation of aluminium phosphate, which is insoluble in carbon dioxide water.

But here another possibility may be noted—that given by the differential absorption of the two ions of a salt. If Pantanelli shows, for example, that, from a solution of potassium sulphate, the chick-pea absorbs 1·82 mg. of potassium and only 0·51 mg. of the sulphate radical, this means that a considerable surplus of free sulphuric acid is left in the soil and will exercise the solvent action of a strong acid. On the other hand the acid radical is more extensively absorbed from some salts, such as calcium nitrate, and this results in an increased alkalinity of the medium. Redfern (1922) has shown that in very dilute solutions of calcium chloride the two ions are absorbed almost equally, though in stronger solutions the calcium is absorbed more readily. Jones and Shive (1922) find that in a complete nutrient solution the acidity tends to diminish under the action of roots, except when ammonium salts are present. We have already referred to other sources of acids in the soil.

Submerged Plants.—The root systems of submerged plants have been described, and it has been found that they are active in absorbing water. That this should be a primarily important function is scarcely credible. The water current from root to leaf in a water plant might be thought a functional relic of a remote land ancestor. In fact, it serves to promote the supply of salts. Snell (1908) has shown that plants of Elodea and Potamogeton allowed to root in the mud grow much more vigorously than plants in the same vessel supported away from the substratum. In one experiment shoots of Elodea showed an average increase in length of 9·85 cm. when rooted in mud, of 4·4 cm. when free, and of 2·6 cm. when rooted in sand. Similar results were obtained by Pond (1905) and by Brown (1913), who, however, interprets them as due to an increase in the supply of carbon dioxide (see pp. 110, 201). The mineral content of fresh water is generally very low, and the advantage of rooting in the rich mud is evident. Further, it has recently been shown by Mayr (1915) that in some submerged leaves, *e.g.* of *Alisma plantago*, only certain cells or cell groups, called *hydropotes*, are easily permeable by salts and water.

§ 6. Exceptional Means of absorbing Water and Salts

The soil is the normal habitat of the land plant and the root is the normal organ of absorption. The great majority of plants make good their requirements of water and salts by this means ; they are efficient, as we have seen, not only in ordinary cases but in extreme conditions ; the submerged water plant thrives best rooted in the mud, and the plants of the most arid deserts draw an exiguous but sufficient supply from an almost dust-dry soil. There are many plants, however, which depend on specialised or peculiar organs, drawing water from rain or even dew and fog, or from reservoirs of their own. These plants are the exception ; but in certain stations they are prevalent, and the peculiarities of their water supply have always excited interest.

Epiphytes.—The biological class which is most conspicuous in this respect is that of the epiphytes—plants living attached to other plants, most often clinging to, or rooting in, the bark of trees. In the tropical rain forests epiphytes reach their highest degree of luxuriance, the stems, branches, and even leaves of the trees being covered by a host of species and individuals. Outside the tropical rain forest true epiphytic communities are found in the temperate forests of Chili and New Zealand. The less highly specialised epiphytes may grow in crevices of the bark, in clefts of the trunk, in the crooks of branches, where accumulations of dust, humus, and other debris furnish them with a limited amount of soil. But the most characteristic cling to their hosts by tendrils or by roots, these frequently forming an extensive network round trunk and branches. Sometimes, as in some epiphytic species of Ficus, special roots descend to and penetrate the soil; many roots are dependent on such supplies of water as they can get on the trees; sometimes they hang freely in bunches or singly in the air, wholly dependent on atmospheric precipitation. These subaerial roots have in general lost their geotropism. They are often strongly haptotropic, negatively phototropic, and hydrotropic. They are usually unbranched, though if they enter the soil they may branch there. The growing zone is often of considerable length.

Roots with Velamen.—In many epiphytic orchids and aroids the roots have a special absorbing tissue lying outside the cortex and termed the *velamen*. This is morphologically equivalent to the epidermis, but it is usually several cells thick. The cells are dead, thin walled, and empty, and are prevented from collapsing by fine spiral or netted thickenings. They are open to each other and to the outside by pores. Within this velamen lies the living cortex, the cells of which frequently possess chloroplasts; its external layer is specialised as an *exodermis*, most of the cells having thickened and suberised walls impermeable to water. Here and there a *transfusion cell* of the exodermis remains thin walled and serves to pass water inwards. When dry, the velamen has a silvery, parchment-like appearance. On wetting, it absorbs water greedily and rapidly, the pores letting the air escape. The amount of water taken up is very considerable—according to Goebel's measurements (1893), from 40 to 80 per cent. of the weight of the root. A distinction between absorption by a velamen and by an earth root is that in the former case the whole surface of the root is active, and in the latter only the restricted root-hair zone. The wet root appears green, the chlorophyll of the cortex being visible through the wet velamen. The water passes more slowly into the cortex and conducting tissues of the roots. When the velamen once more dries up it acts as a covering, protecting the root from drying out to a very appreciable extent. The supply of salts must be rather precarious, depending on the detritus washed down by rain and collecting about the roots. Velamina seem to be possessed by all epiphytic orchids, *e.g. Epidendron nocturnum, Vanda furva, Odontoglossum Barkeri*. They are less general among the Aroids, *e.g. Anthurium egregium* and *A. acaule*.

The velamen is not only common among epiphytic orchids and aroids, it is also widespread among terrestrial forms such as Sobralia and Phajus, though generally less well developed. As there is no sharp line between terrestrial and epiphytic forms, this is not surprising. A velamen has, however, been found in a number of monocotyledons which are strictly terrestrial, by Goebel (1922). These belong to the Liliaceæ, *e.g. Agapanthus umbellatus, Aspidistra elatior*, and Amaryllidaceæ, *e.g. Crinum longifolium, Clivia nobilis*. The occurrence of a well-developed velamen on the roots of these terrestrial South African plants, which have certainly no epiphytic tendencies, is of great interest. These roots produce also root hairs, and it is to be supposed that the velamen functions chiefly in producing a rapid increment in the water available for growth when the rains set in, and perhaps a certain reserve in dry weather.

The occurrence of the velamen in earth roots of orchids and aroids might be taken to be a relic in species descended from epiphytic ancestors; that it is found in roots of typical geophytes can only indicate that it is a structure which may have been evolved in the earth, and that epiphytism has taken advantage of it.

Collection of Humus.—In a number of epiphytes humus is collected by the plant and forms a substratum into which the roots grow. A good example of one of these "nest epiphytes," as Schimper terms them, is the aroid *Anthurium ellipticum*, described by Goebel. A mass of adventitious roots arises near the base of the stem, growing upwards and outwards and forming a massive felt in which humus collects. The West Indian orchid, *Oncidium altissimum*, has a basket-like network of roots as big as a man's head, in which the rubbish falling from the tree-tops collects. In the Javan orchid *Grammatophyllum speciosum*, the root-work surrounds the host tree and may be a couple of yards thick; *Anthurium Hügelii*, a West Indian aroid, collects rain and humus in a funnel formed by its leaves, and into this the roots grow. Here may be mentioned the plants restricted to the humus nests formed by some ants in trees of the forests of tropical South America.

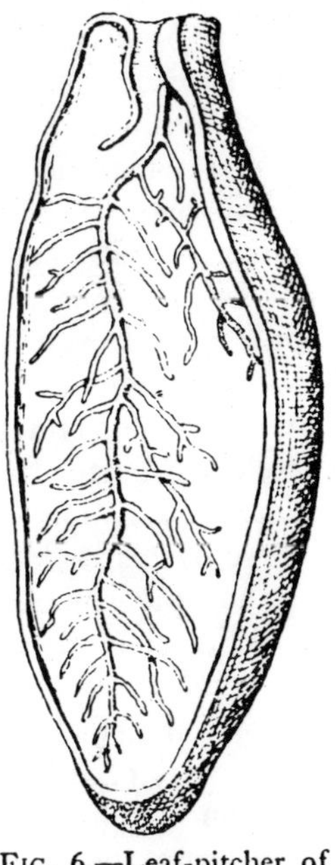

FIG. 6.—Leaf-pitcher of *Dischidia Rafflesiana*; cut open to show the root system which has grown into it. $\frac{2}{3}$ nat. size. (After Treub.)

Yet more remarkable is the arrangement in *Dischidia Rafflesiana*, an epiphytic asclepiad of Java (Fig. 6). The fleshy leaves arise in pairs, and one develops as a pitcher completely closed except for a narrow opening above. Into this opening grows an adventitious root arising beside the leaf; inside it branches vigorously and lines the wall with a network of rootlets. Water and fine organic material find their way into the pitcher and supply these roots. In another asclepiad, *Conchophyllum imbricatum*, the leaves are fleshy, and only slightly hollowed, and are pressed close to the bark of the tree. Adventitious roots grow out under them and are at least kept moist by the sheltering leaves.

Absorptive Hairs.—In all cases dealt with so far the absorption has been carried out by the roots. We may now turn to plants in which the leaves take up water. The most specialised cases of this are found in the Bromeliaceæ, a tropical American family of which the pine-apple is the most familiar example. In *Tillandsia regina* and *Tillandsia unca*, which live in the soil, especially in rocky places, and *Tillandsia bulbosa* which is epiphytic, the sheathing bases of the sword-shaped leaves form a funnel in which a considerable amount of water collects, along with dust, humus, and a whole diversified fauna and flora of algæ, frogs and their larvæ, spiders, and even little snakes. "Like watertight tanks," writes Schimper, "they collect rain-water of which a full litre may descend from one of the larger forms on to a careless collector." This water is used by the leaves, the actual absorption taking place through specialised hairs. These have a stalk of two or three cells sunk in a depression in the epiderm, and an expanded umbrella-like plate or shield of cells for a head. The marginal cells of this head are often expanded so as to form a wing. The outer walls of the shield are frequently much thickened but have no cuticle, or only a very thin one. The shield cells are dead and empty; those of the stalk living. If not supplied with water the hair collapses and closes the depression in which the stalk arises, the empty cells of the shield preventing excessive evaporation. When water is available it is readily sucked in by the shield cells which expand; as the stalk cells absorb water they become turgid and swell, so that the shield rises slightly over the depression, and the supply of water to the stalk cells and so to the mesophyll of the leaf is increased. The inner (upper) side of the basal region of the leaves is thickly beset with these absorbing trichomes, and the water requirements of the plant are fully covered. The poorly developed root system serves only for anchoring; it cannot alone supply the plant's needs. The pine-apple has less highly specialised hairs. Water-absorbing hairs also line similar leaf cisterns in the Javan epiphytic orchid *Eria ornata* (Raciborski, 1898).

Another species, *Tillandsia usneoides*, has a very different habit : it hangs in long shaggy festoons from the branches of trees. " This most remarkable of all epiphytes, often completely covering the trees in tropical and subtropical America, consists of shoots often far more than a metre in length, thin as thread, and with narrow grass-like leaves, and only in early youth fixed to the surface of the supporting plant by weak roots that soon dry up. The plants of Tillandsia owe their attachment to the fact that the basal parts of their axes twine round the twigs of the host "

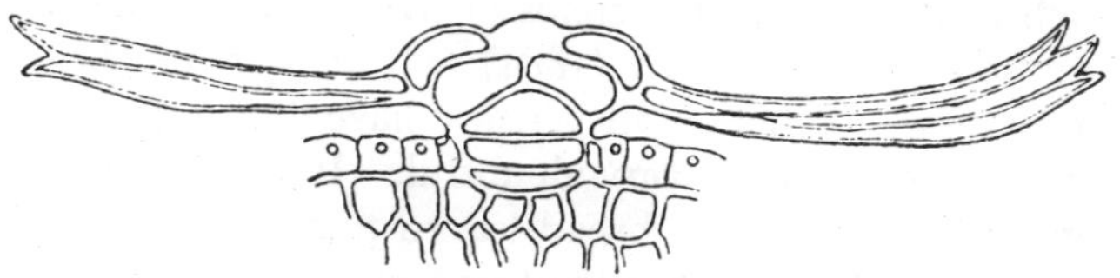

FIG. 7.—Section through water-absorbing hair of *Tillandsia usneoides*. × 280. (After Aso.)

(Schimper). The whole plant is covered with shield hairs similar to those described, but not sunk below the epiderm (Fig. 7). When dry the plant is silvery, like the velamen-covered root of an orchid. Goebel remarks that in South America Tillandsia is often used as a balcony decoration ; the shoots are simply tied on to the rails, where the flourish of blossom is a sufficient proof that such root system as is present can be dispensed with so far as water absorption goes.

It has been asserted by Mez (1904) that certain forms of Tillandsias of the *T. usneoides* type draw their supplies less from rain than from dew. There seems no reason to doubt that dew falling on such a plant may be utilised. Aso (1910) has shown that the hairs of *T. usneoides* are capable of absorbing salts, while those of the related *Ananas sativa*, the pine-apple, cannot do so.

Volkens (1887) assigns an important rôle to dew in the

Egyptian desert, and describes simple absorbing hairs from species of Diplotaxis and Heliotropium. Haberlandt has found absorbing hairs on other plants, *e.g. Centaurea argentea*. Volkens also supposes that the saline incrustations of the glands of various desert plants condense water at night, and that this water is utilised by the plant. The salts certainly deliquesce at night, but, as the plant would require to develop an osmotic pressure of about 400 atmospheres to withdraw the water, and as such pressures are quite unknown even in desert plants, it does not seem likely that this can be a source of water supply.

The hairs which grow in little bunches in the angles of the veins of many European trees, and such hairs as those which form a line on the stems of the chickweed, have been supposed by Lundstrom (1884) to function in absorbing rain or dew. Whether they really have any importance in this respect must be very doubtful. It is true, however, that the leaves of many broad-leaved trees do take up water quite readily. If the leaves of one branch of a forked twig of beech be immersed in water, those of the exposed branch remain turgid for some days; water is therefore absorbed and transported in sufficient quantity to replace that lost by transpiration. It is possible that this faculty of absorption may be of some importance to a plant in enabling it to restore its turgor rapidly in rain, or dew, after a drought. That any plants can actually condense and utilise atmospheric moisture seems doubtful (cp. Wille, 1884).

Summing up, we may say that rain and dew are of extreme importance to many epiphytes which possess specialised absorbing organs; and that rain and dew may occasionally be utilised by ordinary plants.

The Bladderworts.—Finally, we may mention the case of the terrestrial and epiphytic bladderworts. Characteristically these insectivores have a submerged vegetative system. Glück (1905) has described land forms of most of the common species, and Goebel (1893) has fully investigated several American species which habitually live in moss and epiphytically. Thus *Utricularia nelumbifolia* and *U.*

Humboldtii are epiphytes, but only in a restricted sense, for they live in the water cisterns of a giant bromeliad in Guiana. They can be cultivated in damp moss. Several species, *e.g. U. reniformis* and the Singalese *U. bifida*, inhabit moss; *U. Jamesoniana* creeps on the bark of trees. The Utricularias possess no roots, but they send through the moss vigorous runners which bear bladders, and which serve as water-absorbing organs. In *U. bifida*, short stout "rhizoids," probably of leaf nature, are formed; these do not seem to play any important absorbing rôle. The related terrestrial genus Genlisea absorbs water by subterranean bladder-bearing organs of leafy nature. The morphology of the vegetative organs of the Utriculariaceæ will be referred to in another connection.

CHAPTER II

ASSIMILATION AND TRANSPIRATION

§ 1. Significance of Assimilation and Transpiration

Assimilation.—The possession of that complex of four pigments which we call chlorophyll is the fundamental attribute of the plant kingdom. From the unicellular alga to the flowering plant chlorophyll is found without essential change. Such exceptions as the copper beech are apparent only, the green pigment being present, though masked by red sap anthocyans. In the two great groups of red and brown algæ, chlorophyll is associated in the chloroplasts (special organs of the protoplasm) with other pigments which modify the colour. In all the flowering plants so far investigated even the relative amounts of the four constituents are nearly constant.

Chlorophyll absorbs light and, in some way, enables a fraction of the energy thus obtained to be utilised in reducing carbon dioxide in solution. There results a simple compound, the first stage in the building up of organic matter; what this first product is we do not certainly know, though the evidence suggests that the classical hypothesis of Bayer, which identifies it with formaldehyde, is correct. Whatever its nature this first product never accumulates, but changes immediately into higher molecular compounds, and gives rise to sugars (glucose, fructose, and saccharose) and to starch. The exact sequence of the formation of these has not been finally determined. It is possible that the synthesis of the nitrogenous organic compounds, culminating in the proteins, also starts with reactions between the earlier, transient products of photosynthesis and the nitrates which are carried to the leaf cells in the transpiration current. The oxygen liberated by the reduction of the carbon dioxide leaves the plant as free gas.

The whole process is referred to as *carbon* (or better, *carbon dioxide*) *assimilation*, or as *photosynthesis*. The former term regards the raw material, the latter the end products, but both are in common use for the whole process.

The importance of photosynthesis for the plant, and for life in general, needs no emphasis. There are colourless organisms—among the bacteria—which carry on the reduction of carbon dioxide in absence of light, using chemical energy, but their effect is relatively negligible. The organic matter of which the green plant is formed, and from the oxidation of which it derives energy, the organic matter which parasitic plants obtain from green ones, and all the organic food of the animal kingdom is synthesised in the first place from carbon dioxide and water through the action of light on chlorophyll. The major part of organic matter available on the land surface is the product of the flowering plants. Ruskin's " vast family of plants, which, under rain, make the earth green for man, and under sunshine give him bread," must be extended from the grass to the plant in general, and from man to the animal kingdom. Nor must a secondary effect almost as important be overlooked—the elimination from the atmosphere of the carbon dioxide constantly poured into it, naturally by the respiration of living things, and, one might say, artificially, by combustion ; and its replacement by oxygen. The warring processes, as is well known, keep the amount of carbon dioxide in the atmosphere roughly constant at about 3 parts in 10,000. It was in searching for the process responsible for keeping air "good" that the great chemist, Joseph Priestley, made the first discovery of this activity of the green plant in 1771.

Essential for photosynthesis are the absorption of light by the chlorophyll and a sufficient supply of carbon dioxide and water ; the process goes on vigorously only within the narrow range of temperature in which plant life is active. The absorption of light is, of course, best carried out by an expanded surface of chlorophyll. In the most primitive form we see such a surface in the green dust of Protococcus on the bark of a beech tree, or in the thin thallus of a sea lettuce. In more highly organised plants, the evolution of a more or less massive body has gone hand-in-hand with the evolution of special light-absorbing and assimilating organs —the leaves—having a greatly expanded surface capable of forming organic substance in excess of the needs of the actual assimilating cells, and so supplying those parts of the plant which, with other structure and other functions, possess no chlorophyll.

The origin of the leaf is still obscure. It is regarded by some as a specialised appendage or outgrowth of the stem ; others look on it as the primary organ from which, in turn, the stem has been derived ; or it may be that stem and leaf have both been derived by differentiation of a primitive thalloid plant body, such as we see in the brown seaweeds. It is possible that in different groups of plants the origin of the leaf has been diverse. This question need not further concern us here. The leaf is now a highly specialised, yet plastic, organ, the fundamental character of which, in its normal form, is that it spreads a large area of chlorophyll to the light. It is instructive to stand under a well-grown beech and observe the architecture which brings its multitude of leaves to the light, spread over the tips of branches and twigs ; or under a maple with its tiered branches, between which the light strikes to leaves borne far in towards the trunk. There are scarcely any gaps in these leaf screens, and yet little mutual interference.

Of the materials of photosynthesis the water is of course supplied from the soil, but the carbon dioxide comes from the small quantity present in the atmosphere. This is readily demonstrated by simple experiments. Sachs used the formation of starch, which takes place in the leaves of most dicotyledons during photosynthesis, and the presence of which in small quantities is easily shown by iodine, to follow the effect of various conditions. A plant kept in the dark for a day or two loses the starch from its leaves, by conversion into sugars and subsequent removal. Such a plant exposed to light under normal conditions, forms, in the course of a few minutes, sufficient starch to be demonstrated ; but if it be exposed to light in an atmosphere deprived of carbon dioxide, then no starch formation takes place.

Transpiration.—It is essential, therefore, that the chlorophyll cells should be freely supplied with air containing carbon dioxide ; but this, of course, means that gases in general must be able to diffuse in and out of the leaf. Thus oxygen diffuses into the leaf in the dark, supplying the needs of respiration ; by day the excess produced by the more active photosynthesis diffuses out. Much more important consequences for the structure and behaviour of the plant follow from the diffusion outwards of water vapour. The leaf is a water-saturated organ ; it is freely exposed to an atmosphere which has normally a marked saturation deficit of water vapour ; air currents accentuate the deficit by tending to prevent increase of moisture in the neighbourhood of the leaf. Because the leaf is exposed to light, the evaporation from it is increased with rise of temperature in the sun. The consequence, then, of the free exposure of a large surface, which must be an evaporating surface because it is permeable to carbon dioxide and oxygen, is that normal foliage gives off very large quantities of water vapour—it *transpires*; in fact, as we have seen, water lost in transpiration makes up the larger part of the water taken in by the roots.

Photosynthesis and transpiration take place in the same organ. Essential processes of each are subject to the laws of gaseous diffusion. They go on concurrently; yet they may be interfering processes, since excessive transpiration may limit assimilation, both directly by diminishing turgor, and because of the peculiarities of structure shown by plants in arid stations. The two processes are linked together, and for that reason, although their rôles are utterly different, we may best treat them together.

The function of photosynthesis is well understood, but the same cannot be said of transpiration. Two rôles have been assigned to it: (1) It undoubtedly tends to reduce the temperature of a leaf exposed to the sun. The Italian peasant puts his drinking-water in a porous earthenware jar in the sun, if he wishes it cool; the rapid evaporation, using up energy in converting liquid water into vapour, lowers the temperature of the water. The same thing happens with the leaf. This may be important in plants exposed to strong solar radiation, when temperatures so high as to be dangerous may be reached.

(2) Transpiration entails a constant water current from root to leaf through the wood of the plant; this is called the *transpiration current*. It has been supposed that this current is responsible for maintaining a sufficiently rapid supply of the mineral salts, absorbed in small quantities by the roots. It is difficult to get exact evidence for this view, and some of the evidence available is not favourable, as we shall see later. Some American investigators, *e.g.* Barnes (1910), believe that transpiration is an unavoidable evil, without benefit to the plant. Yet it is difficult to imagine an adequate supply of salts being delivered to the leaves of

a forest tree by diffusion alone without the aid of the transpiration stream. We shall return to this subject at the end of this chapter when we have considered the available evidence. Meanwhile, we may say that, though it is certainly unavoidable that the plant should lose large quantities of water through transpiration, it is likely that the process—inevitable in an organism with the structure and function of a flowering plant—has its uses. It may reduce danger of overheating. It may promote the supply of salts to an extensive shoot system. It also keeps up the supply of growth water, and it must be doubted whether this supply could be maintained in an organism of the size of a tree without the special conducting system which primarily meets the necessities of transpiration.

§ 2. The Leaf

No other organ of the plant has a wider variety of form than the leaf; yet we recognise as typical or normal the possession of the thin, expanded, light-absorbing blade or *lamina*, and the slender stalk or ***petiole*** which gives the possibilities of accurate adjustment of position, and of escape by bending from mechanical injury. Through the stalk run the vascular bundles, entering the stem as leaf traces to link up with the general conducting system, entering the leaf and appearing there as the midrib, if one is present, giving off, or breaking up into a network of veins. The larger veins are complete vascular bundles; in the finer there may be only a few elements, or, finally, only one left; the water supply is distributed in the most thorough fashion.

The veins run through a soft tissue termed the *mesophyll*. Towards the upper surface one or several layers of cells are arranged regularly with their long axes perpendicular to the surface of the leaf, rather close together. In a cross-section of the leaf, the columnar appearance of this part of the mesophyll earns it the name of ***palisade parenchyma***. Towards the lower surface the mesophyll consists of irregular cells between which are large air spaces—the *spongy parenchyma* (Fig. 14). The combined volume of the air spaces—the internal atmosphere of the leaf—may be as high as 77 per cent. of the total volume of the leaf, or as low as 3·5 per cent.: commonly, it is about 25 per cent. Over all stretches the epiderm of shallow cells, fitting together to form a continuous covering, their outer walls waterproofed to a greater or less extent by an impregnation of cutin, and an exterior delicate cuticle of the same waxy nature.

The Stomata.—In the developing epiderm some cells undergo special, regular divisions ending in the production of two equal sausage-shaped cells lying side by side, with a slit between, due to the dissolution of the middle lamella of the dividing wall. The slit provides an opening between the atmosphere and the intercellular spaces of the leaf. The whole apparatus is the *stoma*, the slit is the *pore*, the two cells the *guard-cells* (Fig. 10). Through the stomata the main interchange of gases takes place, relatively little going on through the cuticle and wall of the epiderm. The stomata are typically most abundant on the lower leaf surface, and here they lead into the extensive air spaces of the spongy parenchyma. From these diffusion goes on to the palisade parenchyma in which assimilation is most active, partly owing to the more abundant chloroplasts, partly owing to the arrangement of the cells, and their more favourable position for absorbing light. Water passes into the leaf and is distributed along the network of veins, finally diffusing a short distance through a few living cells. The products of photosynthesis diffuse through living cells towards the veins, and so to the general conducting system of petiole and stem.

§ 3. Relation of the Stomata to Gaseous Exchange

In the intercellular spaces of the leaf the air differs from that outside. It has no constant composition, but is always tending towards equilibrium with the atmosphere on the one hand, and with the tension of gas in solution in the water saturated cell walls on the other. It is saturated, or nearly so, with water vapour; in light it is relatively rich in oxygen, in the dark in carbon dioxide. By day there is diffusion of water vapour out through the stomata, and constant evaporation from the cell walls results; the oxygen which, liberated by the reduction of carbon dioxide, saturates the cell water, passes into the air spaces and so outwards; the carbon dioxide in solution is continually used up, and fresh supplies pass into solution from the intercellular air, and diffuse from the atmosphere in through the stomata. At night as assimilation ceases the conditions affecting oxygen and carbon dioxide are reversed; frequently the evaporation of water practically ceases as the atmosphere becomes saturated with falling temperature, and as the stomata close. This continual drift of gases takes place primarily by diffusion and not by any sort of pumping action on the part of the leaf, though possibly mechanical bending by air currents has a certain accelerating effect by alternately increasing and decreasing the volume of the internal air space. Neger (1918) has shown that in many cases diffusion is free throughout the internal air spaces of a leaf, *e.g.* in the holly, ivy, and spindle tree; in most leaves, however, intercellular spaces are not continuous and the leaf is divided into airtight compartments, usually bounded by the major veins, each of which is independent as regards gas exchange: such are the leaves of sweet chestnut, oak, beech, and elm.

Diffusion is governed by physical laws. The molecules of a gas are in continual motion, and in a gas mixture where the different components are unequally distributed, the molecules of each gas drift from the higher towards the lower concentration, the rate of drift depending on the difference of concentration—the steepness of the *diffusion gradient*—on the nature of the gas, and on the temperature. Between the intercellular spaces and the atmosphere, gases can pass by two ways—through the cuticle, and through the stomata. Since the cuticle is more or less impervious and since stomata are actual free openings, diffusion through these is most important. This conclusion is borne out by experiment for the exchange of carbon dioxide, and transpiration of water vapour. The details of this relation are most conveniently investigated by comparing the rate of gas exchange through the upper and lower epiderms of a leaf with different numbers of stomata on the two surfaces, and comparing the results with the relative number of stomata. A striking rough demonstration for the case of water vapour is given by making use of the change of colour of filter paper soaked in a solution of cobalt chloride. When thoroughly dry this salt is a brilliant blue; exposed to air it absorbs water rapidly and changes to pale pink. If a leaf of cherry laurel or ivy be placed between two pieces of this cobalt chloride paper, previously dried to the blue colour, and the whole be protected by two sheets of glass, then in the course of a few seconds the paper touching the lower side of the leaf changes to pink; on the upper side the change takes many minutes. This corresponds to the presence of stomata on the lower surface only. Exact measurements may be made by absorbing the aqueous vapour given off from each surface by a hygroscopic salt (calcium chloride is convenient) and determining the amount by weighing. Using this method, Unger (1862) (quoted from Burgerstein, 1904) obtained results shown in Table XVII.

TABLE XVII

RELATION OF TRANSPIRATION TO STOMATAL NUMBERS

Plant.	No. of Stomata per sq. mm.		Ratio of Stomata.	Ratio of Transpiration.
	above.	below.	above : below.	above : below.
Fuchsia fulgens ..	0	200	— : —	1 : 8·0
Aucuba japonica ..	0	145	— : —	1 : 40·0
Nicotiana Tabacum ..	100	207	1 : 2·0	1 : 4·3
Helianthus annuus ..	207	250	1 : 1·2	1 : 1·25

From this it is clear that the stomata are most important as regards transpiration, but that transpiration does also take place through the cuticle. In Fuchsia and Aucuba water vapour is given off by the upper surface on which are no stomata. The *cuticular transpiration* from the upper surface is one-eighth of the combined cuticular and *stomatal transpiration* of the lower surface in Fuchsia, while in Aucuba it is only one-fortieth ; this is evidence of the effect of the cuticle, which is thick in the latter, and thin in the former. In the sunflower the transpiration ratio is very close to the stomatal ; in the tobacco the two are different. The latter is the common case and indicates that factors such as the relative size and state of the two sets of stomata, the relative thickness of the cuticle above and below, and the different conditions of light, etc., at the two surfaces, must enter into the result. Renner (1910) has estimated the ratio between the total stomatal and total cuticular transpiration for a number of leaves. Some of his results are given in Table XVIII.

TABLE XVIII

RELATION OF CUTICULAR AND STOMATAL TRANSPIRATION

Plant.	Ratio.		
	Cuticular Transpiration	:	Stomatal Transpiration.
Nuphar luteum	1	:	3·2 (9·9)
Hydrangea hortensis	1	:	7·1 (15·0)
Archangelica officinalis	1	:	3·4 (20·5)
Callisia repens	1	:	3·9
Tradescantia viridis	1	:	3·5
Rhododendron hybridum (hort.) ..	Cuticle thick, cuticular transpiration negligible.		

These figures must not be taken as constants for the various leaves, since in different experiments they vary considerably, and there is a marked difference between the ratios found in still air and in wind. The ratios in wind for three plants are given in brackets.

There is, therefore, no constant relation between the

two; but we may draw the definite conclusion that stomatal transpiration is predominant, and that this predominance is much accentuated in moving air—the condition in which the plant most often transpires. A different result, obtained by F. Shreve (1914*a*) for *Pilea nigrescens*, a herb of the Jamaican rain forest, must be noted. He found that the cuticular transpiration was 30 per cent. *greater* than the stomatal. This may be characteristic of rain-forest plants, living in a very moist atmosphere, and having leaves with an extremely thin cuticle.

The path of diffusion of carbon dioxide in respiration and assimilation has been investigated by Blackman (1895), and by Browne and Escombe (1900). Table XIX gives some of Blackman's results for the intake of carbon dioxide in assimilation and escape of carbon dioxide in respiration.

TABLE XIX

PATH OF GASEOUS EXCHANGE IN RESPIRATION AND ASSIMILATION

Plant.	Ratio.		
	Stomata. upper surface / lower surface	Diffusion out of CO_2 in respiration. upper / lower	Diffusion in of CO_2 in assimilation. upper / lower
Nerium oleander	0/100	3/100	—
Prunus laurocerasus	0/100	3/100	—
Hedera Helix	0/100	4/100	—
Platanus occidentalis	0/100	2/100	0/100
Ampelopsis hederacea	0/100	3/100	0/100
Polygonum sacchalinense	0/100	6/100	0/100
Alisma plantago	135/100	121/100	145/100
Iris germanica	100/100	107/100	—
Ricinus communis	39/100	40/100	—
Populus nigra	18/100	26/100	—
Helianthus tuberosus	41/100	36/100	—
Tropæolum majus	50/100	37/100	—

There is a close relation here between the passage of carbon dioxide and the number of stomata. It is clearest in the plants which have no stomata on the upper surface; in these there is practically no diffusion of gas through that surface. It may be noted that the thin cuticle of the leaves of Platanus, Ampelopsis, and Polygonum is as effective in stopping diffusion of carbon dioxide as is the thick cuticle of Hedera, Nerium, and *Prunus laurocerasus.* Where stomata are present on both surfaces the relation is not quite so close, but here we must take into account the different conditions of the stomata on the two surfaces.

Browne and Escombe found that in assimilation the carbon dioxide absorbed by the upper surface, when stomata are present there, was always considerably greater than the stomatal ratio would account for. Thus for *Rumex alpinus*, with a stomatal ratio of 37/100, the gas exchange ratio was 73/100. They believe that this is due not to excessive diffusion through the cuticle but to wider opening of the stomata on the more strongly illuminated upper surface, and perhaps to more rapid utilisation of the carbon dioxide by the palisade tissue. As regards respiration their results confirm those of Blackman. The diffusion of oxygen has not been measured directly; it may be taken to agree with that of carbon dioxide, the determination of which in gaseous mixtures is much more convenient to carry out.

We may conclude that the stomata are the main path of the gaseous exchanges of the leaf. As regards carbon dioxide and oxygen, the cuticle offers an almost complete barrier to diffusion; the stomata are also the chief exits for the water vapour, though appreciable amounts pass through the cuticle.

In submerged water plants, with a fine cuticle, diffusion of carbon dioxide and oxygen in solution takes place directly through the saturated epidermal walls over the whole plant, except, perhaps, where the leaves have the special permeable hydropotes described by Mayr (1915).

§ 4. Distribution and Dimensions of Stomata

It is a well-known fact that stomatal numbers vary greatly in different plants; that their relative numbers are usually different on the two surfaces of the same leaf is clear from examples already quoted. In the ordinary dorsiventral leaf there are generally more stomata on the lower surface; in many cases, indeed, especially in hard and leathery leaves, they are completely absent from the upper surface. In erect leaves, as in grasses, and in succulent or fleshy leaves the numbers tend to be more equal. In the floating leaves of water plants they occur on the upper surface alone. The general tendency is for the stomata to occur on the surface next the best developed aerating system; this is normally the surface best protected from the extreme drying influence of the sun. In the exceptional case of the floating leaf stomata can function on the upper surface only. The stomata are not evenly distributed over the leaf surface; they are in general more frequent along the larger veins, and their density often varies from tip to base. Table XX gives some idea of the range met with.

TABLE XX

DISTRIBUTION AND NUMBER OF STOMATA

Plant.	Stomata per sq. mm.	
	Upper surface.	Lower surface.
Nymphæa alba	460	0
Quercus Robur	0	346
Pirus Malus	0	246
Gentiana lutea	61	123
Mimosa pudica	187	308
Triticum sativum	47	32
Sempervivum tectorum ..	11	4

De Bary gives a compilation from an extensive investigation by Weiss, which shows that, of 157 land plants examined, 12 species had fewer than 40, 42 species had 40–100, 38 species 100–200, 39 species 200–300, 12 species 300–400, 4 species over 400 stomata per sq. mm.: 675 for the olive and 716 for the turnip are the largest numbers. A moderately large leaf with an average density of stomata may possess several millions.

The actual opening is more or less oblong or elliptical in surface view. There is, again, a great range in the area of the pore of the fully opened stoma in different species. Measurements of the pore made by Renner (1910) for a number of plants are summarised in Table XXI.

TABLE XXI

EXAMPLES OF THE SIZE OF THE STOMATAL PORE

Plant.	Pore.			
	Length.	Breadth.	Area.	Depth.
	mm.	mm.	sq. mm.	mm.
Hydrangea hortensis	0·01	0·003	0·0000236	0·013
Nuphar luteum	0·0095	0·003	0·000022	0·0154
Aconitum lycoctonum	0·0129	0·0063	0·000063	0·013
Callisia repens	0·0275	0·0088	0·00019	0·0176
Rhododendron hybridum (hort.)	0·0055	0·001	0·0000044	0·013

Taking the sunflower as a typical case, we have 330 stomata per sq. mm., each with an area when full open of 0·0000908 sq. mm. The combined stomatal area per square millimetre is therefore 330 × 0·0000908, or 0·03 sq. mm., or one thirty-third of the total area.

§ 5. DIFFUSION THROUGH THE STOMATA

The proper functioning of the leaf requires a restricted loss of water vapour and a sufficiently free passage of carbon dioxide and oxygen. It seems obvious that restriction should be the result of the practical limitation of gaseous diffusion to the stomata, which, though so numerous, are so small, and occupy so small a fraction of the total leaf surface. It might well be thought that the access of carbon dioxide would be unduly impeded by these means, and that, in the case of the sunflower, for example, the flow of gas would be only one thirty-third of what might take place into a free absorbing surface of the same area as the leaf. This is not so. It has been shown by Browne and Escombe (1905*b*) that a *perfectly absorbing* surface of sodium hydroxide can absorb from ordinary moving air about 0·128 c.c. carbon dioxide per square centimetre per hour. These investigators also show that under ordinary favourable conditions a sunflower leaf absorbs from 0·029 to 0·044 c.c. carbon dioxide per square centimetre per hour. Other investigators have found higher figures; Thoday (1910) finds assimilation to proceed at a rate equivalent to the absorption of 0·149 c.c., and this is probably the most accurate determination made for a (detached) leaf in ordinary conditions. At the lowest estimate, therefore, the assimilating leaf can absorb carbon dioxide at one fourth the rate of a free absorbing surface of sodium hydroxide, instead of at only one thirty-third that rate; at the highest estimate it is rather more efficient.

Taking the case of transpiration we find a similar discrepancy. Thus Bakke (1914) found that the loss of water from a sunflower leaf might be as much as three fifths, while from the dahlia it might be nine tenths, of that from a freely evaporating water surface. Such high values are not infrequent, though one quarter to one half are more usual ratios.

The existence of cuticular transpiration does not explain this difference; it is much too small. To carbon dioxide exchange the cuticle is, as we have seen, an almost complete barrier. The explanation is given by the physical laws which govern the diffusion of gases through small openings. The fundamental work was done by Browne and Escombe (1900), primarily for the case of carbon dioxide. They also considered the application of their results to transpiration; Renner (1910) has since extended the work on transpiration.

Browne and Escombe determined the rate at which carbon dioxide diffuses through a pore, by a method simple in principle. A nickel covering was sealed to the mouth of a wide glass tube, at the bottom of which was an absorbing layer of sodium hydroxide. The nickel was pierced by a pore of the size it was desired to study, and the rate at which carbon dioxide diffused through the pore from the air was given by the amount of the gas absorbed by the reagent in a given time. The rates for different sizes of pore could then be compared with each other and with the rate for the uncovered tube. The results are reproduced in Table XXII.

TABLE XXII

RATE OF DIFFUSION OF CARBON DIOXIDE THROUGH PORES OF DIFFERENT SIZES

Diffusion of carbon dioxide through small apertures.					
Diameter of aperture.	CO_2 diffused per hour in c.c.'s.	CO_2 diffused per sq. cm. of aperture per hour.	Ratio of *areas* of apertures.	Ratio of CO_2 diffused per hour.	Ratio of *diameters* of apertures.
22·7 mm.	0·2380	0.0588	1·00	1·00	1·00
12·06 „	0·10180	0·0891	0·28	0·42	0·53
6·03 „	0·06252	0·2186	0·07	0·26	0·26
3·216 „	0·03971	0·4852	0·02	0·16	0·14
2·117 „	0·02608	0·8253	0·008	0·10	0·093

From this it appears that the rate of diffusion is proportional to the diameter (or radius, or circumference), and not to the area. Now the diameter is proportional not to the area but to its square root. Therefore if we consider the case of two pores *a* and *b*, of which *b* has an area one half that of *a*, it follows that the diffusion through *b* will be, not one half that through *a*, but seven tenths: that is, the smaller the pore the greater is the diffusion *per unit area*. This is demonstrated by the figures in column 3 of Table XXII. This applies to the diffusion of any gas, to water vapour as well as to carbon dioxide, either outwards or inwards through a pore.

The reason of this relation may perhaps be appreciated by thinking of the way in which water leaves an upturned pipe under low pressure. It *wells* out over the edge of the orifice, and the amount leaving the pipe will depend rather on the length, or circumference, of that edge than on the area of the orifice. In diffusion the gas wells over the edge of the pore, though it diffuses upwards as well as outwards. Browne and Escombe picture it passing through a series of shells, elliptical in section quite near the pore, but soon becoming practically hemispherical: over the surface of each shell the density of the gas is equal (Fig. 8).

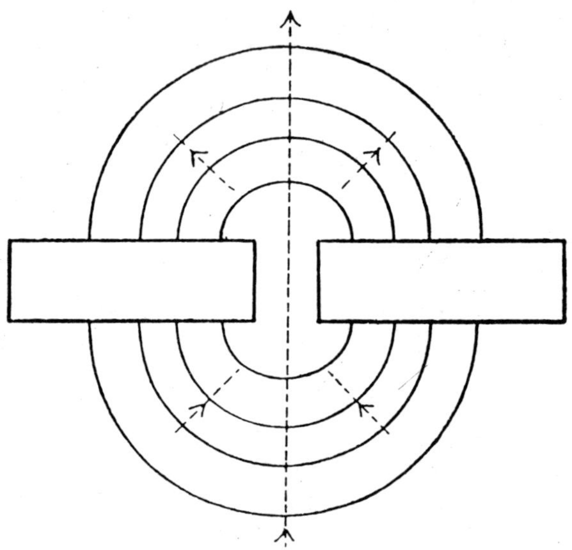

FIG. 8.—Diffusion Shells outside and inside a pore in a septum. The arrows show the direction of flow of the gas.

The amount of gas entering the pore in unit time may be calculated from a formula deduced by Larmoor:

$$Q = d(p_1 - p)4r \quad . \quad . \quad . \quad . \quad (1)$$

where Q is quantity of gas;

d is a physical constant (the *diffusion constant*) depending on the nature of the gas, and the atmospheric temperature and pressure, and expressing the amount which diffuses across unit area, in unit time, under unit pressure;

p_1 is the pressure of the gas in the atmosphere;

p is the pressure of the gas at the pore;

r is the radius of the pore.

For a gas leaving the pore, p_1 and p are reversed. As may be seen, in this formula it is the *radius* and not the *area* which determines the relation of the size of the pore to the diffusion rate.

It is clear, therefore, that if instead of a single pore, the septum through which diffusion takes place is pierced by a number of small openings, much higher rates of diffusion will occur than through a single opening with an area equal to the combined areas of all the pores. Browne and Escombe demonstrated this by using celluloid septa pierced with different numbers of pores set at different distances apart. Their results are given in Table XXIII.

TABLE XXIII

DIFFUSION OF CARBON DIOXIDE THROUGH A MULTIPERFORATE SEPTUM

Average area of tubes employed, 9·39 sq. cm.
Diameter of each perforation in septum, 0·38 mm.

Distance apart of pores in terms of diameter.	No. of pores per sq. cm. of septum.	Per cent. area of pores of area of septum.	Per cent. diffusion through septum of diffusion through open tube.	Theoretical per cent. diffusion with no mutual interference between pores.
2·63	100·00	11·34	56·1	87·6
5·26	25·00	2·82	51·7	63·7
7·8	11·11	1·25	40·6	44·0
10·52	6·25	0·70	31·4	30·7
13·1	4·0	0·45	20·9	21·9
15·7	2·77	0·31	14·0	15·5

From this it appears that when the septum is pierced by about 1000 pores, the combined area of which is 11·34 per cent. of the area of the septum, the diffusion is 56 per cent. of that taking place with no septum present; while with only 26 pores, having a combined area of 0·31 per cent., the diffusion is 14 per cent., that is, nearly fifty times what could be accounted for by the available *area*.

When the pores are close together they interfere mutually. This may be most easily understood by taking the case of transpiration. The water vapour leaving one stoma meets with that leaving its neighbours before it has fallen to the vapour pressure in the atmosphere, and the rate of flow is thereby reduced. The last column shows the theoretical rate of diffusion if no interference were present, and it may be seen that the observed value is as great as the theoretical when the pores are 10 diameters apart; that is to say, there is no mutual interference at this distance; the vapour shell with a radius 5 diameters that of the pore has the same tension as the vapour of the atmosphere.

The epidermis may be looked on as such a multiperforate septum, of which the stomata are the pores. Very generally they are not less than 10 diameters apart; in the sunflower, with a large number of stomata, they lie at an average distance of 8 diameters apart. But the results so far considered require some modification before they can be applied to the leaf.

(1) The stoma is not a circle but an ellipse. It has been found possible, however, to treat it as a circle of equal area, and to use the radius of such a circle in calculation.

(2) The stoma is not a simple pore in an indefinitely thin septum. It is a little tube the depth of which is, as we have seen, often greater than its diameter.

The rate of diffusion *through a tube* depends on the area of the cross-section, the drop in pressure between the two ends, and the length of the tube. It is given by the equation

$$Q = \frac{\pi r^2(p_1 - p)}{L} \quad . \quad . \quad . \quad . \quad (2)$$

where L is length of tube,

r is its radius,

p_1 and p are the pressures at the two ends.

Let us consider the case of water vapour leaving the stoma, and let us simplify it for the moment by supposing that at the inner opening the air is saturated and the pore kept supplied by vapour at maximum pressure, p_1. The rate of diffusion through the stomatal tube is then given by equation (2). But we cannot use this equation to calculate the amount of vapour passing through the tube, because it contains the quantity p (pressure of vapour at outer opening), which we cannot determine. The amount of vapour passing through the stomatal tube is, of course, equal to the amount leaving the opening which is given by $Q = d(p - p_0)4r$, where p_0 is the vapour pressure in the atmosphere ; but here again we have the unknown quantity p. What happens is that the mode of diffusion from the opening prevents the immediate fall of the vapour pressure at the outer end of the tube to the lowest value, p_0, and therefore adds a certain resistance to diffusion through the tube. We can get over the difficulty by supposing that diffusion takes place through a longer tube than the stoma $(= L + x)$, at the outer end of which the vapour pressure is p_0, and such that the extra length x has the same effect as, in fact, the external diffusion shells produce. The rate of diffusion through this system is, of course, equal to the rate through the stomatal tube *or* through the diffusion shells, and we therefore have

$$Q = d(p - p_0)4r = d\frac{\pi r^2(p_1 - p)}{L}$$

and

$$d\frac{\pi r^2(p_1 - p)}{L} = d\frac{\pi r^2(p_1 - p_0)}{L + x}$$

From the first pair we can find a value for L, and substituting this in the second pair we find that

$$x = \frac{\pi r}{4} \quad . \quad . \quad . \quad . \quad . \quad . \quad . \quad (3)$$

The rate of diffusion through the stoma is therefore

$$Q = d\frac{\pi r^2(p_1 - p_0)}{L + \frac{\pi r}{4}} \quad . \quad . \quad . \quad . \quad (4)$$

where all the quantities may be determined.

We assumed that at the inner end of the stoma the air was fully saturated with water vapour. This is not the case. Full saturation will occur at the surface of the evaporating mesophyll cells, and from these a series of diffusion shells of diminishing density will run towards the internal opening of the stoma, resembling those occurring outside the more strongly the larger the air space below the stoma. To allow for these we must simply double the correction x, and the formula for diffusion through the stoma in still air becomes

$$Q = d\frac{\pi r^2(p_1 - p_0)}{L + 2x} = d\frac{\pi r^2(p_1 - p_0)}{L + \frac{\pi r}{2}} \quad . \quad . \quad (5)$$

If, however, the leaf is swept by wind, no diffusion shells will form outside, the value p_0 will occur and be maintained at the stomatal opening, and the correction is again reduced to x (for the internal shells), the formula being given by equation (4).

This equation applies equally to diffusion of carbon dioxide into and out of the leaf, and to the diffusion of water vapour. In assimilation, p_1 represents the density or pressure of carbon dioxide in the atmosphere, and p_0 is zero. In transpiration, p_1 is the pressure of the vapour in saturated air, and p_0 the pressure of the vapour in the atmosphere.

The total diffusion through all the stomata on one square centimetre leaf surface per hour is given by Browne and Escombe (modified) as

$$Q = n \times d\frac{\pi r^2(p_1 - p_0) \times 3600}{L + \frac{\pi r}{2}} \quad . \quad . \quad (6)$$

where n = number of stomata per square centimetre, and 3600 is introduced to bring the value from seconds to hours. Applying this to the actual case of the sunflower, we have :—

n = 33,000,
d = 0·145 C.G.S. units,
p_1 = 0·0003 atmos.,
p = 0,
r = 0·000535 cm.,
L = 0·0014 cm.

From which it may be calculated that Q = 2·095 c.c. CO_2 per square centimetre of leaf surface per hour. In wind the value becomes 2·578 c.c. This means that if the mesophyll cells absorb carbon dioxide instantly and completely, then under ordinary conditions these quantities of carbon dioxide can *diffuse through the stomata*. As a matter of fact, the greatest known rate of absorption of carbon dioxide by the sunflower from ordinary air is that given by Thoday (1910) of 0·15 c.c. per square centimetre per hour. Browne and Escombe point out that the stomatal system amply provides for a full supply of carbon dioxide, since, in fact, at its highest assimilating power in ordinary air, the sunflower leaf uses only 5 to 6 per cent. of the diffusive capacity of the stomata. The reason they give for the low value found is that diffusion through the cell solutions is very slow, quoting Graham's remark, " liquid diffusion of carbonic acid is a slow process compared with its gaseous diffusion, quite as much as days are to minutes." The reason that more carbon dioxide is not used is, therefore, that it cannot travel more rapidly *into the cells*.

Taking the case of transpiration from a sunflower leaf in wind, at a temperature of 20° C., and with a fall in vapour pressure from saturation (0·02 atmos.) inside the leaf to one-quarter of that amount in the atmosphere, Browne and Escombe found that diffusion through the stomata could take place at the rate of 0·1730 grm. of water per square centimetre per hour; the maximum amount obtained experimentally was 0·0276 grm., or one-sixth of the theoretical quantity. Browne and Escombe conclude, therefore, that maximal transpiration can be effected by means of the stomata.

They do not, in this case, offer an explanation of why the actual amount falls short of the theoretical, and it is difficult to see why it should do so, since at the evaporating surface of the mesophyll cells of a turgid leaf the air must be saturated with water vapour. As we have seen, the transpiration from the surface of a sunflower leaf may reach three-fifths of the value of an equal area of free water surface. Renner points out that the value calculated by Browne and Escombe is three times the rate of evaporation from a free water surface of equal area, which is of course impossible. Renner concludes that equation (5) requires to be modified. He assumes that a second series of diffusion shells forms over the *leaf as a whole*, and that this offers a further resistance to diffusion. It is not very easy to appreciate this, when we take into account the fact that the stomata do not interfere mutually. Such a ***vapour dome*** over the whole leaf would, however, be formed in any case by the vapour transpired through the cuticle. The equation, devised by Renner, taking this outer dome into account, is

$$Q = d(p_1 - p_0)\frac{2\pi R^2}{\frac{\pi R}{4} + \frac{\pi R^2}{n\pi r^2}\left(L + \frac{\pi r}{2}\right)} \quad . \quad . \quad (7)$$

where R is the radius of a circle having the same area as the leaf. Using this formula, and taking cuticular transpiration into account, Renner does, in fact, get a much closer agreement between observed and calculated values in quiet air, as may be seen from Table XXIV.

TABLE XXIV

TRANSPIRATION FROM VARIOUS LEAVES BY RENNER'S FORMULA

Name of plant.	Transpiration in grams per minute.				Ratio Transpiration. Wind/Still.	
	Still Air.		Wind.			
	Observed.	Calculated.	Observed.	Calculated.	Observed.	Calculated.
Nuphar ..	0·025	0·0206	0·11	0·141	4·4	6·8
,,	0·0167	0·025	0·083	0·221	4·9	8·8
Hydrangea ..	0·012	0·019	0·021	0·067	2·0	3·5
Archangelica ..	0·033	0·035	0·12	0·344	3·5	9·8
Gentiana ..	0·018	0·025	0·044	0·065	2·4	2·6
,,	0·0143	0·0132	0·045	0·032	3·0	2·4

The effect of wind, according to Browne and Escombe's formula, is not very great; in the example quoted the theoretical absorption of carbon dioxide is increased by about 20 per cent. Using Renner's formulæ the difference is much greater, for in wind the vapour dome for the whole leaf disappears and equation (4) becomes applicable. The discrepancy in Browne and Escombe's experiment is, therefore, not completely explained; and it is probable that their leaves, which were cut, were badly supplied with water. Thoday has emphasised the difficulty of keeping sunflower leaves fully turgid.

Renner's results for wind do not agree well with his calculated values. He himself observes that this may be due to the "quite primitive" method by which the wind was produced, *i.e.* by a fan. It is of course clear that wind of different strengths would affect transpiration differently, and very exact methods would be required for a satisfactory investigation. The ratios obtained, however, do approach the theoretical ratios obtained by using equations (7) and (4), and are far greater than the ratios (1·2 for the sunflower) given by equations (4) and (5).

We have said that Renner's theory of the formation of a "vapour dome" over the leaf as a whole is not quite satisfactory. An alternative explanation of the depression of the transpiration below the values given by Brown and Escombe's formula, may be based on the viscosity of a gas which causes a thin layer to "stick" to a solid or liquid surface. The presence of such a layer, perhaps one-tenth of a millimetre thick, on the surface of the leaf would of course slow down diffusion. The layer would not be removed, though it would be diminished, in wind. This would agree with the fact that observed wind values are almost always too low.

Even apart from such difficulties, these equations cannot be taken as absolutely precise expressions, because the stoma is not a true cylindrical tube, and because the air space below the stoma does not allow of the formation of regular diffusion shells. They do, however, express the fundamental laws of gas exchange between the leaf and the atmosphere, and in practice they give a good approximation to the actual rates of exchange in transpiration. The important thing is that the work of Blackman, Browne and Escombe, and Renner, has firmly established the facts: (1) that the stomata are practically alone concerned in exchange of carbon dioxide, and are most important in the exchange of water vapour; and (2) that diffusion through the stomata takes place at a rate sufficient to satisfy the maximum needs of the plant in assimilation, and according to physical laws capable of exact expression.

It might be asked what advantage in the way of protection from desiccation the leaf gains from its cuticle, if transpiration sometimes takes place in quantities approaching the amount of evaporation from a free water surface. We must take three points into account. (1) The cuticle on the exposed surface may be continuous, so that no transpiration takes place here. (2) The presence of the epiderm with its cuticle and stomata permits of great variety of specific and individual adjustments to special conditions of transpiration. (3) The stomata are capable of closure.

§ 6. Stomatal Movements and their Mechanism

The pore of the stoma is not normally permanently open. Changes in the size and form of the guard cells lead to diminution in the size of the pore, or to complete closure. The changes are possible because of the peculiar structure and properties of the guard cells, and are carried out in response to changes in external conditions and in the water relations of the leaf cells. In surface view the guard cells stand out sharply from the other epidermal cells by their smaller size, their shape, and the fact that they contain chloroplasts. In a transverse section through the stomata, the guard cells are seen to be smaller than the other epidermal cells, and to have walls thickened in a unique manner. (See Figs. 10 and 19.) Typically, the wall next the pore is thicker than that which separates the cell from the neighbouring epidermal cell. At the point of junction with the latter specially thin strips may be present, forming a sort of hinge. The wall next the pore has often a thickened projection above and below, and a bulge in the middle, so that the pore is divided into outer and inner "courts." Frequently the thickening of the guard cell walls is so pronounced that the cell lumen is much reduced. The guard cells contain abundant starch grains ; this is the case even in leaves which do not otherwise form starch, as in most monocotyledons. When the guard cells are flaccid, because of low water content, they are nearly straight and lie with their inner walls touching each other, so that the stomatal pore is closed. With increasing water content, and consequent greater turgor pressure, the walls are expanded. As the outer wall—away from the pore—is thinner it stretches much more than the inner. In doing so it pulls the inner wall with it, so that

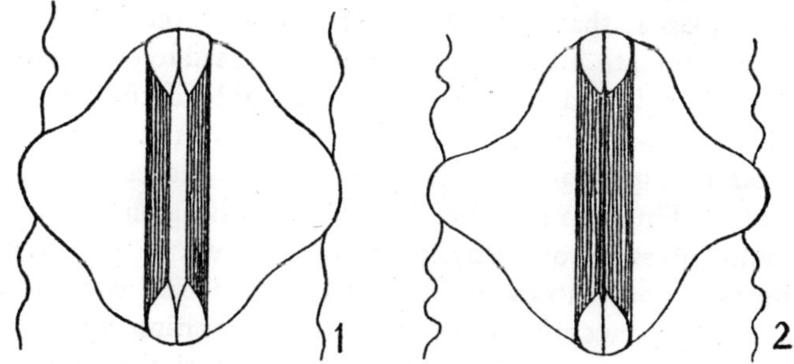

FIG. 9.—Stoma of Cereal (diagrammatic). 1. Open. 2. Closed. The thickened portion of the guard-cell is shaded.

the guard cell becomes bent outwards and the stomatal pore is opened. That the opening of the stoma is actually accompanied by a distension of the guard cells has been shown by Schwendener (1881), to whom much of our exact knowledge of the mechanism of this movement is due ; he found an increase in width of the guard cell of about 10 per cent. The effect of loss of water is strikingly shown by plasmolysing a strip of epiderm with open stomata ; closure at once takes place.

Other types of structure and mechanism occur. The most important of these, characteristic of the grasses and cereals, may be described (Fig. 9). The two ends of the guard cell are thin walled, and are joined on one side to those of the sister guard cell. The wall of the middle portion is nearly uniformly thickened, and to such an extent that the lumen is reduced to a narrow slit. When the cell is turgid, the thin-walled ends expand and the middle portions, which remain straight, are pulled apart. In this open condition the guard cells are dumb-bell shaped. This type of mechanism seems to be less delicate than the other, since the grass stomata show relatively slight activity of movement.

The bulging out of the guard cells takes place against the resistance of the neighbouring epidermal cells. It is therefore possible only if the guard cell can develop a higher turgor pressure than these, and to do this the cell sap must have a higher osmotic pressure. Iljin (1915) has found, for steppe plants, that the osmotic pressure of the guard cells of open stomata may reach the very high value of 90 atmospheres, while that of the neighbouring epidermal cells is only about one-quarter of this. Wiggans (1921) obtained similar though smaller differences for the leaves of mesophytes. Thus the guard cells of Cyclamen had a maximum osmotic pressure of 29·49, those of the beet of 31·5 atmospheres; the values for the epidermal cells were 10·09 and 12·55 atmospheres respectively. Ursprung and Blum (1916, 1918) found the *suction force* of the guard cells about 2 atmospheres higher than that of the epidermal cells in the beech, and about 3 atmospheres higher in the ivy.

The collapse of the guard cells, resulting in stomatal closure, may be due to two causes. Reduction of water content by excessive transpiration may lead to loss of turgor by all cells of the leaf, including the guard cells, when collapse and closure will occur; or a fall in the osmotic pressure of the guard cell may so far reduce turgor pressure that collapse takes place. It will be seen that the first cause of movement depends on the water relations of the leaf as a whole. It is unlikely that conditions favouring high transpiration—low humidity, high temperature, etc.—affect the guard cell independently. Its rate of transpiration may be increased, but, while it maintains a high osmotic pressure, it must also tend to withdraw water from the neighbouring cells, and to maintain its relatively greater turgor. It has been asserted by F. Darwin (1898) that in plants brought into a dry atmosphere the stomata close. Knight (1917) has, however, shown that mechanical shocks, due to shaking of the leaves, may lead to rapid closure, and this is more likely to have been the cause of the closure observed by Darwin than the change in atmospheric humidity.

Action of Light.—The external factor which most markedly influences stomatal condition is light. In plants with a sufficient water supply the stomata normally open through the day and close at night, though, as we shall see, other conditions may occur. The action of light might be

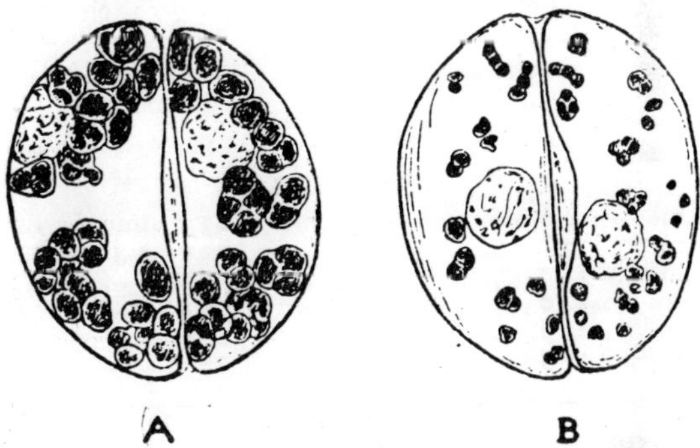

FIG. 10.—Starch content of guard cells of *Fouquieria splendens*; A, starch abundant; B, starch almost absent. (After Lloyd.)

supposed to be due to its effect on photosynthesis. The guard cells contain chlorophyll; in light they should accumulate carbohydrates with a consequent increase in osmotic pressure, giving the condition favourable to opening. This, however, is not the case. Lloyd (1908), Loftfield (1921), and Iljin (1915) have found that the opening in light is accompanied by a great *decrease* in the amount of starch in the guard cells; as the stomata close in the dark the starch is reformed (cp. Fig. 10). This is, of course, not the behaviour of the ordinary assimilating cell. Cases have been observed, *e.g.* by Ursprung (1917), where intense insolation has led to disappearance of starch; but, normally, starch is formed by day and disappears at night. In the case of the guard cell it is quite easy to see that, if the large store of starch is converted into sugar, an increase in osmotic pressure will occur, giving the conditions for stomatal opening. Iljin found, in fact, that the osmotic pressure of the guard cells falls to that of the epidermal cells as the starch increases and the stomata close. Wiggans (1921) determined the osmotic pressure of guard and epidermal cells of several mesophytes, at various times throughout the day. The results for Cyclamen and the beet are given in Table XXV.

TABLE XXV

OSMOTIC PRESSURE OF GUARD AND EPIDERMAL CELLS

Hour.	Cyclamen, 2nd January.		Beet, 4th January.	
	Guard cells.	Epidermal cells.	Guard cells.	Epidermal cells.
7 a.m.	14·6 atmos.	10·2 atmos.	23·5 atmos.	12·5 atmos.
9 a.m.	15·8 ,,	,, ,,	22·0 ,,	,, ,,
11 a.m.	31·0 ,,	,, ,,	31·6 ,,	,, ,,
1 p.m.	22·0 ,,	,, ,,	31·6 ,,	,, ,,
3 p.m.	25·9 ,,	,, ,,	30·2 ,,	,, ,,
5 p.m.	18·5 ,,	,, ,,	25·0 ,,	,, ,,

The difference in behaviour of the two kinds of cell is striking; the guard cells alone show a regular increase in osmotic pressure, and this is accompanied by stomatal opening. In neither of the plants cited were the stomata found completely closed. From other experiments it seems probable that, with complete closure, the osmotic pressure would fall to that of the epidermal cells, as Iljin found it to do. Fig. 11 shows the relation between starch content and stomatal opening in the Lombardy poplar.

There has been difficulty in demonstrating the formation of sugar in the expanded guard cells, though there is little doubt that it is formed. Lloyd saw oil globules appear as the starch disappeared (in *Verbena ciliata*); the function of this oil is unknown.

That the action of light is not due to increased

assimilation is further shown by the fact that opening takes place under blue glass screens which greatly lower the rate of assimilation, and also in light in an atmosphere free from carbon dioxide in which no assimilation takes place. Lloyd and Loftfield suppose that light activates an enzyme of the nature of diastase, which then converts the starch into sugar. Recently Sayre (1923) claims to have shown that light acts by changing the hydrogen ion concentration, and thus favouring enzymatic conversion of starch into sugar; in the dark a reverse change in the acidity reverses the reaction.

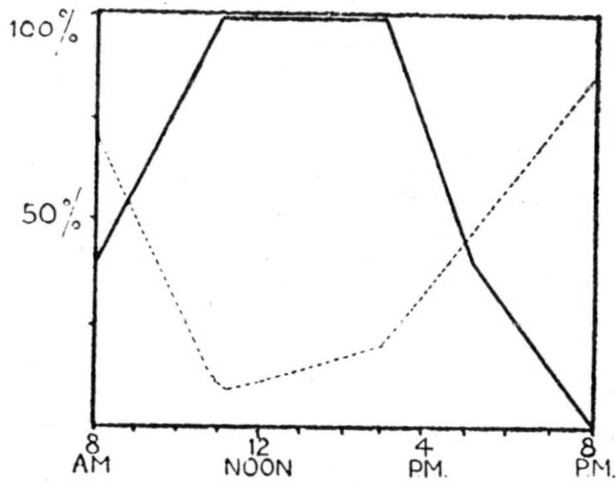

Fig. 11.—Starch content and stomatal opening; the heavy line shows the change in per cent. opening of the stomata of the Lombardy poplar through the day, the broken line shows the changes in starch content. (After Loftfield, modified.)

Experimental alterations in hydrogen ion concentration lead to the same result. Sayre also found that decrease in water content on wilting led to a change in hydrogen ion concentration, and to an increase in starch. How far closure on wilting is due to such changes rather than directly to loss of water remains to be investigated.

Water Content of Leaf.—The behaviour of the stoma as the leaf wilts is not simple; there are many conflicting statements on this point. Knight (1917) has, however, shown conclusively that in the initial stages of wilting, when the leaf is losing more water than is supplied to it, the stomata show a temporarily *increased opening*, thus confirming earlier work by F. Darwin (1898). Closure takes place only when wilting is very pronounced (see Fig 12). This temporary opening probably takes place when the epidermal cells, with their lower osmotic pressure, have lost so much water as to become flaccid ; the still turgid guard cells have

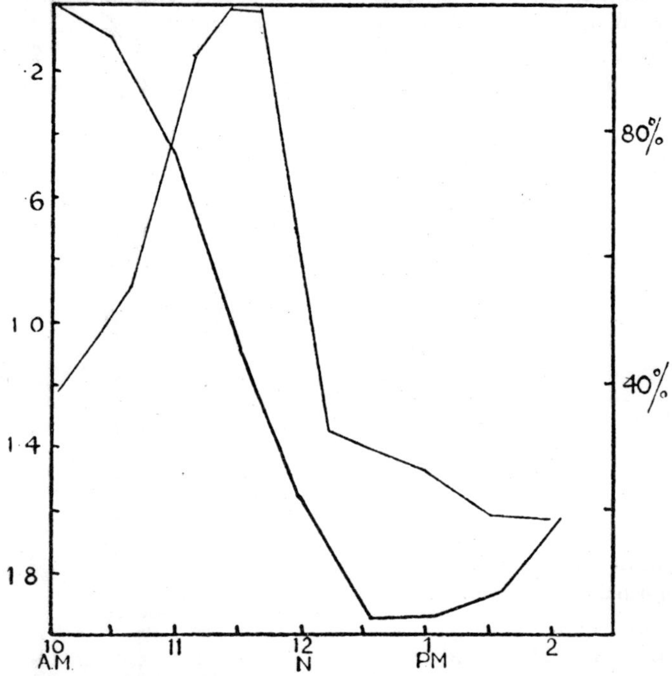

FIG. 12.—Water loss and stomatal aperture; the heavy line shows the net water loss, the light line the per cent. opening of the stomata, during wilting. (After Knight, modified.)

now a smaller resistance acting against them and so bend further apart. Delf (1912) found a very rapid closure to take place within 7–15 minutes after leaves of succulents were detached from the stem. Even complete wilting does not always lead to stomatal closure. Marsh herbs and shrubs, such as the water plantain, brooklime, willow, and alder, were found by Stahl (1894) to have open stomata in the extreme wilted condition, and Darwin confirmed this in some cases, though not in all. Linsbauer (1917) found the stomata to close in all cases.

Temperature.—Loftfield has shown that temperature has an important effect on the rate of movement of the stomata. At temperatures of 1° C., 10° C., 20° C., and 30° C., the times of opening of the stomata of alfalfa in light were 6 hours, 4 hours, 2 hours, and 1 hour respectively; a rise of 10° C. doubles the rate of the movement; this is what one would expect with the enzymatic conversion of starch into sugar as the fundamental reaction.

We have noted the effect of mechanical shock in causing closure.

Rhythm.—The result of the relation to light is that the stoma is normally open through the day. It is possible, however, that the daily opening at the normal hour is partly due to an inherent rhythm. This possibility is suggested by Darwin (1898). Lloyd, using more exact methods, did not obtain conclusive results, though some of his experiments may be taken to support Darwin's view.

Metabolism of Guard Cell.—If the chlorophyll of the guard cell is not directly concerned with the opening of the stomata in light, we may still suppose that, through it, the great amount of starch on which the mechanism depends is built up and maintained; but this conclusion has not been established. The guard cells of white-margined Pelargonium leaves contain plastids, but no chlorophyll. Yet Kümmler (1922) has shown that they contain abundant starch—rather more than normal guard cells—and that they open in light; wide opening only occurs if the conditions of water supply are very favourable. From this it would appear that the presence of chlorophyll is not an essential part of the equipment of the guard cell, though it is difficult to believe, in absence of decisive evidence to the contrary, that it does not function, and so assist in the formation of the carbohydrate reserve in ordinary cases.

The guard cell must be looked on as specialised, not only in its structure, but in its metabolism. In this it is marked off not only from the other epidermal cells, but also from the mesophyll. It may be that special permeability relations enable it to obtain and retain an excessive amount of carbohydrates; the chloroplast may function primarily as a starch-building and not as an assimilating organ. In light the equilibrium between starch and sugar is pushed much further towards the sugar side than is usual. This may be due to changes in the acidity of the cell induced by light. The precise effect of the chlorophyll content is obscure.

When we ask of what use the stomatal movements are to the plant, we find ourselves in a region where opinions have differed greatly. Formerly stress was laid on the stoma's ready reaction to changes of atmospheric humidity, or to changes in the moisture content of the plant; the apparatus was looked on as providing a delicate means of cutting down transpiration when too great a loss of water was threatened, and so averting the danger of wilting. Recently this view has been combated. We will deal with the evidence later. It may be pointed out here, however, that the stoma responds most readily to changes of illumination, not of humidity, and that its normal condition is open through the day and closed at night. Both facts emphasise its close relation to the supply of the raw material of photosynthesis, and tend to discount a primary connection with limitation of transpiration. Nor does behaviour on wilting favour the view that the stomata can save the plant from excessive transpiration. The meaning of the closure at night is not clear. It might lead to an accumulation of respiratory carbon dioxide which would then be available for assimilation in light.

§ 7. Gaseous Exchange of Aquatics

Submerged water plants draw their supply of carbon dioxide from gas dissolved in the water. As they have a very fine and permeable cuticle, water and dissolved substances can pass freely through the external cell walls over their whole surface. Diffusion is the easier as the leaves are in general thin; and it is the more extensive as the leaf surface is often very great. Many submerged plants have long ribbon-shaped leaves, *e.g.* the grass wracks of salt and brackish water; others have leaves divided into many narrow segments, as the water crowfoot. We may note the remarkable assimilating roots, often thalloid in form, of the Podostemaceæ (see pp. 198, 294, and Fig. 23). The oxygen which is formed in assimilation diffuses out in solution through the whole surface. Oxygen is much less soluble in water than carbon dioxide, and as a consequence of this a considerable pressure may exist inside the plant. Angelstein (1911) found in Elodea an excess pressure of about one-sixth of an atmosphere; this was reduced gradually in the dark as the oxygen diffused out, or was used up in respiration. If a submerged plant is wounded, the oxygen formed in photosynthesis escapes from the wounded surface as small gas bubbles. A bunch of Elodea, inverted in water under a test-tube, gives off so much oxygen from the cut stems that a few cubic centimetres may easily be collected. The gas is not pure oxygen—it contains nitrogen as well—but it ignites a glowing splint as oxygen does, a good demonstration of the liberation of oxygen in photosynthesis. This "gas-bubble" method has also been used for determining the relative rates of assimilation under different conditions; the number of bubbles given off from a cut shoot in a given time is used as a basis of comparison. The method must be used with caution, as it is subject to many errors. It has recently been improved by Wilmot (1921).

Water in contact with air absorbs carbon dioxide according to definite physical laws; it comes to an equilibrium with air when it contains about the same percentage of gas as does the atmosphere. If the air were the only source of carbon dioxide for submerged plants they would be badly off, for diffusion into water is slow; the rapid mixing due to air currents in the atmosphere is wanting, and the supply could not keep pace with the plants' requirements. It is supplemented in two ways.

In water overlying mud rich in organic matter, carbon dioxide is supplied by the processes of decomposition which continually go on. The presence of bicarbonates in many natural waters also increases the supply. In fresh water calcium bicarbonate is most important. It is hydrolysed and dissociated into several ions, carbon dioxide being set free.

The amount of free carbon dioxide present in solution depends on the concentration of bicarbonate; but if the water is in contact with air, carbon dioxide passes into the air until the solution is in equilibrium with the partial pressure of the gas in the air. In contact with the atmosphere this point is reached when the water contains about 0·03 per cent. of the gas. As the gas passes into the air the splitting of the bicarbonate goes on, with the result that, in the end, calcium bicarbonate in solution is almost completely converted into the insoluble carbonate which is precipitated. If plants are present they utilise the carbon dioxide, and, as they use it up much more rapidly than it could diffuse into the air, they appear to take an active part in breaking up the bicarbonate. It is this effect which explains the conclusion reached by Angelstein (1911), that the plant actively splits bicarbonate. Wilmot (1921) has shown that, in a solution of bicarbonate of given strength, Elodea assimilates at the same rate as in water containing that amount of carbon dioxide which should, on theoretical grounds, be present in a bicarbonate solution of the strength employed. There is, therefore, no active splitting. Ruttner (1921) has found that the plant can convert the bicarbonate completely into carbonate, and this does not take place spontaneously. The reason evidently is that the spontaneous splitting ceases when the solution has reached equilibrium with the atmosphere as regards the carbon dioxide content, while the plant goes on using up the carbon dioxide till no more is formed and the conversion, in a closed vessel, is complete. The presence of bicarbonates in natural waters is certainly of great importance in increasing the carbon dioxide supply to aquatics. The employment of this source sometimes results, especially in algæ, in a deposition of calcium carbonate in the cell walls.

§ 8. Gaseous Exchange of Succulents

Another group of plants peculiar in their relation to carbon dioxide supply are the succulents. They characteristically inhabit dry situations where economy in water is important ; their gas exchange with the external atmosphere is often limited. When they respire at night the production of carbon dioxide is small, organic acids—malic acid in the Cactaceæ, isomalic acid in the Crassulaceæ, and oxalic acid in the Mesembryanthemaceæ—being produced instead. This is an incomplete form of respiration with reduced energy production, but it also means the retention of carbon compounds in the plant. During the day these plants carry on assimilation partly at the expense of the stored acids, and only partly at the expense of atmospheric carbon dioxide. During the day the amount of acid in the sap may be reduced to one-tenth of its night value ; and the necessity of gas exchange with the atmosphere is reduced. One consequence of this is that the ratio of carbon dioxide absorbed to oxygen given off in assimilation falls considerably below unity, which is the normal value. The metabolism of these succulents is peculiar in other respects, as we shall see later.

§ 9. Energy Relations of Assimilation

The leaf is a green light screen or filter which absorbs most of the light falling on it, reflecting or transmitting smaller fractions. The proportion of light absorbed by different types of leaf must be very different; the thin translucent leaf of a Tropæolum obviously absorbs less than the thick leathery leaves of holly or cherry laurel. Exact measurements have not been made for many plants. Browne and Escombe (1905*a*) found that the percentage of direct sunlight absorbed ranged from 64·7 in *Polygonum Weyrichii*, to 78·7 in *Acer Negundo*. The sunflower leaf absorbed 68·6 per cent. Young leaves absorbed rather less than old ones, as might be expected. The leaves investigated, however, are all of much the same type, so that the relatively small range observed does not really give a good idea of that likely to be found in a more varied selection. The absorption in diffused light has not been specially studied.

It would be natural to suppose the main part of this absorbed light to be taken up by the green pigment; but this is not the case. Comparison of the absorption by white and green portions of variegated leaves of *Acer Negundo* showed that the former absorbed 74·5 per cent. of the incident sunlight and the latter 78·7 per cent., the chlorophyll being therefore responsible for absorbing only 4·2 per cent. of the incident or 5·35 per cent. of the absorbed light.

But even this small fraction of the available energy is not all used in assimilation. Browne and Escombe worked out complete balance-sheets for the disposal of the incident energy by a number of leaves under different conditions. To take one example of their results, they found that *Senecio grandiflorus*, assimilating in bright sunshine, absorbed 65·49 per cent. of the light and transmitted 34·51 per cent. Of the absorbed energy 64 per cent. was expended in vaporising water in transpiration, and 1·22 per cent. in carrying on photosynthesis, the remaining 34·78 per cent. being lost by radiation and convection from the heated leaf surface. Thus, of the total energy incident on the leaf, only 0·8 per cent. was used in photosynthesis. In light of half this strength, 1·59 per cent. was used. These figures are typical; in a long series of twenty-four experiments, under varying light conditions, the amount of available energy used in photosynthesis never reached 5 per cent., and on the average it was about 1 per cent.; about 99 per cent. was always lost—so far as the building of organic matter was concerned—by transmission, re-radiation, and transpiration. These experiments were carried out in ordinary conditions of illumination and at favourable temperatures, with the exiguous supply of carbon dioxide always limiting the amount of assimilation. If the leaf is better supplied with raw material, it makes a better show as an economic factory of organic matter. In an atmosphere with 5 per cent. carbon dioxide, Willstätter and Stoll (1918) measured an assimilation ten times as great as that obtained by Browne and Escombe. If this took place in sunlight, about 10 per cent. of the incident energy would be employed; even under such conditions the waste is enormous.

§ 10. Orientation of the Leaf and Illumination

The leaf is, however, not always exposed to bright sunlight; in such climates as ours bright sunlight is rather the exception. It works late and early, in mist and under dark clouds; owing to the presence of other vegetation it may work habitually in deep shade. We find that it is usually so oriented as to absorb efficiently what light may be available, even though occasionally or frequently the supply may be much in excess of what it can utilise. Its broad expanded surfaces must, however, be taken not only as serving for the absorption of the energy, which is abundant, but also as making for efficient utilisation of raw material, the carbon dioxide, which is scarce.

Even a cursory examination of the leaves of a few shrubs and trees shows that the blades are so disposed as not only to receive good illumination individually, but also to avoid mutual shading. The primary arrangement of the leaves with reference to the shoot axis is usually very regular; this *phyllotaxy* may be referred to a comparatively small number of ground types. A species has a very constant leaf arrangement, though this may change from one type to another during development. The final position assumed by the leaf blades is, however, determined largely by the direction of incidence of light, and may completely mask the nature of the original relation to the shoot. The position of the mature blade is attained by bending and twisting movements of the shoot axis, of the petiole, of a leaf joint or *pulvinus*, or of the leaf blade itself. It is clear that the adjustment of a stalked leaf is a much easier and more accurate process than that of a sessile one.

If we examine the erect shoots of a clump of garden mint growing in the open, more or less evenly illuminated from all sides, and shaded, again more or less uniformly, by neighbouring shoots only, we find a simple type of arrangement. The leaves occur in pairs, and each succeeding pair stands at right angles to the one below (*decussate*); each leaf is nearly horizontal. The shading of the lower leaves by those above is thus largely avoided, for the rays of light, falling on the plant in a slant from above, have a much larger space to pass through than they would have if each leaf pair stood directly above the next lower. Further, the leaves decrease in size upwards, and the greater spread of the larger lower leaves is not shaded by the smaller leaves above. Towards the base of the stem the oldest leaves are smaller; in the grown plant they have already ceased to function, and they wither away.

A more complicated arrangement is seen in the sunflower. The first leaves are again arranged in crossed pairs, but soon a spiral arrangement sets in. Starting with a leaf low down on the stem, and following the spiral upwards, we find that we pass twice round the stem, and only when we reach the sixth leaf in succession do we arrive at one standing directly above the first; this arrangement is perfectly definite, and can be characterised by the fraction $\frac{2}{5}$. The denominator indicates that we can treat the leaves in groups of five, the first leaf of each group standing directly above the first of the group below, the second above the second, and so on. The numerator indicates that if we draw a spiral through the leaf bases, then in each group the spiral passes twice round the stem. We could define the relation otherwise, namely by the angular divergence between two successive leaves, which in this case, is $\frac{2}{5}$ of 360°, or 144°. This $\frac{2}{5}$ spiral phyllotaxy is characteristic of many plants. In other plants other arrangements occur, expressed by the fractions $\frac{1}{2}$, $\frac{1}{3}$, $\frac{2}{5}$, $\frac{3}{8}$, $\frac{5}{13}$, etc. This regular arrangement has been the subject of extensive experimental and mathematical treatment, but no wholly satisfactory explanation has yet been given, though it is certain that it is an expression of the interplay of forces in the meristematic apical region of the shoot. The works of Jost, Thompson (1917) and Church (1904) should be consulted.

It is plain that the $\frac{2}{5}$ spiral of the sunflower is more efficient than the decussate arrangement of the mint, for the leaves are more evenly distributed round the stem and all available space is occupied. A glance at the foliage of the sunflower shows that while the leaves overlap very little, there is also very little space left between adjacent leaves. The petioles rise at an angle from the stem, and the blades droop at an angle from the stalk; the whole shoot is a rounded cone covered with green. The rosette plants show a special case of this kind of arrangement; excellent examples are afforded by the daisy, the dandelion, the primrose, or the plantain. They may be looked on as telescoped stems of the sunflower type, and they show at a glance the way in which space is utilised and shading avoided.

Further complications may be seen in a young maple tree. In the erect terminal shoots the natural decussate arrangement of the leaves is conspicuous. The lower leaves are large with long petioles, the upper are smaller with shorter petioles. Looking down on the top of such a shoot, one notices the absence of mutual shading, associated with complete utilisation of available space. Below the apex the side branches are inclined, and are more or less shaded from vertical illumination. On these the leaf blades are so disposed as to receive oblique illumination; depending on their original position on the stem, the leaves may simply be bent somewhat, or the petioles may be more or less twisted. Finally, in horizontal branches, we find the leaf blades all brought definitely into a single plane, but with their origin in four rows, above, below, and to both flanks, still plainly showing. The petioles of the lowest leaves may be very long, and the spaces between these may be filled by the small leaves of short side shoots. A further complication is seen here, and is of widespread occurrence: the leaves arising on the upper side of the branch remain smaller, and have shorter stalks, than those to the flanks and below. In trees like the maple this *anisophylly* is solely determined by the relation to light. In some plants it is hereditary, and may go so far that the smaller leaves are reduced almost to scales, as in Procris. Fine examples of anisophylly are the hemlock spruce and the silver fir. The needles are borne spirally, but by torsions and bendings they come to lie in one plane, those below being largest, those above smallest, the series graded from one extreme to the other. The angular, lobed leaves of the maple are seen to fit exquisitely into each other, forming what has been well termed a "leaf mosaic." The spirally arranged leaves of the beech and elm show similar relations. The narrow leaves of many willows cast little shadow and are borne more abundantly towards the heart of the tree, than are those of the beech. Much-divided leaves, like those of the hemlock or ragwort, allow abundant light to penetrate to those placed below.

One or two further examples may be given. The periwinkle (Plate III.) has both vertical and horizontal shoots. The leaves are paired in the vertical shoot, the arrangement is decussate, and the leaves stand out from the stem. In the horizontal shoot the leaves lie in one plane and in two rows, so that their origin in four rows is not obvious. If we suppose an erect shoot laid on its side, then one pair of leaves, which occupies the flanks, will reach its new position by a twist of the short petiole through 90 degrees. The leaves of the next pair, however, lie one above and one below. The stalk of the former bends down and is also bent horizontally through 90 degrees, that of the latter bends up and also sideways. In these a complex movement is required to bring the blades to the side of the stem. The same result is achieved by Buddleia (Plate III.) and Philadelphus in a totally different fashion. In these plants the leaves are again decussate, and in the horizontal branches

PLATE III

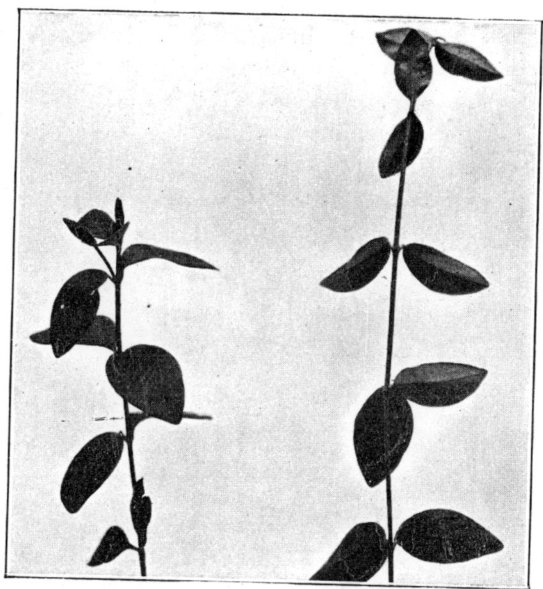

Above, erect (left) and creeping (right) shoots of Periwinkle, the latter with twisted petioles. Below, horizontal shoot of Buddleia with twisted internodes.

they lie in two rows in one plane. They reach this position, however, by the *internodes of the stem* twisting through 90 degrees, alternately to right and left; the short stalks of the individual leaves then twist and bring the blades horizontal.

Leaf adjustment is most striking in plants growing against a wall with strictly unilateral illumination. The garden nasturtium, *Tropæolum majus*, is as good a case as can be found. Its leaves are spirally arranged, but when the plant grows against a wall the petioles curve and twist so that all the leaves come to lie to the front of the stem; the angle which the peltate blade makes with the stalk, and the inclination of the apical part of the stem, bring the blade accurately normal to the incident light.

Quite different is the relation of the grass-like type of leaf to light. It is typically long and narrow, and it stands more or less upright. In the relatively small surface exposed in the upright position, and in its small powers of adjustment, it appears to be less efficient than the broad-leaved type. But two points must be kept in mind. Such plants very often grow in crowded communities, and the narrow leaf, allowing the penetration of light through the mass of vegetation right to the ground, must be an important factor in making this type of community possible. The upright leaf, too, utilises horizontal rather than vertical light. The anatomical structure of the grass leaf, with assimilating parenchyma on both faces, and the stomata more or less equal in number on the two surfaces, emphasises this relation to light. An extreme case is offered by the Iris, with its double rank of bifacial leaves. This arrangement in a broad-leaved plant would be very inefficient; but when the light utilised is mainly horizontal the arrangement takes on a different aspect.

§ 11. Mechanism of Leaf Adjustment

To alter their primary position and come into relation with the incident light, the leaves must of course carry out definite movements. Unequal illumination is the stimulus which causes the leaf to move, and determines the direction of the movement. This response is termed *phototropism*. In leaves, and in leaflets, which possess joints or pulvini, the movement is carried out by the joint, and is due to changes in turgor pressure. The leaves of a scarlet runner, for example, have a pulvinus at the base of each leaflet, and another at the base of the stalk. Each leaflet of a clover or of a wood-sorrel leaf possesses a pulvinus. The pulvini remain capable of movement long after the leaf is mature, and the leaf may at any time readjust itself to new conditions of illumination. A potted scarlet runner placed near a window brings all the leaf blades, by bendings and twistings in the joints, to face the light; if it is reversed, then in the course of a few hours it readjusts the whole light-absorbing surface. The turgor movements in the pulvini are completely reversible, and may be repeated almost indefinitely.

Most leaves, however, possess no pulvini, and then the movement is carried out by differential growth on the different sides of the petiole—particularly near its base and apex. Such growth movements can only take place in the growing organ, and they commonly cease or become very slight as the leaf comes to maturity. In some cases, *e.g.* the garden nasturtium or the house geranium, the petiole may remain capable of growth for a long time. As a rule, however, such leaves assume a *fixed light position*, though this phrase must not be taken in too rigid a sense. Now the direction from which maximum light strikes the plant varies throughout the day, most markedly so in sunshine, and a fixed position cannot, therefore, be assumed in relation to sunlight. It has been found that it is the direction from which comes the maximum *diffused light* that determines, in the main, the leaf position of ordinary plants. It may, of course, frequently occur that this is also the position in which the maximum of direct sunlight falls on the leaf. We have said that the plants of temperate climates have to work for the most part in diffused light, and this conclusion may be extended to cover all plants which grow in communities where much mutual shading occurs. The relation to diffused light is thus a generally important one.

The phototropic reaction is seen at its simplest in organs with radial symmetry. A young stem, when lit from one side, curves, as the result of unequal growth rates on the two sides, till its axis lies in the direction of the light-rays, and so is uniformly illuminated. Most often the shoot, if the zone of growth is long, bends somewhat past the position of equilibrium and then reverses the movement, till finally adjustment in the direction of the light is attained. Such curvatures are frequently seen in potted geraniums and fuchsias grown in windows, and by this means, even without adjustment of the leaf, better illumination of the more shaded leaves is secured.

It is with such *orthotropic* organs (especially with the *coleoptile*, or first sheath-like leaf, of the oat, which is peculiarly sensitive) that the most exact investigations on the mechanism of the phototropic reaction have been made. From these we know that the stimulus may be perceived in one region of an organ and the differential growth reaction take place in another. The conduction of the excitation is in all probability due to the transport of some chemical substance, a mode of conduction which finds its analogy, not in the nervous system, but in the transference of hormones in the higher animals. This is the general mode of the conduction of excitation in plants. For orthotropic organs, at least, it is well established that the essence of the stimulus lies in the *difference in intensity* of the illumination on the two sides, and not in the oblique *direction* of the rays; although the response results in the assumption by the organ of a definite relation to the direction of the rays. The blue end of the spectrum is active in inducing phototropic response; red light has practically no effect.

The reaction of the leaf is much more complex. As we have seen its final position is often attained by torsions, and the changes in growth rate leading to a torsion are obviously more intricate than are those leading to a simple curvature; they are little understood. Again, the change in position is brought about, frequently by a combination of reactions, at the base of the blade, in the stalk, and even in the shoot. Finally, the leaf blade takes up its position at an angle to the direction of the controlling illumination—it is a *plagiotropic*, not an *orthotropic*, organ. Although the reaction is carried out in response to light it is influenced by gravity, so that some leaves, at least, assume a horizontal position even in absence of light.

The case of the sunflower showed us the petioles inclined rather steeply above, and the blades steeply below, the horizontal. The position of the leaf is thus attained by the movement of petiole and blade (or the point of junction of blade and petiole) in different directions. Is, then, the stimulus perceived by one or by both organs, and how is the ultimate adjustment brought about? These questions must be answered differently for different plants. Haberlandt (1905) has shown that there are three main types of behaviour, connected by intermediate types:

(1) In *Begonia discolor* and *Monstera deliciosa* the leaf blade alone perceives the stimulus, which is transmitted to the petiole, in which the appropriate movements are carried out.

(2) In the simple primary leaves of the scarlet runner the petiole and the pulvinus, which is here present, perceive the stimulus directly, and bring the leaf blade into the favourable position. The blade is little sensitive, although it may have a certain influence.

(3) In most plants both petiole and blade are sensitive. Either, illuminated by itself, brings the blade into the proper position; where the blade is illuminated the stimulus must of course be transmitted. The petiole, when illuminated alone, can bring about an approximate adjustment only. Illumination of the blade alone produces more perfect adjustment. The indirect stimulus through the blade can overcome the direct stimulus to the petiole. Haberlandt makes the acute suggestion that the direct stimulus of the petiole secures a "coarse" adjustment, while

the stimulus to the blade secures a "fine" adjustment. Among plants showing this mode of response are the garden nasturtium, the hop, the mallows, and the Virginia creeper.

The recognition of the *rôles* of blade and petiole does not take us far towards an understanding of these very complicated movements, but at present our knowledge stops here. Mention must, however, be made of Haberlandt's ingenious theory of the mode of perception of the light stimulus by the leaf blade. He has shown that in many cases the epidermal cells, by virtue of the curvature or structure of their external walls, act as concentrating lenses, so that, if they are brightly lit from above, a spot of light is thrown on the back wall of the cell. So perfect is the lens action that photographic images of external objects may be obtained with these structures. He supposes that the leaf is in equilibrium only when the light spot falls on the centre of the cell. This is not the case when the illumination is oblique, and then the leaf moves until the light falls normally to its surface and brings the spot to the central position of equilibrium. Ingenious as this theory is, and supported by a mass of observation, it has not received much experimental support. It has been shown, for example, that leaves with the epiderm removed still respond, as do those which have been covered with a layer of liquid paraffin, which converts the cell walls into dispersing lenses. The most highly developed type of lens was found in *Fittonia Verschaffeltii*, where it is a reduced trichome, consisting of a small cell resting on the epidermal cell beneath; but in two very closely related species, the leaves of which are just as sensitive in their adjustment to light, these cells are entirely wanting, and this tells distinctly against Haberlandt's theory.

§ 12. Orientation of Strongly Insolated Leaves

Not all leaves are oriented so that the maximum amount of light falls on the blade. Where insolation is very strong the leaf may take up a position in which a minimum surface is exposed to the sun's rays. This may be seen in our native wood-sorrel, and in other species of Oxalis. Through the day, in its natural shaded habitat, the three leaflets of Oxalis are spread out at right angles to the petiole, but if the plant is exposed to direct sunlight they droop and fold together, assuming the position that is taken up normally in the dark every night. These leaflets move by pulvini, and the drooping in the sun is connected with loss of turgor, Similar movements are carried out by the leaflets of Acacias and of many other plants. Robinia, the false acacia, is a good example of this behaviour ; in the morning the leaflets are horizontal, but, as the sun rises about 30 degrees above the horizon, they move up and assume a vertical position. In many plants, particularly those of very hot, dry climates, the leaves assume a fixed light position of this nature, hanging vertically, and with the edges turned in the direction of maximum insolation. This *profile position* is shown by *Eucalyptus globulus* and many other members of this Australian genus. The famous " compass plants " also show this relation. Of these the European *Lactuca Scariola* is the best known. Its leaves are borne in $\frac{3}{8}$ spiral phyllotaxy, and, according to Neger, in shady places, and in high latitudes as in Norway, they maintain this arrangement. Where they are exposed to intense sunlight, they twist so that they come to stand in two rows, with their edges pointing north and south. So constant is this arrangement in some compass plants of the American prairies, *e.g. Silphium laciniatum*, that they are said to have been utilised by the plainsmen in finding their way. The same feature is shown by the leaves of many tropical forest trees, *e.g. Amherstia nobilis*, which when young and tender hang vertically, and later, as they harden, take up a horizontal position. The flat segments of some cactuses, *e.g.* Opuntia may stand north and south. We might also regard the grass type of leaf, especially in extreme forms such as that of the Iris, as escaping direct insolation. The spruce bears its needles erect on exposed branches, and horizontal when shaded.

In all such plants the leaf takes up a position which avoids the strongest illumination, but it is quite unlikely that the plant is benefited by receiving less light. It is much more probable that the profile position has its use in preventing overheating and also excessive transpiration. An exact investigation for *Lactuca Scariola* has been carried out by Karsten (1918). He found that leaves in the horizontal position attained temperatures as much as 7·6° C. higher than those standing vertical, both being fully exposed to the sun. A vertical leaf, receiving the sun at right angles to its blade, showed temperatures of 6·3° C. higher than the same leaf when edge on to the sun—in profile position. His results for transpiration are not very easy to interpret, since the data for temperature are not complete. He compares the transpiration for the equal periods from 10 a.m. to 1.30 p.m., that is when the sun strikes the edge of the leaves, and from 1.30 to 5 p.m., when the rays strike more and more at right angles to the blade. The average of the water losses from five plants for a sunny day were, for the first period 5·55 grm., and for the second 6·49 grm. In the first period the air temperature varied from 20·5° to 27° C.; in the second, from 27° to 22° C. The loss was therefore greater in the second period, but it is not possible to say whether this was due to the different orientation to the sun of the leaves, or to the higher air temperature. With an air temperature of 27° C. about midday, however, it is clear that the leaf exposed at right angles to the sun's rays would take on a very high, and probably dangerous, temperature.

Wiesner (1907) has applied the term *euphotometric* to those leaves which are so oriented as to receive the maximum of diffuse light, and *panphotometric* to those which are placed so that they avoid maximum direct insolation. Reviewing the occurrence of the latter, we see that they form a very characteristic feature in the vegetation of regions of great heat and drought, being specially prominent in the Australian bush. Their occurrence, often in less pronounced form, is frequent in exposed positions in more temperate climates.

It is legitimate to regard all narrow, more or less erect, leaves, such as those of many grasses, as showing this feature to a certain extent. The panphotometric condition may be assumed temporarily by pulvinate leaves. A leaf arrangement which does not receive full illumination is a great deal more common than would at first sight appear. This aspect of the vegetation of our heaths and pastures awaits, and would repay, investigation.

§ 13. Leaf Structure and Assimilation

The four pigments which we term collectively chlorophyll are localised in plasmatic bodies of definite structure and individuality which are called chloroplasts. There are good grounds for believing that the chlorophyll is present in these in the form of a colloidal solution. Among the algæ we meet with great variety in the form, size, and number of these specialised plastids. From the mosses upwards, with rare exceptions, many small, thin, elliptical chloroplasts occur in each cell. In the typical dorsi-ventral leaf they are most numerous in the palisade parenchyma, next the upper surface. In leaves of the bifacial type, more or less typical palisade tissue abuts on both surfaces. Haberlandt has described the various types of palisade tissue. Its chief feature is the regular elongation of the cells at right angles to the surface of the leaf, so that the light strikes down through a series of tubes lined with green plastids. These cylinders are interspersed with narrow intercellular spaces. Below the palisade lies the spongy parenchyma with fewer chloroplasts and larger air spaces, through which the carbon dioxide passes to the palisade tissue. Frequently there can be made out a definite relation between groups of the palisade cells and cells of the spongy parenchyma, leading to efficient transport of the sugar formed in photosynthesis, or produced at night by the hydrolysis of the starch stored in the plastids. The chloroplasts line the walls of the palisade cells, and more particularly the vertical elongated walls. When a cross wall is bounded to the outside by an intercellular space it also shows chloroplasts, but the cross walls between neighbouring cells may be bare of chloroplasts ; this is the case for the walls marching with the upper epiderm. We see in this an arrangement which brings the chloroplasts into the most favourable position for gas supply, lying as they do against the walls through which they may receive carbon dioxide by direct diffusion from the intercellular spaces. The same features may be seen on the long walls of the palisade cells. There is a distinct tendency for the chloroplasts to be most abundant on the walls bounding the air spaces. In Sempervivum the palisade is arranged in plates, with long narrow spaces between, and it is on the walls next these spaces that the chloroplasts are most abundant.

In relation to light absorption this chloroplast distribution does not appear so favourable at first sight. It might be thought that if the chloroplasts lay on the transverse walls, a better utilisation of light would be obtained. We must, however, look at the leaf as a whole. With such an arrangement the upper cells would seriously shade the lower, and they would also absorb far more light than, with the very limited supply of carbon dioxide at their disposal, they could utilise ; the actual arrangement permits a much fuller utilisation of light. We must also remember that light does not strike straight through the leaf, even in the case of a leaf at right angles to the direct rays of the sun. As it passes through cells, with walls lying in various directions, and with heterogeneous contents of varying refractive powers, it is reflected and scattered in all directions. In actual fact the chloroplasts do not really lie edge-on to the light ; they receive the scattered rays from all directions. At the same time any injurious effect of direct insolation is minimised. The palisade arrangement is therefore favourable for the individual chloroplasts, and for the leaf as a whole, while at the same time the supply of carbon dioxide is also dealt with advantageously. The absence of chloroplasts from the epiderm may well be connected with carbon dioxide supply, since the cuticularised epiderm, despite the fact that it is directly exposed to the atmosphere, is very poorly supplied with carbon dioxide.

The chloroplasts lying in the peripheral plasma are not always fixed in position. Especially in the lower plants movements under the influence of different degrees of illumination are common. Among flowering plants movements of the chloroplasts in the fronds of Lemna, the duckweed, are best known. In light of moderate intensities the chloroplasts lie on the upper and lower walls of the cells ; mutual shading is here not serious, as the frond has only two layers of chlorenchyma. In direct sunlight the chloroplasts move to the side walls, and so take up a profile position to the sun's rays. At night the chloroplasts are more scattered. In the palisade of ordinary leaves the chloroplasts do not move, but movements occur in the spongy parenchyma. In direct sunlight there is a tendency for the chloroplasts to gather in masses in the shade of the overlying palisade, and this may lead to an alteration in the colour of a leaf. Thus an elder leaf growing in the shade has a deep green colour ; if it is exposed to direct sunlight it becomes lighter green.

§ 14. Chlorophyll and the Absorption of Light

The green colour of the chloroplast is due to the presence of a group of four pigments, the composition of which has been the subject of a great deal of research. Recently, the work of Willstätter (1913) and his collaborators has greatly advanced our knowledge and put it on a sound chemical basis. There are present two bright green pigments, named *chlorophyll a* and *chlorophyll b*, and two yellow pigments, named *carotin* and *xanthophyll.* The two latter occur widely, apart from chlorophyll, in fruits and flowers of yellow and orange colour. Alone they are incapable of carrying on assimilation ; it is not known whether they play a definite part in the process in conjunction with the chlorophyll proper, though many attempts have been made to show that they do. The four pigments are present in the chloroplasts of all flowering plants, and in fairly constant proportions. We need not here go into the theories of the way in which the chlorophyll complex acts, but we must consider its relation to light absorption, and some related points.

The green colour indicates that when white light passes through chlorophyll both the blue and the red rays are absorbed, allowing the green to pass on. Spectroscopic examination shows that chlorophyll possesses four absorption bands in the orange end of the spectrum, the strongest lying between the B and C lines, and three at the blue end. A familiar demonstration experiment of timing the bubbles emitted by Elodea, exposed to sunlight behind red and blue screens, shows that under these conditions the red light is much more active in assimilation than the blue. The experiment is very faulty, because no account is taken of the relative strength and purity of the light of the two colours, but, in fact, it is certain that in sunlight the red end of the spectrum is most effective. To determine accurately the effectiveness of the different portions of the spectrum, and to relate them to the absorption by chlorophyll at the corresponding points, and to the energy of the light, is so difficult a matter that, despite repeated attempts, no satisfactory result has yet been attained. Pfeffer found maximum assimilation in the red at the C absorption band; Engelmann (1883) found maximum assimilation near the same point, with a second lesser maximum in the blue; Kniep and Minder (1909) found red and blue light of equal energy to produce the same rate of assimilation, while green produced none. Most recently Ursprung (1917), in an investigation carried out with all physical precautions, found that assimilation occurred in every region of the spectrum, in proportion to the energy of the light and to the extent of its absorption by the chlorophyll; unfortunately, his estimate of the amount of assimilation was based only on a rough colorimetric estimate of the starch formed. From such results we can draw no definite conclusions; we can only say that the balance of the evidence is in favour of an important utilisation of blue as well as of red light.

This is of great interest in relation to the kind of light which is at the disposal of the plant. Stahl (1909), in a stimulating book, has treated of the question of the relation of the colour of the pigment to the quality of light available.

The green colour is very constant, the quality of light is not. Every one who has taken photographs knows the difference in the quality of sunlight at noon and in the evening, and the much longer exposure which is required later in the day. As the sun nears the horizon its rays pass through a much thicker layer of atmosphere, through gas molecules, minute globules of water, and flying dust. These interfere with the short wave-length blue rays much more than with the longer wave-length red rays. The blue rays are more extensively scattered, and the red rays pass on in greater proportion. The thicker the atmospheric layer the greater is the loss of blue light, and so, in the evening, the sunlight is relatively much less active on a photographic plate. The extensive scattering of blue rays in the atmosphere is responsible for the blue colour of the sky and of the diffuse light received from it, for the blue of hazy weather, and for the blue of distant hills, seen through the atmosphere in which blue light is scattered. Thus, with the sun at its zenith, the intensity of radiation at the A line in the red is 1·28 times that at the F line in the blue; with the sun at 11 degrees above the horizon the total light intensity has fallen to one-half, and the intensity at the A line is 3·5 times that at the F line. In diffuse light from the sky the intensity at the F line is 6 times that at the A line, 4 times that at the B line, and 3·5 times that at the C line.

The relative intensities of rays of different wave lengths which the plant receives from the sun thus change continually throughout the day. But, as we have insisted, the plant receives not only sunlight; it often depends mainly on diffuse light. A Virginia creeper on the north wall of a house may, in the brightest weather, receive *only* light from the sky. This light, as well as diffuse light in the shade and from clouds, is relatively richer in blue rays. Not only is this the case, but the plant growing in the open receives, as well as sunlight, light from the sky. As Blackman and Matthaei (1905) have shown, the " sky " light is a very important fraction of the whole. In a horizontal leaf illuminated by the sun and by a cloudless sky, the ratio of sunlight to " sky " light is, with the sun at 60 degrees elevation, 1·58, with the sun at 45 degrees, 0·83, and at 15 degrees, 0·24. The plant growing in the open, even more than the shade plant, is subject to illumination varying constantly not only in strength but in quality, and like the shade plant it depends very largely on diffuse light.

Now, as Stahl points out, the green leaf pigment is able to utilise both blue and red light, as it absorbs both very markedly. It is unnecessary to make the carotin and xanthophyll responsible for the absorption of the blue rays, as Stahl does, for the chlorophyll itself shows, as we have seen, strong absorption of the shorter wave lengths, as well as very strong absorption of the orange-red. The question might be asked, Would not a grey or black leaf absorbing all light be even more efficient ? Stahl denies this, for the transmission of the extreme red and ultra red, and of the yellow and green, is advantageous, since the danger of overheating which would occur if these rays, so strong in direct sunlight, were absorbed, is partly obviated. This is a good general explanation ; though the most strongly heating red rays are in fact absorbed, the transmission of a part of the radiant energy must lessen the danger.

Wiesner (1907) has criticised Stahl's views. He argues that, if chlorophyll is so closely related to the colour of the light, we should find variations in its tone corresponding to the different qualities of light falling on plants in such different situations as an exposed mountain top, a steppe, a meadow, or a woodland, and that such differences are not in general to be observed. But this criticism is not well founded. For the point is just that chlorophyll represents a stable compromise which can make the best of all sorts of light conditions, and that in this, and not in any marked degree of plasticity or specialisation, lies the reason of its extraordinary success. It is a very remarkable thing that, while in every other respect the organisation of the plant shows the widest sort of variation, in this one point, of pigmentation, there is extraordinarily little change throughout the vegetable kingdom.

This interpretation of the utility of the green colour of chlorophyll is supported when we consider that there is every reason to believe that the pigment complex we know to-day has persisted throughout the history of the plant kingdom from its origin. It has been a successful feature of plant-life under general conditions of illumination different from those of the present day; for it is likely that in former epochs an atmosphere much more nearly saturated with water and much cloudier than that of to-day, made diffuse light the normal kind of illumination. This again emphasises the importance of diffuse light. We may look on chlorophyll as one of the most successful and most conservative of the products of life, with relations so generalised that a very close fit to any particular set of conditions is not to be expected.

§ 15. Chlorophyll in its Relation to Assimilation

The relative amounts of the four chlorophyll pigments is fairly constant throughout the flowering plants, but the total chlorophyll content is subject to considerable variation. In the life-history of an individual leaf it, of course, changes in a definite and well-known fashion. The young leaf contains relatively little chlorophyll, although it may contain much yellow pigment; on exposure to light the amount of chlorophyll increases to a maximum which, as Wiesner has shown, occurs about the time the leaf attains its full size. This is the case at least for deciduous leaves; evergreen leaves may not attain their full colour till the second year of their life, or later. After this maximum is reached, intense insolation may weaken the green colour; in the natural course of events the leaf yellows in the autumn before its fall. The chlorophyll proper is broken up, and it is likely that the important amounts of nitrogen and magnesium it contains pass back into the storage organs of perennial plants. The yellow tints of the autumn leaf are due to the carotinoid pigments remaining, perhaps not in their normal condition; the reds are due to anthocyanin, the function of which is even more obscure here than when it is produced by actively functioning organs, as in the copper beech or in many young shoots. Purple and mauve and orange tints arise from differences in the acidity of the cell sap and from combinations of anthocyans and carotins. The increase in chlorophyll in the developing leaf is familiar in the change from the foliage of spring with its brilliant yellow-green tints to the full green colour of summer. It is more strikingly seen when etiolated plants are exposed to light. Seedlings grown in the dark are drawn, and white or pale yellow; illuminated, the formation of chlorophyll begins in a few minutes, and progresses rapidly till the maximum is reached, under favourable conditions, in a few days.

Apart from this change in chlorophyll content in the history of a single leaf, great differences exist between the content in normal mature leaves of different species. This may be due to differences in the concentration of the pigments in the plastids, or to different numbers or sizes of plastids in equal amounts of tissue. A comparison of sections through leaves, for example, of a house-leek and of a cherry laurel shows a very much greater number of chloroplasts in the latter.

A good many investigations on the relation between the amount of chlorophyll and the assimilating capacity of different leaves have been carried out. Haberlandt attempts a correlation between assimilation activity and chlorophyll content. Table XXVI gives his results.

TABLE XXVI

RELATION OF PLASTID NUMBER TO ASSIMILATION

Plant.	Relative assimilating energy.	Relative number of chloroplasts.
Tropæolum majus	100·0	100
Phaseolus multiflorus ..	72·0	64
Ricinus communis	118·5	120
Helianthus annuus	124·5	122

The assimilation activity values are based on results of an early investigation by C. A. Weber (1879). The chloroplast numbers are obtained by actual counts. The correspondence is remarkable, but it is likely that it is partly due to chance, for Weber's results were based on determinations of dry weight increase after forty-eight days' growth under greenhouse conditions, and can give no information as to the maximum assimilating capacity of the plants. There is no evidence that chlorophyll content is proportional to plastid numbers. The relation is really one to plastid material rather than to pigment. Lubimenko (1908) found the chlorophyll content of broad-leaved trees higher than that of conifers, and related a higher assimilating capacity of the former to this.

Willstätter and Stoll (1918) have studied this question with more reliable methods. They have made estimations of the actual amount of chlorophyll present, and have measured the rate of assimilation under conditions of high carbon dioxide supply (5 and 10 per cent. carbon dioxide) and strong illumination. They express the relation of assimilating power to chlorophyll content as the ratio of carbon dioxide in grams assimilated per hour, to 1 gram chlorophyll—that is, the amount of carbon dioxide assimilated per hour by the amount of leaf substance containing 1 gram of chlorophyll. This ratio is called the *assimilation number*. Table XXVII reproduces some of Willstätter's results for leaves assimilating in light of the strength of sunlight, in an atmosphere containing 5 per cent. carbon dioxide and at a temperature of 25° C.

TABLE XXVII

RELATION OF CHLOROPHYLL CONTENT TO ASSIMILATION

Plant.	Chlorophyll in mg. per 10 grm. fresh leaf.	CO_2 assimilated per hour in grm.		
		per 10 grm. fresh leaf.	per 100 sq. cm. leaf surface.	A.N.
Aesculus Hippocastanum ..	24·7	0·159	0·033	6·4
Ampelopsis quinquefolia ..	28·8	0·178	0·028	6·2
Tilia cordata	28·1	0·188	0·028	6·6
Sambucus nigra	22·2	0·146	0·034	6·6
Ulmus sp...	16·2	0·111	0·022	6·9
Helianthus annuus ..	16·5	0·230	0·080	14·0
Cucurbita Pepo	17·5	0·213	0·063	12·1

It will be seen that these plants fall into two classes. The first five have an assimilation number of about 6, the last two of about 12. All the plants examined by Willstätter fall into one of these two groups. The conditions of experiment were such that neither an increase of light nor of carbon dioxide supply could increase the rate of assimilation, so that the figures may be taken as expressing the plant's maximum capacity *for a temperature of* 25° *C*. Taking the plants of either group, we see then that there seems to be a very close relation between the assimilating capacity and the quantity of chlorophyll present. The plants of the second group, however, are able to employ their chlorophyll to much better purpose than the others. As Willstätter points out, they are plants which are remarkable for rapid and luxuriant growth. The amount of chlorophyll available is, therefore, not the sole factor in producing vigorous assimilation. The same amount can allow in Cucurbita double the rate shown by Ampelopsis. In the latter, at least, it is therefore in excess.

This is further demonstrated by the behaviour of young leaves. In these the assimilating capacity is greater, related to the amount of chlorophyll, than in old leaves; as the leaf matures the amount of chlorophyll increases, and so, in general, does the assimilating power, though not to the same extent. This may be illustrated by Willstätter's results for *Quercus Robur* given in Table XXVIII.

TABLE XXVIII

CHANGE IN ASSIMILATING CAPACITY OF THE OAK

Date.	Material.	Chlorophyll, mg. per 10 grm. fresh leaf.	CO_2 grm. per		
			10 grm. fresh leaf.	100 sq. cm.	A.N.
11th May	94 leaves (=5 grm.)	6·6	0·072	0·013	10·9
20th May	19 leaves (=5 grm.)	8·6	0·136	0·024	18·8
9th June	11 leaves (=5 grm.)	21·6	0·194	0·038	9·0
20th June	8 grm. of leaf	25·0	0·196	0·041	7·8

The rate of assimilation for unit quantity of chlorophyll increases up to a point, and then decreases till maturity is reached. This does not cover the complete history of the leaf. It would be necessary to begin with a stage having no chlorophyll. This has been done for seedlings of the barley by Miss Irvine (1910), and of Phaseolus, the French bean, and other plants by Briggs (1920, 1923).

The remarkable fact was brought out that in Phaseolus, for example, for about ten days after the first leaves have unfolded, assimilation has a low value even when the leaf is fully green. By ingenious experimental methods Briggs maintained the chlorophyll content of the leaves at a constant low value for eleven days, and showed that, during this time, the rate of assimilation constantly rose; that is, assimilation increased though chlorophyll did not. Some factor in the process, other than chlorophyll, is therefore increasing during this early stage of development. The increase in this factor depends on the supply of food substances, for in the sunflower, the marrow, and the maple, in which the food-store of the seed exists in the cotyledons, which are also the first assimilating organs, assimilation starts in these at a high value as soon as chlorophyll is formed. In all seeds in which the food is not stored in the first assimilating organ assimilation lags behind chlorophyll formation. This is the case in the French bean, where the food store is in the cotyledon and the first assimilating organ is a foliage leaf; in the castor oil plant, where the food is stored in an endosperm and the first assimilating organ is the cotyledon; and in the cereals, where the food is stored in an endosperm and the first assimilating organ is a foliage leaf. In all these cases a transfer of food must take place from the storage organ to the assimilating leaf, and photosynthesis is more or less delayed. Briggs points out the interesting fact that, in so far as getting a start with the work of assimilation is concerned, the least specialised type of embryo, that in which the same organ functions as food store and assimilating organ, is the most efficient.

Willstätter believes that the factor which develops more slowly than chlorophyll is an enzyme, and is concerned with the later stages of the assimilating process carried on by the protoplasm, as distinct from the early stages in which light and chlorophyll are effective. Briggs, however, brings forward evidence to prove that the factor in question affects both the "light" (or chlorophyll, or photochemical) stage, and the "dark" (or protoplasmic, or chemical) stage of the process. He suggests that "The conception of the process of photosynthesis which seems best to fit the facts is that the seat of the process is the *surface* of the chloroplast, for here the proportion of light absorbed will tend to be greater, and this will be the portion of the chloroplast with which the carbon dioxide will first come into contact. To give more detail to the picture, we may postulate that the photochemical phase of the process consists of an activation, by means of light energy, of the molecules of the surface of the chloroplast either before or after combination with carbon dioxide, and that the chemical phase is an interaction of such activated molecules." This "reactive surface" of the chloroplast is not necessarily the actual surface area; it may be the "internal" surface of a colloid. It is the development of the full reactive surface which is required before assimilation can attain its maximum rate; the presence of chlorophyll alone is insufficient. At the same time the degree of development of the reactive surface affects both stages of the process, for a more extensive surface means a greater number of molecules activated by light, and it also means a more intense interaction of the activated molecules.

Similar relations were found by Willstätter in ageing leaves. In leaves which turn yellow, the decrease in the chlorophyll is accompanied by a decrease in the rate of assimilation. The assimilation numbers tend to rise for a time, and then to fall to a value lower than that of the mature leaf; that is to say, the smaller amount of chlorophyll at the end is less advantageously used; or, to put it otherwise, the protoplasmic factor is failing. This is still more marked in leaves which remain full green even after autumn frosts. In these, although the chlorophyll content may be quite high, assimilation may almost or altogether cease. Yet such leaves, if kept in a warm moist atmosphere for a few hours, recuperate. Willstätter found, for bright green leaves of *Ampelopsis Veitchii* gathered on the 17th of November, an assimilation of 0·006 grm. of carbon dioxide per hour; after a day at 25° C. they assimilated six times as vigorously; the assimilation numbers were less than 0·8 and 4·6 respectively.

Of interest, too, are the results obtained with normal and golden varieties of the same tree; thus, "aurea" varieties of oak, elder, and elm possess one-thirteenth, one-thirtieth, and one-tenth respectively the amount of chlorophyll of the normal green varieties, yet their assimilatory activity is one-half, two-thirds, and equal to, that of the green varieties. The assimilation numbers corresponding are enormous, *i.e* 55, 113, and 77, compared with 7·8, 6·4, and 8·4 for the green leaves. Here we have plants with only a trace of the normal amount of pigment showing assimilating capacities approaching that of the full green leaf; in conditions of intense illumination they are probably as efficient as the normal varieties, and indeed one may readily observe that the growth of the "aurea" elders is about as vigorous as that of the green forms. This is true only for high light intensities; for low intensities the assimilation of the yellow leaves is relatively much poorer—they are less fitted for work in poor light, and it is possible that here the chlorophyll is working at maximum capacity and limits the rate.

This gives us some idea of the enormous capacity for assimilation which the chlorophyll present in the ordinary plant would develop if the reactive surface of the plastid were greater. In fact, the chlorophyll content of the ordinary leaf may be regarded as enabling it to make the best of the low light intensities with which it must often work; it is far greater than is necessary in good illumination.

The similarity of the assimilation numbers within one class of leaves seems to indicate, however, a close connection between amount of chlorophyll and assimilating powers; against this is the existence of two classes of plants with exactly the same sort of chlorophyll, yet the one with assimilation numbers twice as large as the other. It is possible that the similarity within a class is due to the association of a definite amount of assimilating plasma (plastid substance) with a unit amount of chlorophyll. The chlorophyll would then be an indication of the amount of plasma, or enzymes, active in assimilation, and this would be the relation expressed by the assimilation numbers. The difference between the two classes might be due to a fundamental difference in the type or extent of reactive surface developed in the individual plastid.

§ 16. External Conditions and Assimilation

The chlorophyll acts as an absorber and transformer of light energy; whether it takes any part in the ensuing chemical changes we do not know. It is clear, in any case, that it is not the chlorophyll alone that is effective; the protoplasm or its products must also be active. We have seen how these two agents come into play, though we do not know any details. As the assimilation process goes on its rate is affected, at one stage or another, by such factors inside the plant as the rate of diffusion of raw materials, and of finished products, and the flow of water, and by such external factors as the supply of carbon dioxide, and of light, and by the temperature.

Our knowledge of the principles underlying the regulation of assimilation—and of many other functions—by external conditions is due to F. F. Blackman, who has published, along with his collaborators, a series of papers on this and allied subjects. The principles involved are discussed in Blackman's paper on "Optima and Limiting Factors" (1905). Extensive data and discussions are given by Blackman and Matthaei (1905), Blackman and Smith (1911), and Matthaei (1904).

There are according to Blackman "five obvious controlling factors in the case of a given chloroplast engaged in photosynthesis:

(1) the amount of carbon dioxide available;
(2) the amount of water available;
(3) the intensity of the available radiant energy;
(4) the amount of chlorophyll present;
(5) the temperature in the chloroplast."

Of these, two may be called "internal"—the amount of chlorophyll, with which we have already dealt, and the water available; for the water in the chloroplast is regulated by the water relations in the internal leaf cells. To these factors we must add, in view of the recent work of Briggs (1923), the supply of nutrient salts.

Water Supply.—The question of the influence of water supply is a difficult one. Probably the actual assimilating mechanism in the chloroplast is always abundantly supplied with water; but we do not know this, and its relations must be complex. The chlorophyll exists in the chloroplast, a body of colloidal nature, supplied with its water from the cell sap either directly or through the protoplasm. The condition of the plastid colloids must be altered by the changes in the amount of water and of dissolved substances present in it. Now, as the water in the cell sap decreases—through a failure of supply completely to cover evaporation—it becomes a more concentrated solution, and will tend to remove water from the plastids, perhaps to an appreciable extent; we do not know. Further, during assimilation, the concentration of sugars in the plastid may change, and this must mean a normally recurring alteration of the water relations. The exact effect on the rate of assimilation is obscure; but it is known that increase in the products of assimilation tends to inhibit further assimilation, and it is possible that the water balance of the plastid is affected. Thoday (1910) has shown that as the sunflower leaf loses turgor the rate of assimilation falls, till it reaches zero in the wilted leaf. Referring to work of previous investigators, he concludes that this may be a frequent occurrence in hot weather with some plants, *e.g.* the beet. The effect may be due to stomatal closure, but it may be related to diminution of water content in the assimilating cells.

Temperature.—The other four factors we may call "external." We may first take the relation to temperature. The temperature of the chloroplast must be taken as the temperature of the leaf; this may be accurately determined by thermo-electric methods. In dull weather it is practically the air temperature; the amount of heat liberated in respiration is small and cannot appreciably raise the temperature of a thin freely radiating and transpiring organ like the leaf; the heat lost in evaporation must be quickly made good by absorption from the air. In sunlight, on the other hand, the leaf temperature may be considerably higher than that of the air.

It is a fact of common knowledge that, within a certain range, metabolic processes are more active at higher than at lower temperatures. The favourable range of temperature was looked upon as specific for given organisms and for given functions, such as assimilation, although liable to modification; a certain point in it was called the *optimum temperature*, it being supposed that there the function in question took place most vigorously. The relation of temperature to assimilation, growth, respiration, etc., was commonly expressed by a graph having a characteristic form, with a branch rising to the optimum and another descending from it, and tending to zero at the *maximum* temperature—at which the process ceased. Now Matthaei showed that the assimilation rate of the cherry laurel, expressed in mgs. CO_2 per 50 sq. cm. leaf surface per hour was 2·0 at 0·4° C., 3·6 at 9·2° C., 7 at 15° C., 10·1 at 23·7° C. The coefficient of increase for a rise of 10° C. is 2·1. For Elodea the coefficient for 10° C., calculated from the observed values at 7° and 13° C., is 2·05. The primary relation between temperature and assimilation is that assimilation increases with rise of temperature, the coefficient for 10° C. being just over 2. This is the relation which Van 't Hoff had shown to hold between rise of temperature and an ordinary chemical reaction. The temperature coefficients are generally much lower for purely physical or photochemical reactions (about 1·1). That the coefficient for photosynthesis is so high, again points to the importance of the protoplasmic factor, which must be concerned with chemical reactions. Willstätter found for elm and elder coefficients of about 1·5. He considers that this lower figure indicates limiting effects by physical processes, such as the rate of diffusion of carbon dioxide.

At any temperature from 25° C. downwards, the rate of assimilation remains constant for prolonged periods ; successive hourly determinations during six or seven hours give the same value. But from 30° C. upwards this is not the case. At 30° C. it was found that the rate fell from 15·7 mg. carbon dioxide per 50 sq. cm. leaf surface for the second hour to 12·0 mg. for the fifth hour ; at 37° C. the corresponding figures were 23·7 mg. and 10·9 mg. ; at 40° C. they were 14·9 mg. and 4·8 mg. It will be seen that at 40° C. the fall is more rapid than at 37° C., and at 37° C. more rapid than at 30° C. It is clear that the determination for the second hour—the earliest that could be made in practice—does not give us the *initial* values for these temperatures. These initial values can, however, be obtained by extrapolation. The values so found agree with the values obtained by calculation from the coefficient for 10° C.

Blackman's interpretation of this result is that the *primary* relation between temperature and assimilation is that between temperature and a chemical reaction ; that, if we could measure the rate at the moment the high temperature takes effect, we should find it to have the value indicated even at the critical temperature—probably between 50° C. and 60° C.—where almost instantaneous death ensues. Above 30° C. temperature has a second effect ; in some way it has a progressively deleterious action on the protoplasm. We do not know precisely the nature of this deleterious action, and Blackman has called it simply the *time factor*, a term which indicates its increasing effect with time, but does not tie us to any theory as to its nature.

It is clear that the old idea of an " optimum " temperature is purely fictitious ; for the temperature at which assimilation *appears* to have a maximum value will depend on the interval which elapses between the raising of the temperature and the making of the determination ; the shorter the interval the higher will lie the apparent optimum. This conclusion, as well as the whole mode of analysis of the action of temperature, is fundamental not only for photosynthesis, but for metabolic processes in general.

Light and Carbon Dioxide.—The relations between photosynthetic activity and light and carbon dioxide supply are simpler ; here we are dealing with the necessary energy and the necessary raw material. We might expect that if we doubled the available energy or the available material, then the rate of assimilation too would be doubled, and this is what Blackman and Smith (1911) found. The graph connecting assimilation with either of these factors is an inclined straight line ; the relation is arithmetical. This only holds within certain limits, for, with very high light intensities or very high carbon dioxide concentrations, a " time factor " is again introduced, and the rate falls off. For Elodea the time factors set in only when the light intensity exceeds nine times that of direct sunlight, or when the carbon dioxide concentration is over three volumes per cent. For land plants, constant rates of assimilation have been obtained with as much as 10 per cent. carbon dioxide. It is known that very high light intensities have an injurious effect on the protoplasm and on the chlorophyll, while carbon dioxide acts as a narcotic. But in natural conditions, neither light nor carbon dioxide content can ever retard assimilation.

Limiting Factors.—The second fundamental advance made by Blackman is his analysis of the way in which these three factors interact. He states as an axiom: "When a process is conditioned as to its rapidity by a number of separate factors, the rate of the process is limited by the pace of the 'slowest' factor." This is often referred to as the theory of *limiting factors*. The effect of this in practice may be illustrated by considering the case of a plant assimilating in ordinary air under increasing illumination. As the light gets stronger the rate of assimilation rises until a certain point is reached, at which no further rise takes place however much stronger we make the light. Now this does not mean that we have reached a point at which increase of light, as such, has no further effect; it means that at this point the carbon dioxide supply is being completely used up. Carbon dioxide supply is now a limiting factor and increase of light is without effect, because the materials of assimilation are lacking. If we artificially supplemented the carbon dioxide supply, we should find that increase of light would again give higher values.

In this account we have neglected temperature, but in practice it must of course be included. Suppose our object of study is a plant in the open air, and we are following its rate of assimilation in the early morning. We might well find that increase of light and increase of carbon dioxide both failed to give higher rates of assimilation, in which case the temperature would be acting as the limiting factor. The rate at which a plant assimilates (assuming that it is well supplied with water and has its stomata open) is always limited either by temperature, by carbon dioxide supply, or by illumination. Whichever of these three is least favourable will determine the rate of assimilation. No matter how bright the light may be, or how warm the day, the plant cannot assimilate more rapidly than is possible with the supply of carbon dioxide available. We may further note that that factor is limiting, an increase in which leads to an increase in assimilation rate (Fig. 13).

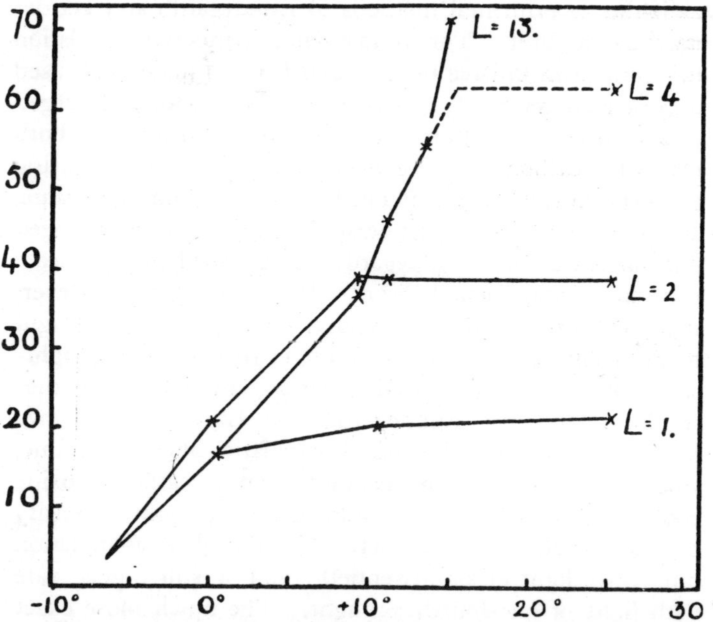

FIG. 13.—Light, temperature and assimilation; each graph represents the course of assimilation (mgs. CO_2 per 50 sq. cm. per hour) with rising temperature for a given light intensity (L = 1, 2, 4, 13): in each, assimilation rate rises till light becomes limiting, and then the graph takes a horizontal course: the exact point of inflection for L = 4 is conjectural, and this part of the graph is dotted: light = 13 is not limiting for the highest temperature (15° C.) for which a value is given. (After Matthaei, modified.)

These conceptions of Blackman's have the breadth and simplicity characteristic of so many important generalisations. They have been the subject of some criticism which has not shaken their position. Recently two pieces of work have appeared which, if they receive confirmation, may necessitate some modification of the limiting factor hypothesis. In the first place, both Lundegardh (1921, 1922) and Harder (1921) claim to have shown that assimilation rate is not directly proportional to light intensity; the graph representing these relations is not a straight line but a curve. Blackman's figures, which are supported by results of Willstätter, are definitely at variance with this. It seems as if the experimental methods of Lundegardh and Harder may be at fault. The fundamental temperature relation does not seem to have been established. Lundegardh used a closed air space, and Harder, working with submerged plants, used a solution of sodium bicarbonate; in both cases the carbon dioxide concentration must fall off during the experiment, though it need not reach a limiting value, and in the latter there must be other changes, *e.g.* in hydrogen ion concentration. Lundegardh's published figures indicate very large experimental errors. It may be noted, however, that for "aurea" plants Willstätter found the graph for light to be a curve, and that Warburg (1919) found the light-assimilation graph to be a curve for certain algæ. The other result obtained by Lundegardh and Harder is that two factors may influence the rate of assimilation at the same time. Lundegardh found, for instance, that a carbon dioxide concentration which was limiting for a light intensity of one-fortieth sunlight, yet gave a higher assimilation rate with light of one-twentieth, and a still higher rate with light of one-fourth sunlight. The much more exact results of Matthaei and of Blackman and Smith do not agree with this, though perhaps the number of experiments bearing on this point is too small and the results not sufficiently uniform to give definite refutation. Smith (1919) writes: "It is conceivable—and indeed probable—that when, so to speak, two factors are close to the limiting value a change in the one not limiting may have some appreciable effect on assimilation. This will show itself about the inflexion of the curve when the limiting factor is changing. For example, when carbon dioxide is limiting, increase of temperature may cause a small increase of assimilation by increasing the rate of diffusion of the carbon dioxide." This is, of course, a very different position from that taken up by Lundegardh, who holds that, for each higher light intensity the graph connecting assimilation rate with carbon dioxide supply rises at a steeper angle. Lundegardh's results may, of course, be due to some secondary effect, such as the degree of stomatal opening. It must, however, be emphasised that Blackman has established a method of experiment and of analysis to which all work on photosynthesis must conform if it is to be exact.

Nutrient Salts.—Briggs (1923) has shown that, if plants are grown in culture solutions lacking either potassium, magnesium, phosphorus, or iron, the rate of assimilation does not reach the value shown by plants grown in a complete culture solution, which again does not reach that shown by plants grown in soil. The lower rate is exhibited when calculated either for unit leaf area or for unit leaf weight, and is not therefore an effect of lessened development of leaf surface, which was indeed not markedly different in the different plants. Briggs supposes that the diminution is due to a failure to develop the full reactive surface in the chloroplast. Lack of nutrient salts, therefore, directly affects plant growth through its action on photosynthesis. An indirect effect will also be produced, since the diminished assimilation ultimately means a diminished assimilating leaf surface.

§ 17. Assimilation in Natural Environment

We may now consider the actual effects of changes in these conditions on the rate of assimilation in nature. The action of light and of heat are closely linked because they tend to vary together. Their relations have been studied by Blackman and Matthaei (1905), and by Thoday (1910). Blackman's experiments were carried out on leaves of the cherry laurel, and of the artichoke, a plant with leaves very similar to those of the closely related sunflower. The leaves were allowed to assimilate in an atmosphere containing 2 to 3 per cent. carbon dioxide, an amount which never limited assimilation, and their assimilation under the natural changes of illumination and temperature was determined. From former work the maximum assimilation at different temperatures was known—that is the assimilation with excess of light and carbon dioxide, and with temperature limiting. Thus for the cherry laurel the maximum assimilation at 20° and 25° C. is 0·0085 and 0·0115 grm. of carbon dioxide per 50 sq. cm. leaf surface per hour.

By comparing the rate of assimilation actually observed in natural conditions with these figures it could be determined whether, in any given case, the light or the temperature was limiting the rate of assimilation ; for if the observed figures fall below the maximum value for a given temperature, obviously the light is limiting, while if they are approximately the same, then the temperature is limiting. Some of Blackman's results are given in Table XXIX.

TABLE XXIX

ASSIMILATION BY CHERRY LAUREL IN NATURAL ILLUMINATION

Time.	Illumination.	Leaf temp. °C.	Assimilation grm. CO_2 per 50 sq. cm. leaf surface per hour.	Assimilation per cent. of maximum for temperature.
9.30–10.30 a.m.	Dull, occasional sun	23·5	0·0092	86
10.30–11.30 „	Dull and faint sun	23·2	0·0095	91
11.30–12.30 p.m.	Sun and then dull	23·8	0·0102	97
12.30–1.30 „	Dull	21·4	0·0071	75
1.30–2.30 „	Bright sun	26·9	0·0108	85
2.30–3.30 „	Bright sun and occasional clouds	26·3	0·0093	76
3.30–4.30 „	Sun with dull period	21·8	0·0061	64
4.30–5.30 „	Sun and thin cloud	21·5	0·0054	57
5.30–6.30 „	Sun and thin cloud	20·2	0·0041	48

The second and third readings approach the maximum, and in these temperature may be limiting. The light was not full sun, but in the middle of the day it is at its strongest and, even when clouded occasionally, apparently allows of maximum assimilation at a temperature of 23° C. At 2 p.m. the bright sun again permits a high rate of assimilation, but the temperature is also high, and the possible maximum is not reached. Towards the evening, even although a good deal of sun is available, the natural decline in light intensity never allows the possible maximum to be reached, and light is always limiting. In this experiment the leaf faced south throughout the day, and the change in the incidence of the sun's rays materially affected the rate of assimilation. In a companion experiment, under very similar weather conditions, the leaf was continuously moved so as to lie at right angles to the sun's rays, and here it was found that the maximum was reached, except under very heavy clouds, and that the temperature, ranging from 20° to 25° C., was limiting from 9 a.m. to mid-day. This shows how the position of the leaf affects the play of the external factors. It was also shown that in the sun at noon at the summer soltice the maximum assimilation at a temperature at 29·5° C. is obtained with 0·36 of the available light for the cherry laurel, and with 0·69 for Helianthus, the assimilation of the latter being about twice as vigorous.

These experiments were under natural conditions, except as regards carbon dioxide supply, which was supernormal. Thoday's work (1910) supplements them in this respect. His values are calculated from the increase in dry weight of the leaves of the sunflower during assimilation under completely natural conditions, with the exception that he used cut leaves; this enabled him to control the water supply and maintain the leaves in a condition of turgor, by supplying water under pressure. The sunflower wilts very readily when the leaves are attached to the plant.

Thoday reached the important conclusion that in bright sun the maximum assimilation by sunflower leaves is 17 mg. increase in dry weight per *100 sq. cm.* per hour; this is exactly the value obtained by Sachs, so that it may be taken as accurate. It is equivalent to an assimilation of 27·5 mg. of carbon dioxide. He points out that, for the artichoke, Blackman found the maximum rate of assimilation at 22·3° C. (with light and carbon dioxide in excess) to be 26·2 mg. carbon dioxide; taking this figure as applying also to the sunflower, he draws the conclusion that in bright sun the assimilation of the sunflower leaf is limited by temperature up to 23° C., and that below this the normal supply of carbon dioxide is more than sufficient to cover the possible assimilation. Willstätter, however, has obtained a value of 80 mg. for the sunflower with temperature at 25° C. limiting, and this would indicate that temperature is limiting only below 11° C. 'If Willstätter's result is correct, and it is probably too high, we have the very important fact that the carbon dioxide supply, small though it is, is yet sufficient to maintain assimilation at the highest possible value for all temperatures below 11° C. (or 52° F.) as long as the light is sufficiently strong; while, if Blackman's figure is taken, the carbon dioxide of the atmosphere can maintain maximal assimilation up to 23° C. (or 73° F.). It will be noted that the sunflower is one of the plants with high assimilating capacity. The more numerous plants of the first group in Table XXVII will be even better situated as regards sufficiency for their needs of the carbon dioxide supply, especially in temperate climates.

Looking over these results, we have a vivid view of the way in which the different controlling factors cross each other as the day waxes and wanes and the weather changes. In the cool hours of the morning the supply of carbon dioxide is ample, but temperature or dull light is limiting. The sun mounts the heavens, and temperature rises with light intensity; assimilation also rises, limited from moment to moment by one or the other of these two factors. The sun passes behind a cloud, and the heated leaf no longer receives sufficient light to maintain its full efficiency; it shines for a period, and the carbon dioxide supply is no longer sufficient to supply the leaf's capacity in brilliant light and high temperature. Evaporation increases, the leaves flag, and the stomata close, carbon dioxide supply is cut off, and, it may be, assimilation ceases for that day.

One particular factor is always limiting, but that factor may change a dozen times with the varying conditions of an English summer day.

Shade Plants.—These experiments all deal with plants which receive at least intermittent sunlight. It would be a matter of great interest to know how plants living in constant shade behave. A good deal of attention has been devoted to this question, but, unfortunately, even in the most recent work, that of Lundegardh (1921, 1922), the experimental methods are not completely adequate ; in particular, the importance of establishing the fundamental relation to temperature has not been realised. Lundegardh finds that at normal carbon dioxide content the supply of the gas limits assimilation in a shade plant, Oxalis, when the illumination reaches a value equal to one-tenth sunlight, while for a sun plant, Nasturtium, the rate increases up to one-quarter sunlight. He attributes the difference to lesser efficiency in the diffusion of carbon dioxide into the chloroplasts of the shade plants. The chloroplasts of Nasturtium, for example, are more numerous and smaller and the cells are much larger than those of Oxalis. There is a much greater area of cell surface (through which carbon dioxide diffuses) per unit volume or unit area of chloroplast in the Nasturtium leaf, and the area of chloroplast surface is much greater in relation to the chloroplast volume. Further, the Nasturtium leaf has more numerous stomata, and though these are smaller, the diffusive capacity of the Nasturtium leaf must be at least $2\frac{1}{2}$ times that of the Oxalis leaf. With equal supply of carbon dioxide in the atmosphere, the supply to the assimilating surface will be superior in Nasturtium, which will therefore be able to make use of a higher intensity of light.

Lundegardh and Boysen-Jensen (1918) both find that in thin-leaved shade plants, *e.g.* Oxalis and *Lychnis dioica*, the respiration is much less than in thicker-leaved sun plants, *e.g.* Atriplex and Sinapis. The light intensity at which assimilation, with normal carbon dioxide supply, just balances respiration is always much lower in the shade plant.

Lundegardh found that in the shade plants he examined it lay between 1/120 and 1/140 sunlight, and in the sun plants between 1/40 and 1/60. Boysen-Jensen found this light intensity for shade plants about 1/5 that for sun plants. The shade plant is therefore able to accumulate a carbohydrate surplus at a lower light intensity than the sun plant.

One interesting fact is brought out—that the carbon dioxide content of the atmosphere near the ground in a forest is about 25 per cent. higher than that over open ground, and that this is due to a much greater evolution of carbon dioxide from the soil. Whether this is of great importance to the plants of the forest floor where conditions of temperature or light are usually limiting is much less certain than Lundegardh seems to think; it would be interesting matter for exact investigation. Lundegardh also attempts an evaluation of the actual amount of assimilation by Oxalis throughout the day, and finds that it barely covers the loss by respiration for the twenty-four hours, and indeed may fall below it. It rises above the respiration loss considerably, only when the leaf is illuminated by sun spots penetrating the forest canopy, and this calls attention to the importance for shade plants of this variable and uncertain type of illumination. Lundegardh concludes that plants like Oxalis can build up a surplus of carbohydrate only when light travels freely through the leafless trees in spring and autumn, and that in summer they just cover wastage without increasing dry weight. This investigation emphasises the interest of the conditions of existence in the forest. The question is one which might well be investigated in the tropics, where the shade vegetation is much more luxuriant than in our latitudes, a fact that points to an important influence of temperature, as well as of light, in the assimilation of shade plants. Differences similar to those between the leaves of sun and shade species also occur between leaves of a single species.

Sun and Shade Leaves.—The structure of the assimilating tissue of the leaf—the number of layers of palisade, the depth of its cells, the relative proportions of palisade and spongy parenchyma—is characteristic for a given species. It is, however, subject to modification in different environments. This is well seen by comparing leaves growing in shaded positions with those exposed to direct insolation. In many plants there is a sharp distinction between "sun" and "shade" leaves. A familiar example is the common harebell. In shaded positions, in crevices of walls or in deep grass, we may see it bearing large rounded thin leaves; in sunny stations, where it sends up flowering shoots, the leaves on these are narrow, elongated and thick. Less obvious, but just as distinctive, differences are to be found in the leaves of many trees. The sun leaves of the maple possess two layers of palisade, and the upper layer of palisade is nearly twice as deep as the single palisade layer of the shade leaf; the ratio of the thickness of palisade to the thickness of spongy parenchyma, is, in the sun leaf, 3·9 : 1; in the shade leaf, 1·1 : 1. The total volume of the intercellular spaces is greater in the shade leaves by about 50 per cent. The beech (Fig. 14) shows similar differences. Willstätter and Stoll (1913) found that the shade leaves of the beech contain more, those of the elder less, chlorophyll than the respective sun leaves. In both trees the ratio of green pigments to yellow pigments is greater, in the beech much greater, in the shade leaf than in the sun leaf. The figures relate the chlorophyll to weight of fresh leaf substance,

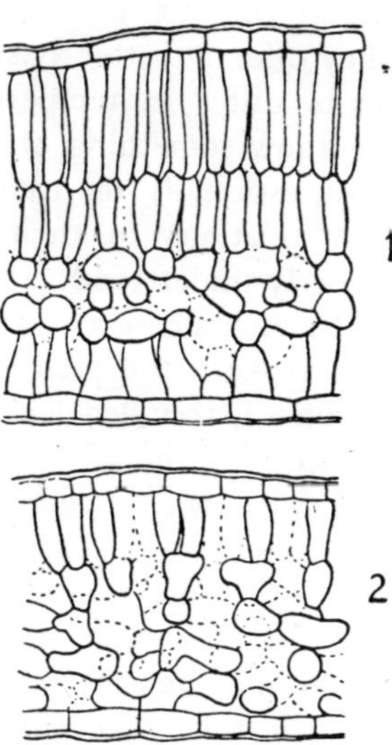

Fig. 14.—Sun and shade leaves of beech: 1, sun, 2, shade leaf. The sun leaf has two layers of palisade tissue at the upper surface, and one layer at the lower surface. × 240. (After Nordhausen.)

not to area. The average area of the leaf is small in the sun, and again small in extreme shade, with a maximum at an intermediate light intensity. Stomata are much more numerous in the sun leaf: the sun leaf of the beech has 413 stomata per sq. mm., the shade leaf only 113 (cp. Nordhausen, 1903, Schramm, 1912). These sun and shade leaves of a single plant are contrasted in the same way as the sun and shade plants we have just considered. They offer a better basis for comparison. Their environments differ in temperature, humidity and illumination. We know something of their assimilation, but much exact work, especially on transpiration, will be necessary before their relations to their respective environments can be properly understood.

In such a case as that of Campanula we can trace a relation between the plant's requirements and the leaf types; the broad, rounded leaf of the shade station offers a large surface for the absorption of the low light intensities available; the narrow upright leaf of the exposed station might serve as an example of a panphotometric leaf. We may also relate the structure of sun and shade leaves of trees like the beech, maple, or elder to the type of illumination they receive in nature. The thin palisade of the shade leaf may be supposed to absorb efficiently light of low intensity, and the available leaf substance is spread over a greater area giving a greater absorbing surface. The sun leaf, on the other hand, with deep palisade, may be regarded as absorbing more efficiently light of higher intensity; the increased energy supply thus available is paralleled by an increased supply of carbon dioxide through the greater number of stomata. As in the thin leaves of shade plants, the respiration of the shade leaf of a tree must be small per unit area, and a low light intensity will permit assimilation to balance respiration: Harder (1923) found this critical illumination for the shade leaves of the ivy to be one-half that for the sun leaves, and Boysen-Jensen obtained a similar result for the elder. A good deal of experimental work has been done on the relative efficiency of sun and shade leaves, not always with very satisfactory methods. The results, however, agree well with the interpretation given above. Thus Müller (1904) found, for the walnut and elder, that, in the shade, sun leaves and shade leaves assimilated at equal rates for equal areas, while in the sun the sun leaf was superior. For equal weights the shade leaf was slightly superior in the sun, markedly in the shade. For the lily-of-the-valley Hesselman (1904) found the shade leaf superior in sun and shade. For the pine and the spruce Stalfelt (1922) found the shade leaves superior for equal weights. For the elder Boysen-Jensen found the sun leaf more efficient, for equal areas, in the sun. We may refer to Blackman and Matthaei's demonstration that equal areas of very different types of leaves assimilate at equal rates when light is limiting. This agrees with Müller's result.

The difference between the sun and the shade leaf is due to the direct effect of different factors only to a limited extent. Nordhausen (1903, 1912) and Schramm (1912) have shown that the particular type of leaf is already determined in the bud, so that it may be regarded as showing after effects of the conditions of a former vegetative period. Trees from an exposed position, transplanted to a shady one, may show the sun type of leaf for several years, and *vice versâ*. They have further traced in detail an important parallel between the leaves formed early in the development of the individual and the shade leaves on the one hand, and between the leaves of more mature growth and the sun leaves on the other.

It is an almost universal phenomenon that the leaves of a plant pass through a definite series of changes in form and structure as the plant grows. We may speak of the earlier and later leaves as *youth* and *adult* forms. Such differences also exist, though less markedly, between the basal and apical leaves of a single shoot. Our example of the harebell shows this difference well. The round leaves are youth forms. In shady places they may be retained throughout the plant's life, though such a plant does not flower. In open stations they are soon lost and the narrow adult type is produced. In the trees the sun leaf is a typical adult leaf, and the shade leaf a typical youth form. The correspondence may be seen from some of Nordhausen's measurements given in Tables XXX and XXXI.

TABLE XXX

STRUCTURAL CHARACTERS OF SUN AND SHADE LEAVES OF THE BEECH

Origin of leaf.		Thickness of leaf.	Number of layers of palisade.	Depth of upper layer of palisade.
Adult tree	Sun ..	210 μ	2	60 μ
	Shade ..	108 ,,	1–2	28 ,,
Young tree	Sun ..	117 ,,	1–2	39 ,,
	Shade ..	90 ,,	1	24 ,,

TABLE XXXI

COMPARISON OF APICAL AND BASAL LEAVES OF BEECH SHOOTS

Illumination.	Position of leaf.	Thickness of leaf.	Number of layers of palisade.	Depth of upper layer of palisade.
Bright light ..	Apical	180 μ	2	60 μ
	Basal	140 ,,	1–2	48 ,,
Shade ..	Apical	132 ,,	2	36 ,,
	Basal	92 ,,	1	36 ,,

In Table XXX we have exhibited the characteristic difference between extreme sun and shade leaves in the grown tree. In the seedling there is much less difference between the two; the characters of the shade leaf are even more pronounced, while the sun leaf comes very near the shade leaf of the adult tree. As Table XXXI shows, on a single shoot of the adult tree the basal leaves are always relatively of the shade leaf type, though it is only in well-shaded basal leaves that the character of the seedling shade leaf is approached.

These results show that the production of the sun or shade type of leaf under appropriate conditions is less a direct structural modification than the resumption or exaggeration of a certain developmental stage. This is well shown in the harebell. In open stations it produces a few rounded leaves and then forms only the linear adult type, these alone appearing on the flowering shoot. By keeping the plant shaded, or by placing an adult plant in the shade, the continued or renewed production of the youth form is forced on the plant. In this case the plant remains immature, for it does not flower.

The plasticity of such trees as the beech and maple is much more limited. As might be expected, it is not possible to modify the leaves of the seedling so much as those of the adult. Moreover, in the adult, it is easier to force the basal leaves of a shoot to assume shade, or youth, characters, and the apical leaves to assume sun characters, than to reverse these processes. A particular leaf, by virtue of the age of the plant bearing it, or of its position on the shoot, inclines to one or the other type; it is always easier to exaggerate that type than to reverse it.

The conditions in which the young plant finds itself are likely to be more or less shaded and moist. Even the harebell sprouting in an open place will be shaded by surrounding grasses, and so also with tree seedlings. Normal development brings the plant gradually to stronger light and drier air. So we have a relation between the environment of the youth form and of the adult shade form, as we have between their structure.

The influence of intense illumination has been studied by Bonnier (1895). A comparison of plants at high altitudes with those of the same species in the plains, showed that the former tend to produce smaller, hairier leaves of an extreme sun type with thick cuticle and numerous stomata. The very high degree of insolation, and particularly the excessive proportion of ultra-violet light, must play an important part in causing these changes, but many other factors are also at work.

Alpine Plants.—Henrici (1921) has recently investigated the assimilation of alpine plants. The lower temperature limit of assimilation is extraordinarily low—down to − 16° C. air temperature; but at low temperatures and light intensities, though assimilation takes place, no starch is formed. If a plant is placed in a high temperature and the assimilation in different light intensities is determined, it is found that the rate of assimilation increases to a point at which starch makes its appearance; thereafter a decrease occurs, followed at still higher light intensities by a second increase. The same thing takes place at high light intensities when temperature is the limiting factor and is gradually increased. A comparison of sun and shade plants of the same species showed that for the latter a time factor set in with the temperature as low as 15° C., and for the former at 31° C. For light a time factor set in at the very low intensity of one-twentieth sunlight for shade plants. We may note here that Matthaei (1904) found that assimilation by the cherry laurel decreased markedly in April. As Lewis and Tuttle (1920) have shown for a number of evergreens that sugar content of the leaves falls, and starch is formed in spring, it seems that Matthaei's result is analogous to that of Henrici for alpine plants.

Tropical Plants.—Mention may here be made of the work of McLean (1920) on the assimilation of the coconut palm in the Philippines. The rate rises rapidly with rising light and temperature in the morning, reaching a maximum about 9 a.m. It falls off somewhat till about 2 p.m., rises to a second and lower maximum about 5 p.m., and falls rapidly to zero between 6 and 7 p.m. The rapid morning and evening changes correspond to the short twilight of the tropics. The cause of the mid-day fall was not elucidated. McLean suggests that it is due to too intense insolation; it seems more likely that it is connected with transpiration and a possible closure of the stomata, the condition of which should be examined. Giltay (1898) found that the production of dry matter by plants in the tropics was not greater than under favourable conditions in Europe. This is explained by Blackman's results; but as the conditions, especially of temperature, may be favourable for much longer periods in the tropics, the total growth is much greater, more rapid and luxuriant.

Such results must be looked on as preliminary ; we are in a position to appreciate how varied are the factors involved in governing assimilation ; we scarcely yet know how variation in leaf structure enters into the matter. Still less do we understand the effect of possible changes in the minute structure of the chloroplast. The investigation of the relation of different plant types to efficiency in assimilation is practically a virgin field.

§ 18. Extent of Transpiration

We come back now to some considerations regarding transpiration, which also have an important bearing on photosynthesis. We have already referred to the extent to which transpiration takes place in ordinary plants, and have seen that the amounts of water passing through the plant are very large. Ganong has reckoned that the transpiration of ordinary greenhouse plants ranges through the day from 15 to 250 grm. per hour per square metre of leaf surface, with an average of about 50 ; at night the average falls to about 10. We may get a more vivid idea of what this means by taking Hales' values—the first obtained—for the transpiration of the sunflower. He found that a large plant with a leaf area of 5616 sq. in. or 39 sq. ft., gave off on an average 20 ounces of water in the 12 hours of a hot day, that is to say, about 1 pint ; while a cabbage with one-half the leaf surface (2736 sq. in.) gave off as much as 1¼ pints. Von Höhnel reckoned that a large birch, bearing about a quarter of a million leaves, might give off over 400 kilos. of water on a sunny day, or about 90 gallons. Such figures serve to emphasise the very considerable water turnover of the leaf in a short time, and the extent to which the root must draw on the soil. Now it is not uncommon in hot sunny weather to see such plants as the sunflower or the potato drooping in our gardens and fields, or the red campion and foxglove in natural stations. And this means simply that the root, under circumstances of extreme demand, temporarily fails in a sufficient supply. In such cases the plant as a rule makes good the deficit through the night, when transpiration is reduced; no permanent damage results. But even so, the stomatal closure which follows wilting means a decrease in the assimilating power of the plant. Prolonged periods of drought have more serious effects; with us they are not common, yet in some years the pastures are "burnt up"—the herbage drying and withering. Under such conditions the mere closure of the stomata on wilting has been insufficient to prevent a ruinous loss of water by the plants affected. It is clear that in more arid conditions, conditions culminating in the desert environment, our plants would rapidly perish, and that the vegetation which does survive must have special powers of water conservation and supply. We have seen that as regards the root system this is so, and we now turn to the general question of the manner in which the water content of the plant is conserved in temporary and more prolonged periods of water shortage.

§ 19. Water Balance, Wilting, and Water Storage

We must first consider the question of the amount of water the leaf may lose without seriously disturbing its functions. When the leaves of a mesophyte flag, how much water, relative to their full content when turgid, have they lost? Knight (1922) found that fully turgid leaves of such trees as the ash and apple wilted after they had lost under 1 per cent. of their water content. This figure is probably representative for mesophytes in general, and it suggests that these plants work with a very small *water balance*. Only when absorption and conduction can cover, or very nearly cover, the loss by transpiration is the plant safe from wilting; it has little reserve to fall back on. This is confirmed by Thoday's experience with the sunflower; he found that on a fine summer day the leaves could be kept turgid only by forcing water into them under pressure. Livingston and

Brown (1912) found daily variations in the water content of leaves of Physalis and Nicotiana of 1 to 6 per cent.

In plants of a more xerophytic type different relations are found. For Parkinsonia, a leguminous tree of the Arizona deserts, a daily variation in the water content of as much as 10 per cent. was found by E. B. Shreve (1914). The small hard leaves do not show wilting, and we have here a leaf that can suffer a loss of water, without serious injury, and in the normal course of events, that would certainly be fatal to the sunflower. Livingston found daily variations of as much as 20 per cent. in desert plants.

Water Storage.—In many plants the leaves or other organs are provided with special water storage tissue which very greatly increases the water balance on which they work. A cross-section of the leaf of Tradescantia shows the epiderm on each side to consist of huge colourless cells between which lies a thin band of assimilating tissue. In some species of Ficus the epiderm cells are large, in others the epiderm is several layers thick. In Peperomia there is a water-storage tissue many cells thick of epidermal origin. In Rhizophora the storage tissue is subepidermal. In the ice plant, *Mesembryanthemum crystallinum*, the water is stored in modified hairs, huge bladder-cells studding the leaf surface and giving the plant its curious glistening appearance. Water storage tissue in the form of enlarged epidermal cells is probably the most frequent; it is common in the orchids, perhaps even in some of our native species.

In succulents the water storage in the mesophyll of the leaf, or in parenchymatous tissues of the stem, is extensive, and these plants have a large water balance. The necessity for this is imperative in such plants as the cactuses, which are exposed to arid conditions for months on end, and have root systems capable of absorbing water only when the surface layers of the soil are moist. We have exact information for the cactuses of Arizona from the work of Macdougal and Spalding (1910). These plants are exposed to two dry seasons of three or four months' duration, in early summer and in autumn, when no rain falls. They transpire continuously, though the amount seems to bear little relation to the weight or surface of the plant. An Echinocactus, a globular type, for instance, weighing 42 kg., was taken up and kept in the laboratory for 16 months without water. It transpired at the rate of from 1 to 29 grm. daily, losing in all 4 kg. of water before it died. Another plant, weighing 17 kg., lost 5 kg. between November and May; it was then taken up and placed in the soil, where it made up its loss and showed growth, weighing 20·5 kg. in October. In the following May it had fallen to 13·5 kg., but was still quite healthy. Here we have a plant losing 30 per cent. of its weight, or 40 per cent. of its water content, without injury. MacDougal states that Carnegeia, a columnar type, may lose as much as 63 per cent. of its water content without permanent injury. The immense quantities of water stored in these plants may be judged from the estimate that a Carnegeia 6 ft. high may contain over 30 gallons of water; while a full-grown branched individual, with a maximum diameter of 2 ft. and a height of 40 ft., may lose and gain, in the course of a season, a matter of 100 gallons. The loss of water is accompanied by very marked contraction.

For succulents of the Sedum or Mesembryanthemum types the balance is probably much smaller, though for these exact data are wanting. Plants with epidermal storage tissue may lose 10 per cent. of their water before flaccidity sets in.

Where a specially differentiated water storage tissue exists, it can be of service only if the assimilating cells can remove water from it, that is, if they have a higher osmotic pressure. Observation has shown this to be the case. The water tissue of Peperomia collapses while the chlorophyll containing cells are still turgid. Haberlandt and Schimper observed that in *Rhizophora mucronata*, one of the mangroves, the younger leaves remained fresh for several days after the older leaves had wilted. The water tissues of these leaves increases from one-half to two-thirds the thickness of the leaf as this matures. Here we have not only a storage tissue in each leaf, but a withdrawal of water from the older leaves by the younger.

Some of the suction force measurements of Ursprung and Blum (1918*a*, *b*) have an application in this connection. In a detailed examination of the different tissues of the leaf in the ivy and beech, they found that the suction force of the epidermal cells was less than that of the mesophyll. Thus in the beech, the upper epiderm had a suction force of 7 atmos. and the palisade of 16 atmos., while the values for the lower epiderm and the spongy parenchyma were 6 and 11 atmos. respectively. For the ivy, values for the upper epiderm and the palisade were 8·3 and 15·6. This is regarded as showing that in these leaves, too, the epiderm functions as a water store, which may be drawn on by the mesophyll. If this is so, then the storage function of the epiderm may be general. In view of the small fall in water content which occurs at wilting in such leaves, it cannot be said to provide a large working margin (cp. also Haberlandt).

We see, then, that the possession of water storage tissue provides many plants with a considerable water balance, with the help of which they can withstand prolonged spells of drought. How assimilation is affected in such plants by a loss of water short of the danger point, we do not know. In the ordinary plant water storage is at a minimum, the available margin is small, and a short period of drought is sufficient to cause wilting or more serious consequences. With the onset of wilting assimilation may practically cease.

We may here note the water storage capacity of the trunks of deciduous trees, which is useful in another connection. Craib (1918) has shown that the water content of the trunk of the maple increases greatly during the winter, a store being formed which supplies the enormous demands of the expanding, and rapidly transpiring young leaves in spring.

§ 20. Atmospheric Conditions and Transpiration

The transpiration of a leaf exposed to the air is greatly influenced by external conditions. Just as rise of air temperature, fall of atmospheric humidity, insolation and wind increase the drought which dries the washing, so they increase the transpiration from the leaf. The leaf is not, however, as simple a thing as a towel hung out to dry; these factors have not only a *direct* effect on its transpiration, they may also influence it *indirectly* by affecting the condition of the leaf. The leaf reacts in various ways to changes of illumination; it behaves differently when turgid and when flaccid. Changes in transpiration rate due to such alterations may be looked on as *indirect* effects of the external conditions, or as a process of *regulation* by the leaf. The atmospheric conditions may have a further indirect effect on transpiration from their influence on soil moisture, and thus on water supply.

Methods of Investigation.—In investigating the effect of external factors on transpiration, two methods may be employed. Records may be made of temperature, humidity, wind, etc., and plotted along with the graph representing transpiration so that their relative effects are exhibited. It is, however, more convenient to use for comparison a record of the *evaporation* from a standard water surface. This sums up the influence of the various atmospheric conditions, and enables us to see any differences between the effects of these on transpiration and on evaporation. The evaporation record may be supplemented by suitable data for particular factors. The water surface generally employed is the saturated surface of a porous porcelain filter candle. This is filled with water, and a glass tube, passed through the cork which closes it, dips in a bottle of water acting as a reservoir. The evaporation may be determined by weighing or, with a suitably graduated bottle, by direct reading. Such an instrument, devised by Livingston, who has also worked out methods for its standardisation, is called an *atmometer*.

If the amount of transpiration is divided by the amount of evaporation, both being referred to unit area, a value is obtained which is called *relative transpiration*. In it the direct evaporating effect of the atmosphere is supposed to be eliminated, and there is exhibited the physiological behaviour of the leaf—that is, the difference in reaction shown by the leaf from that of a physical evaporating surface. This is not always true, for Knight (1917*b*) has shown that different strengths of wind do not affect the plant and the atmometer in the same way, *even directly*; that is, evaporation from different types of physical surface does not keep a constant relation in still and moving air. If the wind velocity is constant throughout an experiment, however, the relative transpiration graph does reflect the physiological behaviour of the plant as related to changes in such factors as temperature and light. The same result is secured if, instead of

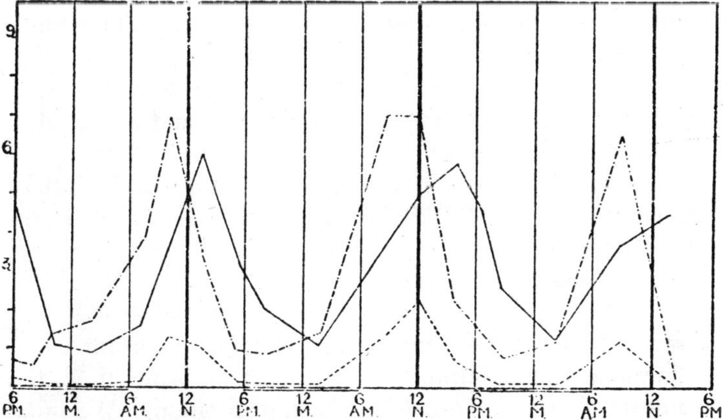

FIG. 15.—Transpiration (broken line) of *Euphorbia capitellata* through three days, compared with evaporation (continuous line) and relative transpiration (dot-dash line): the scale of the ordinates is different in the three graphs. (After Livingston, modified.)

reducing transpiration and evaporation to unit areas, we simply divide transpiration from any (constant) area by evaporation from any (constant) area. The value so obtained may be referred to as the T : E ratio. It is, of course, proportional to relative transpiration. Only occasionally has the condition of constant wind velocity been observed in determinations of relative transpiration, and results must therefore be interpreted with caution.

Example of Relative Transpiration.—In Fig. 15 are

given the graph representing transpiration of a Euphorbia during three days, and the graph of evaporation from an atmometer, as determined by Livingston (1906). It will be seen that there is a general resemblance between the two ; the changing conditions of the passing day have a similar effect on both. Looking closer, however, we see that there is a very important difference. The highest rate of transpiration occurs each day between 10 a.m. and 12 noon, and transpiration then decreases steadily. Evaporation, on the other hand, increases till between 2 and 4 p.m., where the maximum occurs. This means that external conditions are such as to increase evaporation till late in the afternoon, but that some hours earlier a change in the reaction of the plant has led to a diminution of water loss. The same thing comes out in the graph of relative transpiration which is also given. If transpiration and evaporation were affected by external change in the same way and to the same extent, this graph would be a straight line. Actually, it shows that transpiration is increasing more rapidly than evaporation till the maximum is reached, and that subsequently the increase in evaporating power of the air is not reflected in the transpiration.

The fact demonstrated by such an experiment is that the conditions of the atmosphere have an important effect in influencing transpiration, but that, at the same time, the condition of the leaf modifies the transpiration. We must now consider in what way this modification is effected by the leaf..

§ 21. Stomatal Regulation

It was long thought that regulation depended on alteration in the aperture of the stoma, the mechanism of which we have already studied ; but until recently no exact experimental data were available, because no proper method of measuring stomatal aperture, frequently and accurately, was available. We have now two suitable methods. The first is based on the fact that if a piece of epiderm is removed from the leaf and instantly plunged in absolute alcohol, the stomata retain their size and shape as in the fresh state, and can be accurately measured (the average of a considerable number being taken) when convenient. This method is due to Lloyd (1908). Darwin and Pertz (1912) devised an instrument, called the *porometer*, which has the advantage that it may be modified to give continuous readings, or records (Laidlaw and Knight, 1916). It consists in essence of a small glass funnel, the wide end of which, about 1 cm. in diameter, is glued to the surface of the leaf. The other end is connected to a vertical tube of water. The weight of the water column draws air through the stomata, and the rate at which this happens can be ascertained from the time it takes the water column to fall a given distance. Since, in the porometer, air is *drawn* through the stoma, the rate of its passage is proportional to area and not to diameter, as with diffusion. For this reason the *square root* of the time is taken as inversely proportional to the aperture of the stomata. The amount of transpiration can be very accurately determined by weighing; if a potted plant is employed, the pot and earth are covered with a suitable waterproof case. Self-recording apparatus has also been employed here. The rate of absorption of water from a glass tube (*potometer*) has also been employed as a measure of transpiration, but, except in conditions where absorption just balances loss, is obviously less accurate.

The results obtained by the use of these methods have, on the whole, been unfavourable to the hypothesis that stomatal movement is the most important mode of regulation of transpiration by the plant. Following the rate of transpiration of the leaves and the changes in stomatal aperture for *Verbena ciliata* through the 24 hours, Lloyd (1908) showed that, under normal laboratory conditions, the transpiration rate, which was low at night, began to rise at about 2 a.m., reaching a maximum at between 10 a.m. and 11 a.m. Before noon it began to fall, and this fall continued, with a slight check in the afternoon, till midnight. The stomata begin to open about 2 a.m., and become full open at 8 a.m.; they commence to close about 12.30 p.m., and are completely closed at 6.30 in the evening. Fig. 16 shows graphs for the two processes, and at first sight there is a striking similarity, indicating a fairly exact correspondence. When we compare them in detail, however, we find (1) that after the stomata are full open at 8 a.m., the transpiration rate increases for two hours; (2) that the

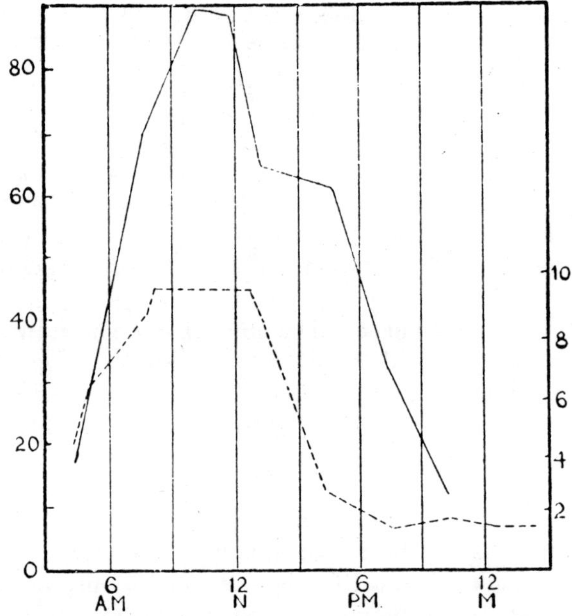

FIG. 16.—Transpiration and stomatal movement; the continuous line represents transpiration of a shoot of *Verbena ciliata*, the broken line represents extent of opening of stomata in microns. (After Lloyd, modified.)

transpiration rate begins to decrease a full hour before the stomata begin to close, and has already fallen to 70 per cent. of its maximal value when closure begins; (3) that the check in the fall of the transpiration rate does not correspond to any check in the closure of the stomata; (4) that the transpiration rate continues to fall after the stomata cease to close, and later starts to increase with the stomata very nearly closed. This particular experiment is typical of the normal march of transpiration and stomatal movement through the 24 hours. Its analysis shows that while the two agree in their general course, there can be no question of a delicate regulation of transpiration by stomatal movement; or, to put it in another way, stomatal aperture is not in general the factor limiting transpiration. In fact, it seems more likely that the loss of water about noon is responsible first for restriction of transpiration, and then for stomatal closure. Lloyd measured transpiration by the potometer.

Artificial changes in the conditions made the discrepancies clearer. To quote only one experiment: a shoot of Verbena kept overnight in a dark room was placed in the sun at 7.15 a.m.; at this time the stomata were scarcely 50 per cent. open, at 7.45 a.m. they had opened to 75 per cent., and at 8 a.m. had closed to 60 per cent. Yet during this period the transpiration rate had steadily increased to four times its original value.

Trelease and Livingston (1916) followed the daily march of transpiration by the cobalt paper method (it is possible to time the rate of colour change in a standardised paper, and so to obtain accurate relative values), by which they determine an "index of transpiring power," which is proportional to relative transpiration in still air. They compare this with the changes in stomatal aperture determined by the porometer. The graphs they give for five series of experiments show that, in two, variation in stomatal aperture agrees closely with change in transpiration rate, while in three it does not. F. Darwin (1916) concludes, from the results of a series of experiments with the porometer, that regulation by the stomata is important.

Knight (1917), in the most exact work so far done on the subject, finds that sometimes the graph of the T : E ratio runs parallel with the graph of stomatal aperture, but that usually this is not so. Graphs for one of his experiments are given in Fig. 17.

Loftfield (1921) has carried out a long series of investigations with American crop plants, which have yielded results of the greatest interest, the details of which should be studied in his important memoir. We may take a few illustrative examples. He studied his plants in the humid spring climate of Minnesota and the hot dry summer of Salt Lake City. He found that they could be divided into three groups as regards stomatal behaviour.

(1) Cereals. In dry conditions the stomata of wheat, barley, and oats remain closed for long periods, showing

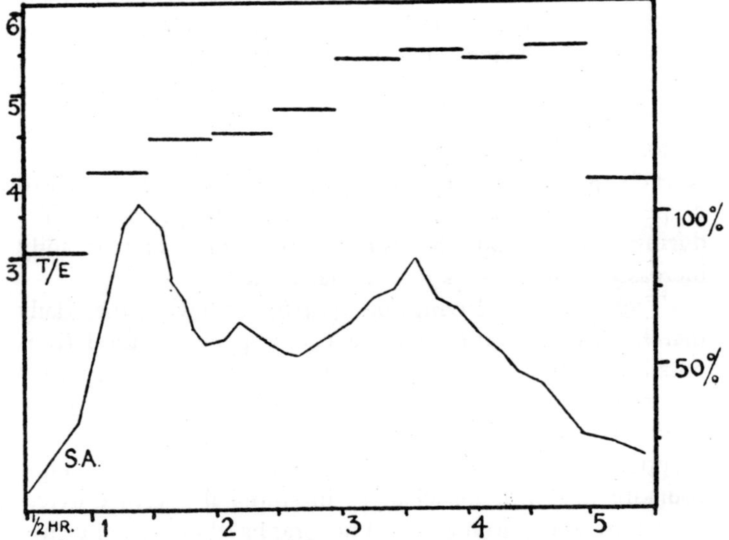

FIG. 17.—Transpiration and stomatal movement; the thick lines indicate transpiration (T/E) for the half-hour periods; the continuous line indicates the per cent. opening of the stomata. (After Knight, modified.)

only a slight and transitory opening in the morning. In greenhouse conditions, with higher atmospheric humidity and better water supply, they show a daily opening of several hours, but only of limited extent. We have already noted that the grass stoma is relatively inactive. In our own climate the opening is probably of longer duration. It would be of interest to know the assimilation relations of such plants in arid conditions. The maize has a partial opening between 9 and 10 a.m., a closure round noon, a second opening at 2 p.m. followed by a second closure, and a third slight opening in the evening. The stomata finally close at 9 p.m. The more considerable opening of the maize stomata may be related to the fact that it is better suited to dry climates than the other cereals.

(2) In the potato growing in conditions of good water supply and low evaporation, the stomata close only for the three hours before midnight, and are open for more than twenty hours. As the water supply fails they close earlier, and may be closed from 4 p.m. till 1 a.m. If along with

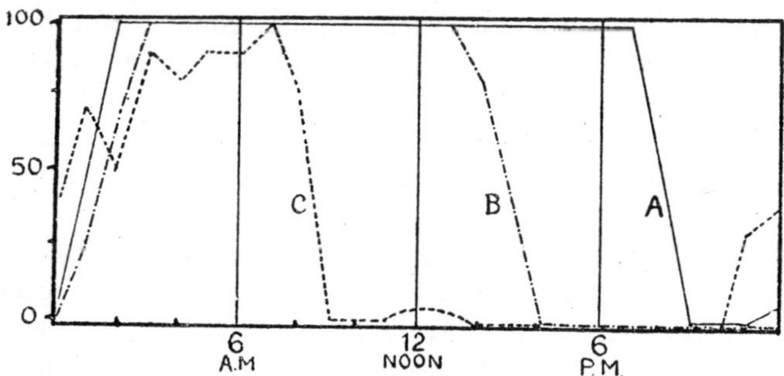

FIG. 18.—Daily stomatal movement in potato; 100 = full open; A, water supply good, atmosphere humid; B, water supply good, atmosphere dry; C, water supply low, atmosphere dry. (After Loftfield, modified.)

poor water supply there occurs high evaporation, closure takes place as early as 9 a.m. and persists throughout the day, the opening movement starting at 10 p.m. This behaviour, which is illustrated by Fig. 18, is characteristic of plants with rather fleshy leaves like the tulip, onion, and cabbage, and of some with thin leaves. It may be noted that F. Darwin (1898), on the whole confirmed earlier work of Leitgeb's (1886) and Stahl's (1897), which showed that the leaves of marsh plants and of plants showing sleep movements are characterised by possessing stomata which remain open day and night, a condition which is approached by that of the potato in a moist environment. E. B. Shreve

(1916) found the stomata of Cactuses more widely open at night, a condition like that of the potato in arid conditions.

(3) Most thin-leaved plants examined, such as alfalfa, clover, tomato, dandelion, showed, under favourable conditions, a gradual opening in the morning, complete from about 11 a.m. to about 3 p.m., and then a gradual closure complete from 9 p.m. to 5 a.m. This we might regard as the normal behaviour of the mesophyte. With lower water supply, a temporary mid-day closure appears, which, with increasingly arid conditions, extends both ways until the stomata remain closed all day. With this there is associated an increasingly prolonged opening at night. The assimilation relations under such conditions would again be an interesting study. In a number of trees examined, cherry, peach, pear and poplar, no mid-day closure was ever found. Loftfield inclines to relate this to the potential water balance in the trunk.

Loftfield gives many graphs showing the relation of the march of transpiration to these movements. In some cases a close relation is evident, in others discrepancies occur. Loftfield's conclusion is that stomatal regulation is important; his own words may be quoted: "Although factors concerned in evaporation have great influence upon transpiration, this influence is definitely controlled by the stomata. When the stomata are wide open, or nearly wide open, transpiration is the result of the action of the factors of evaporation alone, since the stomata in no wise interfere with the action. As the stomata close the influence of the factors is lessened, but until closure has reduced the apertures to 50 per cent. or less, stomatal regulation is still largely overshadowed by the control exerted by these. When closure is almost complete, the regulation of water loss by the stomata is very close, and the effect of the factors overshadowed by the effect of even very small changes of the opening."

This seems a fair statement of the case so far as our present knowledge goes. It means, however, that with the stoma over 50 per cent. open a dangerous depletion of water may take place, a loss over which the stoma has no control. Taken in connection with the fact that at the beginning of wilting the stoma actually opens wider, this indicates that the stomatal mechanism is no guard against wilting. The stoma may remain open while the water content of the leaf is decreasing rapidly.

§ 22. Regulation by Internal Leaf Changes

We have seen, however, that with increasing evaporating power of the air there may be a decrease in transpiration, and Loftfield's statement as to the controlling effect of the evaporating factors must be modified in this respect. If now this decrease is not due to stomatal closure we must seek for some other explanation. Livingston and Brown (1912) consider that it is directly due to the decrease in water content of the leaf. They suggest that a slight drying out of the walls of the mesophyll cells of the leaf, by removing water from the outer surfaces and withdrawing it within the ultra-microscopic pores of the wall substance, increases the force with which water is held, and so decreases the rate of transpiration. They refer to the daily fall in water content of the leaf in support of their hypothesis. Determinations on Physalis show that the water balance of the leaf begins to be drawn on (that is the water content falls) about 9 a.m. and continues to do so till 3 p.m. At noon the relative transpiration begins to fall, indicating the beginning of regulation by the leaf, for evaporation increases till 3 p.m. In the evening the water content of the leaf begins to rise, and the balance is restored about 7 p.m. Thus, after a certain depletion of the water content, regulation sets in, limiting transpiration to a value less than that corresponding to the evaporating power of the air. E. B. Shreve (1914) supports this view.

An experiment of Knight's has a bearing on this theory. A shoot was transpiring rather more rapidly than it absorbed water, so that its water content fell slightly ; its transpiration also fell till it reached a rate which just balanced absorption, and then remained constant. The conditions of the experiment were then altered by placing the apparatus in still air, so that transpiration was much depressed; absorption now predominated, and the water content of the shoot increased; the original conditions were restored, and transpiration at once increased, not to the previous constant value, but to a much higher one. In this case, therefore, an increase in the water content of the shoot was directly responsible for an increase in transpiration, and *vice versâ*, Knight (1922) has shown that the stomatal opening which occurs at the beginning of wilting is accompanied by an increase in transpiration. The maximum transpiration is however reached, and a decline sets in, before the stomata reach their maximum opening. This decline in transpiration must therefore be caused by the decreasing water content of the leaf. As we have seen, the decrease in water content which is effective may be very small. The slight loss of water which may lead to diminution in transpiration has been called by Livingston "incipient wilting."

It may be noted that in all this work no attention seems to have been paid to the cuticular fraction of transpiration. Yet we have seen that this may be an appreciable part of the whole in mesophytic plants, and the work of F. Shreve (1914*a*) has shown that in some rain-forest plants it may be greater than the transpiration from the stomata. It is quite possible that at the critical points in the march of transpiration the cuticular fraction may be an important factor, and its study should not be neglected.

Another possible factor in the regulation of transpiration is change in the concentration of the cell sap. In the plants studied by Livingston the daily fluctuation in water content was, on an average, about 4 per cent. This means an increase in the concentration of the cell sap, which might be further increased by an accumulation of sugars. The consequent lowering of vapour pressure would, however, hardly produce an appreciable retardation of evaporation. It is possible that in the plants with very high osmotic pressure discovered by Fitting (1911), a certain protective effect is obtained; it is unlikely that it is very pronounced or general.

E. B. Shreve (1920) found that the xerophytic summer leaves of the desert composite *Encelia farinosa* contained in their cell sap a brown substance, which appreciably diminished evaporation rate, and which was absent from the more mesophytic type of leaf produced in the rainy season. Such cases are probably rare.

Conclusions on Regulation.—We must conclude, then, that a slight fall in the water content of the leaf acts as an automatic check on transpiration, but that, despite this internal regulation, wilting may set in. On wilting, stomatal closure, after a preliminary temporary opening, follows, and ultimately cuts down water loss to a low value, conserving the supply or even allowing absorption to make up the loss. The mesophyte grows in conditions where such regulation is normally sufficient to obviate serious damage, but the plant which exists in more extreme conditions can do so only by virtue of special safeguards.

§ 23. Limitation of Transpiration by Form and Position of Stoma and by Cuticle

We have assumed so far that the stoma is of the "normal" type in structure and position. Frequently, however, the structure is such that diffusion is slowed down, and the stoma may be so placed that direct impact of extreme atmospheric conditions is avoided. Most simply this happens when the stomata are confined to the less exposed lower surface of the leaf, as is the rule in our broad-leaved trees.

A common mode of protection is seen where the stomata are sunk in pits or grooves below the surface of the epiderm. Grooved leaves are very common among the grasses and sedges; they are prominent in our two dune grasses, *Psamma arenaria* and *Elymus arenarius*. In such cases the stomata are almost or entirely confined to the bottom and sides of the grooves. Grooves are also a common feature on the stems of switch plants, in which the leaves are much reduced, as in Ephedra, or fused with the stem, as in Casuarina, or small and short-lived, as in *Spartium junceum*. A similar feature is found in the leaves of Empetrum and of many Ericaceæ. The stomata are confined to the lower surface, and the inrolling of the leaf margins places them in a partially enclosed space.

The formation of pits is arrived at in a variety of ways. In *Nerium Oleander* the pits are depressions in the leaf surface in which groups of stomata occur. In Pinus and Juniperus, single stomata are sunk by the overarching of the neighbouring cells of the epiderm and hypoderm. This is the case, too, in the Australian *Hakea suaveolens*. The guard cells are, as a rule, smaller than the epiderm cells. and, if the disproportion is great and they are close to the inner edge of the latter, a pit is formed as in *Spartium junceum*. Perhaps the most frequent case is that in which the pit is formed by the presence of an abnormally thick cuticle. A thick cuticle is the commonest feature of xerophytic foliage, and is responsible for the glistening appearance of leaves like those of the holly or cherry laurel. The thick cuticle depresses cuticular transpiration. Renner reckons that in leaves such as those of the rhododendron cuticular transpiration is practically nil. It is commonly associated with a hard leathery leaf, with much sclerenchymatous mechanical tissue, which prevents flagging, and a close structure with reduced intercellular spaces; this may reduce evaporation from the internal surfaces. Combined with a distribution of stomata on the under surfaces of the leaves only, these features give a type of leaf highly resistant to drought. It is the characteristic foliage of evergreen trees, especially of those which retain their leaves through a hot dry summer, as, for example, the evergreen oaks of the Mediterranean and California, and many trees and shrubs of the tropical thorn forests and scrub. This vegetation is described by its leaf type as *sclerophyllous*. In such shrubs as the holly the cuticle forms a layer not quite so thick as the epidermal wall, but frequently it attains a much greater thickness, and in the neighbourhood of the guard cells may take on the form of special overarching outgrowths, forming more or less deep pits often of complicated form. In *Dasylirion filifolium* it forms a double chamber over the stoma; in *Euphorbia Tirucalli*, where it attains no special thickness in general, it builds up a deep well over the stoma.

The general effect of this mode of protection, whether due to individual pits, to grooves, or to an inrolling of the whole leaf surface, is to interpose between the diffusing stoma and the dry, outer air a space in which a high degree of humidity prevails, so that the rate of diffusion is slowed down. For, instead of the whole diffusion gradient from saturation to the minimum humidity being accomplished in the short distance from the mesophyll cell to the external orifice of the stoma (in moving air), it is extended for a considerable distance, and consequently the rate of diffusion falls. The conditions of diffusion for such stomata have been worked out by Renner with such exactitude as is possible for openings so irregular. For *Agave americana*, a leaf in which the pit is of a fairly regular shape, and is formed by cuticular extension, we may represent its effect as that of a wide tube in the bottom of which diffusion shells are formed in the usual way, and through which diffusion proceeds to the outer air, where again diffusion shells are formed. This can be treated by an extension of the formulæ we have already studied, but into the details of which we need not go. Renner calculates that the superposition of the pit leads to a depression of transpiration by 31 per cent. The more complex, almost globular pit, of *Hakea suaveolens* gives a depression of 37 per cent. In Nerium, with its stomata grouped in common hollows, the depression is as much as 77 per cent. Further complications arise when the stomatal pore is not straight. We have noted that, to a slight extent, this is the common case, the actual pore being divided into fore- and after-courts by slight cuticular protuberances. This irregularity may be much exaggerated, giving rise to such an extreme case as the sinuous channel seen in *Nipa fruticans*, where the effect on diffusion must be important. Experimental confirmation of Renner's results on the actual leaf is impracticable, but Renner has substantiated his calculation by evaporation experiments with physical models in which the observed depression is in good agreement with that calculated. We have thus every reason to believe that the protected stoma gives up

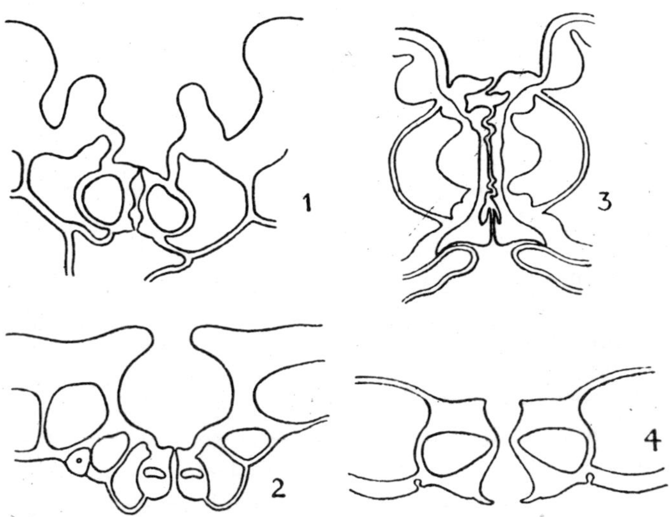

FIG. 19.—Structure of Stomata: sections through, 1, *Agave americana*, × 450; 2, *Hakea suaveolens*, × 500; *Nipa fruticans*; 3, *Verbena ciliata*, × 600. (1 and 2 after Renner, 3 after Bobisut, 4 after Lloyd.)

much less water than one freely exposed. Recently Gradman (1923) has shown that the pit is specially important in protecting the stoma from the drying effects of wind. Diffusion of carbon dioxide is not depressed to the same extent as is transpiration.

Depression of transpiration also occurs if the stoma does not open directly into the mesophyll, but is separated from it by internal cuticularised cells or by sclerenchyma. Occasionally prolonged drought, or the death of the guard cells, leads to a complete stoppage of the passage by thick-walled outgrowths of the neighbouring parenchyma, or by masses of resinous material.

The protection given by grooves is increased by the very general tendency of such leaves to roll in edgewise in drought. This is well seen in Psamma and Elymus, where the leaf rolls completely into a narrow tube in dry weather. Less pronounced cases are those where the two leaf halves fold together as in *Festuca glauca* and *Aira cæspitosa*. Examples are common in grasses of steppes or dry pastures. In such cases the stomata occur in grooves on the upper surface only, and it is this surface which comes to lie on the inside of the rolled or folded leaf.

The mechanism of this movement has not been quite cleared up. Along the bottom of the furrows of grass leaves

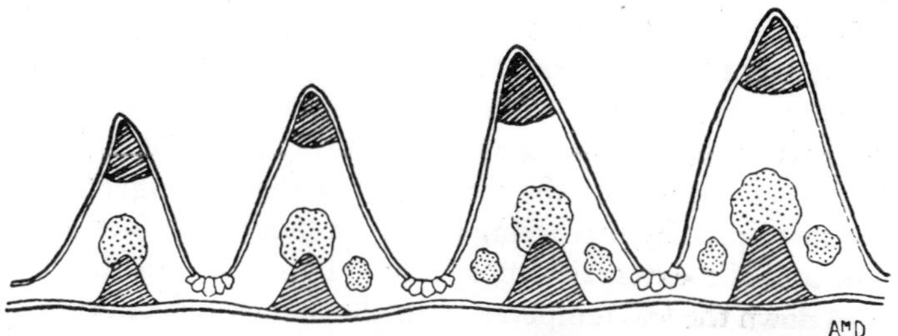

FIG. 20.—Leaf of *Aira cæspitosa*: transverse section showing the deep furrows at the bottom of which are the thin-walled cells which are concerned in the folding in of the leaf: mechanical tissue shaded. × 40.

there lie larger thin-walled water-storing epidermal cells. It is likely that these, as they lose water, have their walls drawn together, and so cause a local contraction of the upper surface. Whether the strong sclerenchyma, which clothes the lower surface, takes an active part in the movement or acts merely as a resistance is not known. Haberlandt's account should be consulted (cp. Fig. 20).

§ 24. Hairiness and Transpiration

Further protection of the stomata may be gained where the leaf is clad with hairs. This is often combined with the pit arrangement. The furrows of Casuarina bear hairs, and the grooved lower surfaces of the leaves of Loiseleuria and Erica are hairy. The production of hairs is a very widespread occurrence, and ranges from sparse and short bristles, or fringes along the leaf edge and about the veins (as in the beech), to dense tomentose coverings which give the leaves—of the Edelweiss, to take a famous example—a grey or white appearance. The hairs are epidermal outgrowths and show a wide range of form and structure ; they may be unicellular or multicellular, simple or branched, straight or wavy ; or they may assume such strange forms as the scales of Elæagnus, the oleaster, or the bifid prongs of the wallflower. Conspicuous hairiness is a common feature of xerophytic plants ; it is a mark of many of the plants of the garigue and macchia of the Mediterranean regions, *e.g.* of *Cistus albidus*, the lavenders, the rosemary, and others ; it is prominent in many mountain plants growing in exposed positions. When the hairiness reaches the tomentose stage, that is when the hair forms a thick covering or felt, it is likely that it acts in much the same way as the sinking of the stomata in pits, by preserving a layer of moister air between them and the external atmosphere. It is also held that it keeps down the leaf temperature in the sun by reflecting and absorbing light. A certain amount of evidence is available to show that " shaved " leaves attain a higher temperature than those with their natural covering. When the hairs are sparse the effect must be very much smaller. Renner (1909) holds that with sparse hairs the most effective type of hair is that which lies close to the epidermal surface, as in the wallflower. It is likely that such a covering may tend to prevent movement of the surface layer of air. In many plants where the degree of hairiness varies, it is marked in dry stations, while in humid surroundings the leaf may be almost glabrous. A striking example of this is afforded by *Polygonum amphibium*, which is known in two distinct habitat forms. As an aquatic it has floating leaves, oblong-oval in form, with a polished upper surface, and quite glabrous. On the land the leaves are smaller, narrower, and pointed, and markedly but not excessively hairy. There is undoubtedly a causal connection between dry atmosphere and hairiness, the mechanism of which we do not know.

Yapp (1912) has analysed the conditions for *Spiræa Ulmaria*, the meadow-sweet, in which variation of hairiness is found not only in plants of different habitats but in the younger and older leaves of the same plant, which, in meadow vegetation, exist under quite different conditions. The leaves may be thickly clad on the lower surface with fine hairs (pubescent), or hairless, or partly hairy ; the differences between the bare green leaf surface and the silvery felt are very marked. The rather complex changes may best be given in Yapp's own words : "This pubescence is subject to a kind of periodicity, and appears only under certain definite conditions. The chief rules governing the appearance of the hairs are as follows : (*a*) The seedlings, also all leaves formed during the first year, are glabrous. (*b*) On the erect flowering shoots of adult plants there is a regular succession of glabrous, partially hairy and completely hairy leaves. The earliest radical spring leaves are glabrous, the cauline leaves hairy. (*c*) The non-flowering shoots of adult plants produce only radical leaves. The earliest of these are glabrous as in (*b*). Subsequently the successive leaves exhibit increasing hairiness up to June or July, after which they are decreasingly hairy until glabrous leaves are once again produced in autumn. (*d*) The distribution of pubescence on the partially hairy leaves is interesting. The terminal leaflet is invariably hairy, and there is a regular decrease of hairiness from above downwards. Individual partly hairy leaflets generally possess a marginal band of hairs, with sometimes additional bands running inwards between the main veins."

In addition to the differences in hairiness the upper leaves show the "sun leaf" characters, the lower and radical leaves the "shade leaf" characters which we have already studied. The shade leaves are larger, thinner, with fewer, but much larger, stomata than the sun leaves ;

in the upper leaves there are two layers of typical palisade, and in the lower the palisade is much reduced ; sun leaf character goes with hairiness, and *vice versâ*.

We can relate hairiness to arid conditions in two ways. In the first place it is most pronounced in leaves unfolding in summer ; this is most marked in the succession on adult non-flowering shoots referred to in Yapp's summary. Records taken with atmometers showed that the period of maximum evaporation, June, July, and August, corresponds exactly with the production of the hairiest shoots. In the second place the position in which the hairy leaves are borne on the stem is where evaporation is most rapid. This brings us to speak of an important relation which has been worked out for marsh vegetation by Yapp in another paper (1909). In any plant community with fairly tall, close vegetation, the conditions as regards evaporation at the upper level of the plant shoots are very different from those lower down, where a considerable degree of sheltering takes place. Yapp investigated this by taking atmometer readings 4 ft. 6 in. above the soil (just clear of the vegetation tops), 2 ft. 2 in. above the soil, and 5 in. above the soil (at the bottom of the vegetation). During eight days in July the evaporation in the three stations was in the ratio 100 : 37·6 : 10·1. During eight days in August and September the ratio was 100 : 28 : 3·2. Temperature records showed a marked lowering at ground level. The main part of the difference in evaporation was probably due to shelter from wind in the lower strata. It is obvious that leaves near ground level are living in an entirely different environment from those 4 ft. up as regards humidity, wind action, temperature, and also light. This difference in environment is reflected in the presence of different layers of vegetation. Plants like Hydrocotyle, Symphytum, and Scabiosa, and the radical leaves of Spiræa inhabit the ground layer ; Carices, and the greater part of the leaf surface of Angelica and Lysimachia occupy the middle layer ; the upper leaves of these and of Spiræa form the upper layer. This is usually referred to as a *stratification of vegetation.* The difference in conditions is striking in a small space in the case of the marsh society, but it occurs equally in any close community; the range of the different factors of the environment being different, for example, in a forest and a heather moor.

Returning to the case of Spiræa, we see that the lower and radical leaves occur in an atmosphere in which evaporation is much reduced, while upwards, as hairiness increases, the evaporation is successively greater. Even if we take a single leaf we see that the most exposed apical leaflet is the one which tends to be most hairy. Even in the individual leaflet the margin first becomes hairy, and this is the region most liable to drought—in many leaves it withers first in keen winds.

Yapp was not successful in experimentally modifying, to any considerable extent, the regular succession of hairiness. No degree of drought could make the spring radical leaves hairy; not even a combination of low light intensity and high humidity suppressed the pubescence of leaves normally hairy; at the most the number and size of the hairs were reduced.

All this forms a very striking parallel to the case of the sun and shade leaves already dealt with, all the more so because the hairy and glabrous leaves are also anatomically sun and shade leaves. We might say that hairiness is here an additional sun leaf character. We have the same general relation between the character and the environmental conditions in which it *normally* occurs; and we have the same limited plasticity shown by a leaf developing under the action of the set of conditions, opposite to that in which it would normally grow. It is supposed by Yapp that diminution of water supply stimulates certain cells to elongated growth—the epidermal cells which produce hairs, the root hairs, and the palisade. Some experimental evidence can be adduced in favour of this view, though it cannot be said to advance much our understanding of the mechanism of hair production. The fact that large changes in the external conditions have so little effect makes the causal connection between the hairiness and the humidity, or light rather hopeless to trace. In other plants, however, Yapp found it easier to modify hairiness ; in *Epilobium hirsutum* and *Mentha aquatica* hairiness was completely suppressed by growth in a damp atmosphere. These plants and others such as *Lycopus europæus*, *Scabiosa succisa*, and *Lysimachia vulgaris* showed under natural conditions the same general succession of leaf type as Spiræa.

We are evidently dealing here with another special case of the very general succession of adult to youth forms. The adult form is the hairy leaf, and it is related to normal adult conditions ; in so far as it shows a diminution of hairiness in poor light and higher humidity, it is exhibiting a reversion to the juvenile form rather than a modification in any one feature in response to special circumstances. In some cases the regular course of development is very definitely rigid, as in Spiræa ; in others, as with Mentha, it is influenced markedly by the impact of the external complex of conditions.

Hairiness is, then, a very general xerophytic character ; it may, in a single species, show marked exaggeration in more arid stations, or at periods of development in which the plant is exposed to more arid conditions. Thus the American desert composite, *Encelia farinosa*, was found by E. B. Shreve (1920), to form large glabrous leaves in the wet, and small hairy leaves in the dry, season. Hairiness is marked in many plants, as is indicated by the frequent occurrence of such words as *tomentosum* as specific or varietal names ; it may be assumed, although direct proof is scanty, that it reduces transpiration and reflects light.

§ 25. Effect of Ethereal Oils

In certain dry regions the occurrence of plants with a strong aromatic scent is very characteristic ; the scent of the macchia of Corsica can be perceived far out at sea. The lavenders, rosemary, rue, thymes, balms, lemon, oleander, eucalyptus, wormwood, and pine are all shrubs or trees which occur naturally in dry stations. So great is the amount of ethereal oil given off from the rue that it is said to be possible to set it in flame on hot days. It has long been supposed that this feature has some biological significance. Dixon (1914) has shown experimentally that diffusion takes place more slowly in an atmosphere laden with such vapour. It is also known that less heat is radiated through these vapours. Experiment has so far failed to show, however, that transpiration is appreciably diminished by this means.

§ 26. Number of Stomata

Reduction in transpiration may also occur through reduction in the number of stomata. The variability of stomatal number in a single plant may be great, and is often marked between two individuals in different stations. Neger has collected some instances which show that, comparing related species, the number tends to be less in those growing in dry stations, *e.g.* in the sedges and poplars. For a single species, however, the exposed leaves have more numerous stomata. Yapp found the number much greater in the sun leaves of Spiræa, but the stomata were also much smaller. As we have seen, sun leaves have in general more numerous stomata than shade leaves. In alpine plants the stomatal number seems to increase with elevation and exposure. An exhaustive study taking account of stomatal size and of conditions of assimilation has still to be made. The exact effect, too, of the light and heat reflecting action of polished leaves has yet to be studied.

§ 27. Reduction of Leaf Surface and Transpiration

Leaf Fall.—We have dealt so far with various features which tend to reduce diffusion through the stomata and the epiderm. Transpiration may also be reduced by the reduction of the evaporating surface and such reduction may be permanent or temporary. A temporary reduction is shown by all our deciduous trees and shrubs, and by perennial herbs which die down in autumn, persisting as underground roots, rhizomes, bulbs, or corms. This fall of the leaf is not related to transpiration only; protection against frost is also important. Yet if we think of the leaf structure of the evergreen trees and shrubs we find that it shows the features we associate with reduced transpiration. The glistening leaves of ivy and holly indicate a thick cuticle; the heaths show protection of stomata, as do the evergreen conifers. It is interesting to compare with the latter the needles of the deciduous larch in which the cuticular development is small and the stomata are flush with the surface.

Transpiration is, of course, normally low in winter; but trees in foliage through this period run a danger when, in periods of temporary high transpiration—warm sunny days—the water supply from a cold soil, or through a frozen trunk, is very slow. The sun is often brilliant and may radiate considerable heat in frosty weather. This danger is real. A conifer is often seen with the needles on one side dead, brown, and burnt; this is generally regarded as due, not to the direct effect of the heat or light of the sun, nor to "frosting," but to transpiration so high, under conditions of insufficient water supply, that the leaves are actually desiccated. Neger (1915) considers that it is due to the action of late frost on young needles. But it is certain that winter leaf-fall is partially related to the danger of excessive transpiration.

A much more definite relation occurs in those plants in which the leaves die during the hot season. This is a characteristic feature of the bulbous plants of arid regions. In the Mediterranean countries and in South Africa the bulbous plants sprout in the early spring or in the wet winter season, complete their flowering, and die down as the dry season commences. The bulbous plant is, indeed, characteristic of semi-arid regions. Our few representatives may have been derived from such types. Their vernal period of vegetation may be a relic of this habit which, in our latitude, fits in with the necessity of obtaining a supply of light before the woods, where for example the wild hyacinth lives, are in full foliage.

Trees with leaf-fall in summer are characteristic of many tropical and sub-tropical forests. Of the four types of tropical forest distinguished by Schimper two, the monsoon forest and the savannah forest, are more or less leafless in the dry season. The teak forests of East Java are an example of the former, and in the season from June to October the trees are completely bare. In the savannah forests of Venezuela and Brazil the dominant leguminous trees, especially species of Cassia, shed their leaves in the dry season ; there also occur evergreen trees with hard leathery leaves. The " caatinga " of Brazil and the thorn scrub of Mexico are very open communities of a bushland type in which leguminous shrubs, shedding their leaves in the dry season, are prominent. Similar conditions prevail in the woods of the Abyssinian highlands where the dominant Boswellia casts its leaves in drought.

Leaf Movement and Profile Position.—Temporary reduction of leaf surface of another kind is shown by those plants with pulvinate leaves which assume a definite sun position. We have already mentioned the vertical position of the leaflets of Robinia and of Oxalis in the sun, and how this position avoids extreme insolation and overheating. The most important consequence is probably a reduction in transpiration rate, but the relation is not simple, for the stomata of Robinia are confined to the lower surfaces of the leaves and thus become exposed. We might also regard the leaves which take up a permanent profile position from this point of view. The actual transpiring surface is not reduced, but the heat-absorbing surface is, and this is a very important point.

Types of Small Leaves.—Leaf surface is permanently reduced in many characteristic xerophytes, and a number of distinctive leaf types may be recognised. The pines have characteristic thick narrow leaves which stand out so that the rays of the sun pass through the foliage: the needles of fir and spruce are flat and broader. The xerophytism of the conifers has already been related to their winter green character. It is equally suited to life in moderately arid conditions, as in Asia Minor and the Mediterranean countries. It is likely that it is primarily related to the fact that conifer wood is a much less effective conductor of water than that of the angiosperm, since it possesses only tracheids, narrow and short compared with the vessels of broad-leaved trees. The resistance to water-flow is much greater. Farmer (1919) has measured the "specific conductivities" of a number of woods; by this he means the rate of flow of water through a standard length and area of wood in unit time. He finds the specific conductivity of the pine to be 13, of the larch 14, and of the yew 12, while for the oak and the beech the figures are 75 and 65 respectively.

This once more emphasises the fundamental point that it is not only the amount of water available in the soil that determines the plant's condition, but the relation between the rate at which the supply can be drawn on and the loss. In trees with so high a resistance to water-flow it is probable that, with external conditions favourable to supply, the internal resistance is frequently the limiting factor. If we look on this as the factor primarily related to possession of a xerophytic needle-leaf by the conifers, we can readily see that these trees would thus be fitted to maintain a winter green foliage, and also to form communities in more or less arid situations.

These considerations also apply to the conifers which possess the scale type of leaf seen in Cupressus, Chamæcyparis, and Thuja. The leaves are small and adhere to the stem over part of their surface; only the tip is raised and free. These types are characteristic of arid stations. Examples of gymnosperms with expanded leaves are the cycads, Agathis, and Welwitschia with leathery xerophytic leaves, and Ginkgo, which is deciduous.

The needle and cupressoid scale types of leaf are not

confined to the gymnosperms. Thus *Hakea sulcata*, and other species of this Australian genus, have long needle leaves.

Scale leaves of the cupressoid type are seen in the succulent saltworts (*Salicornia herbacea*), and in the tamarisk (*Tamarix gallica*). A remarkable case of convergence is shown by the New Zealand *Veronica cupressoides*, in which the shoots are so entirely cupressoid in appearance that it is not possible to say from a cursory inspection of a vegetative shoot that the plant is not a cypress.

Marked xerophytism is also exhibited by the pure linear and setaceous leaf forms of grasses, sedges, and other plants, such as *Linum catharticum*, the purging flax, and *Linaria vulgaris*, the toad-flax. Taken in conjunction with stomatal protection, this type also secures reduced transpiration. There is no sharp line of demarcation between the linear leaf and typical broad shapes. Indeed transitions may be traced in a single individual as in Campanula ; or inside a genus, as in different species of Galium. The narrower the leaf, and the more vertical its position, the more transpiration is reduced. The leaves of some rushes are cylindrical.

Another distinctively xerophytic leaf type may be termed " ericoid " as it is seen in many species of Erica. It is also found in *Loiseleuria procumbens*, the creeping azalea of the Scottish mountains and arctic countries, *Empetrum nigrum*, the crowberry, *Dabeocia polifolia*, *Passerina sps.*, and other South African plants (cp. Thoday, 1921). In shape it is variable, rather narrow or even needle-like ; it is marked by its small size, leathery texture, and evergreen habit, and by characters already noted—the protection of the stomata, which lie in grooves often clad with hairs, and the tendency of the leaf margins to roll in, which may be accentuated in drought. Between such leaves and the broad sclerophyllous foliage of *Vaccinium Vitis-Idæa*, or of a Rhododendron, there is no sharp demarcation. The small-leaved, hairy Rhododendrons—*Rhododendron ferrugineum* of the Alps—come near the Loiseleuria type, as does *Myrica Gale*.

According to Farmer the specific conductivity of the wood of trees and shrubs with sclerophyllous foliage (both of arid and of temperate regions) is small; for *Quercus Ilex*, the holly oak, the figure is 32, and for the holly it is 9.

§ 28. Moorland Xerophytes

Such plants occur in the most diverse stations. The rock-roses grow in dry exposed gravels and chalks. Loiseleuria is an alpine in this country, and arctic in distribution elsewhere; it is sometimes subject to strong insolation, but its xerophytic character must be partly related to its evergreen habit. The heaths are typically a South African genus; some species occur in dry stations in the Mediterranean region; in central and northern Europe they are plants of the heaths and moors. Calluna, in Britain, is characteristic of the dry moors on thin peat, really a type of heath, where in summer extreme drought may prevail. But it grows also in the wet soil of true moor on deep peat, and is there accompanied by *Erica Tetralix*, by *Myrica Gale*, and by species of sedge, cotton-grass, and xerophytic grasses. In fact, the vegetation of our wettest moors is markedly xerophytic, and this is the case in the moors of temperate climates in general. The problem of reconciling the xerophytic character of the vegetation with the wet soil conditions has always exercised the ingenuity of botanists. Though we cannot yet be said to have a full understanding of the subject, we now know that a whole series of factors is involved.

The earliest clear explanation was that offered by Schimper, who distinguished soils bearing a xerophytic type of vegetation into those which were *physically* and those which were *physiologically* dry. The latter include moorland soils, saline soils, and cold soils. By physiological dryness he meant simply the introduction of some factor which made the absorption of water difficult although there was plenty available; in the case of cold soils the low temperature, in saline soils the high osmotic pressure of the solution, was the factor in question. In moorland water the difficulty of absorption was supposed to be due to the presence of the large quantities of humus acids, derived chiefly from the dead and partly modified remains of the bog-moss, Sphagnum, present in the peat, acting as toxins and so reducing absorption directly, or rendering minimal absorption necessary as a condition of absorbing small quantities of toxic matter. Dachnowski (1908) has shown that oats and other plants grown in bog water show signs of stunting, particularly in the root system. Treatment with chalk or lamp-black counteracts this effect. His conclusion is that the lowered growth is due to the presence of toxins, chiefly of bacterial origin. Rigg (1916) also found toxic substances in water in which plants had decayed. Burgerstein (1876) had earlier shown that humus extracts reduced transpiration. On the other hand, Montfort (1921) has shown that the passage of water through maize plants —determined by the rate of excretion of water-drops from the leaf tips of the plants in a moist atmosphere—is not diminished when the roots are placed in bog water, but that later on a toxic action sets in. Stocker (1923) has shown that the root system of moor plants is just as extensive related to their aerial organs as that of ordinary mesophytes—that there is no stunting, and that the transpiration for equal weights of root system is greater. His conclusion is that the moor xerophytes are in reality ever-green winter xerophytes, an opinion which had been previously expressed by Gates (1914). Farmer (1915) has shown that the specific conductivity of an arborescent heath is only 8 compared with an average figure of 70 for broad-leaved trees. The efficiency of the conducting system of the heath must evidently make water economy important. Thatcher (1921) has shown that if a plant is able to produce a healthy root system in peat it transpires in that medium more freely than in ordinary soils, and that it does so at any water content from saturation to the wilting point. She found that healthy root systems were formed by *Salix pentandra*, and *S. cinerea*, *Acer pseudoplatanus*, and

Epilobium hirsutum; but not by cabbage, Pelargonium, or sunflower. She ascribes the action of peat in this respect chiefly to lack of aeration. In his memoir on aeration, after a detailed account of the extensive literature on the subject, Clements (1921) concludes that the moor soil is unfavourable because of its extreme poverty in oxygen.

A different point of view is taken by Warming with regard to the Carices, which he thinks show xerophytic characters inherited from ancestral forms, and which are not always now in accord with their environment.

There is no doubt that moorland plants are xerophytic, and, in view of the evidence available, it seems highly probable that this habit is a consequence of the action of diverse factors singly or in combination in different cases, though it may turn out that one of these is of more general importance than the others.

§ 29. Leaf Reduction: Transference of Assimilating Function

Spines.—Another form frequently assumed by reduced leaves is that of the spine, an excellent native example being the whin. Here the first two or three seedling leaves are normal and trifoliate—a youth form evidently indicating the ancestral type—but these soon give way to the adult form, a narrow, flat, rather soft, mucronate spine. It is in the axils of these that the angular, branched, and much harder, short-shoot spines arise, on which again leaf spines occur. The well-known needle-like spines of the cactuses may be of leaf nature. Stipular or leaf spines may be formed as well as foliage leaves, as in the false acacia and barberry. In many cases spines are modified shoots, while thorns are epidermal and cortical outgrowths. We may here note that spines and thorns of all kinds reach their most unpleasant degree of development in arid regions of the tropics and sub-tropics, where their development goes along with a greater or less degree of leaf reduction. So notable is this that several communities of both the new

PLATE IV

PHYLLANTHUS GLAUCESCENS AND *RUBUS AUSTRALIS.*

Above, Phyllanthus with leaf-like shoots (cladodes), bearing flowers and fruit; below, Rubus with very small leaflets, the stalks of which, and the elongated petioles, are responsible for light absorption.

and old worlds are named by Schimper, thorn forest and thorn scrub.

It can be shown experimentally that spine formation is favoured by high light intensity and low humidity; whin seedlings growing in a saturated atmosphere form broader, soft leaves, and long, " drawn " shoots. Except in those cases where, as in the whin, the spine is also a reduced leaf, the relation to these conditions is not clear. Spines may be a protection against grazing animals, but this does not help us to understand the causal relation of drought to their formation. At this stage of leaf reduction we meet with cases of stems assuming important light-absorbing functions. Most young stems, before cork formation, have chlorenchyma (tissue with chlorophyll) and can thus assimilate. In mesophytic plants the stem surface is always negligible in proportion to the leaf surface, but in the whin it is greater, and the stem is the more important assimilating organ.

The climax of this tendency is reached when the leaf disappears altogether and the work of assimilation is entirely taken over by the branches. The broom is a case in which this is approached. It bears small trifoliate leaves, which in dry stations may fall early. They never persist through the winter. The characteristic switch shoots, bright green and prominently ribbed, are the chief assimilating organs. In *Spartium junceum*, of the Mediterranean region, the leaves are still fewer and more fugacious; the rush-like stems are the assimilating organs. In Casuarina the leaves are fused throughout their length with the axis, only the extreme tip being free; the appearance is that of a typical switch plant. Ephedra is an example from the Gymnosperms; here the opposite leaves are reduced to mere scales. That the substitution of the stem for the leaf as an assimilating organ is effective in reducing transpiration has been shown by Bergen (1903) for *Spartium junceum* growing at Naples. In the season when the plant bears leaves the transpiration from these is 2·6 times that from the stem if equal areas are compared.

The occurrence of chlorophyll in the stem is widespread in desert plants with small and fugacious leaves, or with none. Its distribution has been studied by Cannon (1908) for desert plants of Arizona. Its most important position is in a band in the parenchyma of the cortex, which may persist through the life of the shoot or may be cut off by cork formation. It sometimes occurs in the epiderm, and sometimes even in the parenchyma of the wood. It penetrates to a depth much greater than is customary in leaves. The smaller the leaf surface the more important the chlorenchyma of the stem becomes.

As assimilating organs the stems suffer from their small surface, even though this be expanded by ridges or wings, but as regards reduction of transpiration this is important. They avoid direct insolation even more effectively than leaves in the profile position, as may be seen admirably in switch plants like the broom. At the same time they are illuminated from all sides, and this is advantageous for assimilation.

Phyllodes and Cladodes.—A further stage in the transference of the leaf functions occurs when the stem or the leaf-stalk assumes leaf form and structure. Leaf-like stems are termed *cladodes*, leaf-like petioles *phyllodes*. Of native plants the butcher's broom, *Ruscus aculeatus*, has cladodes (Fig. 21). The leathery shoots are quite like the sharp-pointed upper leaves of the holly ; the position of the flowers in the middle of the upper surface indicates their morphological nature, as does the fact that they spring from the axils of scale leaves. The species of asparagus frequently grown in greenhouses as " ferns " for decorative purposes, seem to have needle-like leaves, but examination shows that these are shoots coming from the axils of scale leaves. Phyllodes are found in many leguminous plants. Their nature may be ascertained by comparison with related species. In many cases, too, the primary leaves have the normal form and give place, often through a series of transitions, to the mature phyllodes. This is well seen in *Acacia nereifolia* (Fig. 22). We may here mention the theory, recently proposed by Arber (1918) and supported by a mass of anatomical evidence, that the leaf of the monocotyledons is in all cases a phyllode, corresponding not to the leaf blade but to the petiole of the dicotyledon leaf. This view is adversely criticised by Goebel.

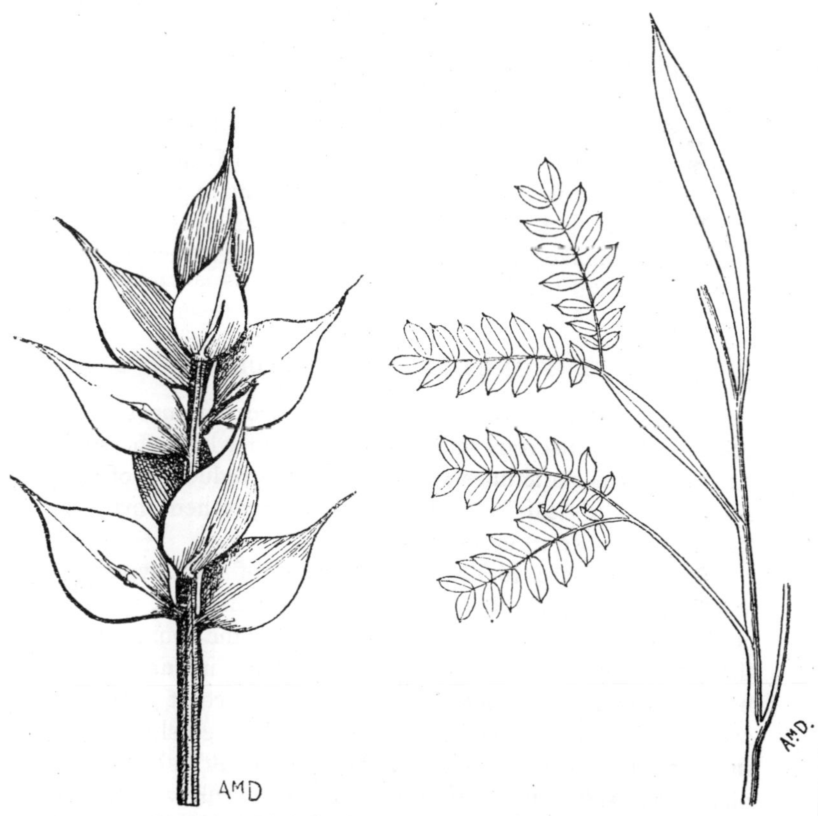

FIG. 21.—Cladodes of butcher's broom (*Ruscus aculeatus*). Nat. size.

FIG. 22.—*Acacia nereifolia* with compound leaf, phyllode, and transition form. $\frac{2}{3}$ nat. size.

What advantage the plant secures by replacing a leaf by a leaf-like stem or petiole it is difficult to see; unless we assume that the leaf, presumably present in some distant ancestor, was of mesophytic type, and that its disappearance, combined with the evolution of a xerophytic leaf-like stem, has been a modification easier, in some cases, than an alteration in the structure of the leaf itself. The change may have originated as a chance mutation.

Succulents.—The leaf function has been taken over by the stem in a great many succulents. The finest examples of this are offered by the succulent cactuses of America, and euphorbias and stapelias of Africa. The leaves of the cactuses are to be seen as small blunt projections on the young shoots, *e.g.* of Opuntia. These soon fall off, and the spines, which may also be modified leaves of side shoots, appear. There is no approach to leaf form in the stems, except in the flattened segments of Opuntia. Our native herbaceous Salicornias or glassworts are also stem succulents with small leaves.

§ 30. TRANSPIRATION OF SUCCULENTS

More familiar in temperate climates are leaf succulents, such as the stonecrops and house-leeks, in which the succulent habit is sometimes associated with a reduction of leaf surface. The leaf succulent Mesembryanthemums are frequently grown in gardens.

Succulents are highly characteristic of two types of station—desert and salt marsh. There seems, also, to be a general tendency for seaside plants to become more fleshy than individuals of the same species growing inland, or for fleshy varieties of a species to occur near the coast; *Matricaria inodora*, var. *salina*, is for example distinguished from the type by its more fleshy habit. Such British seaside plants as *Salicornia herbacea*, *Suaeda maritima*, *Salsola Kali*, *Cochlearia officinalis*, *Glaux maritima*, *Arenaria peploides*, *Cakile maritima*, *Mertensia maritima*, *Plantago maritima*, *Aster Tripolium*, and *Statice Limonium* are all more or less markedly fleshy. Saline soils were distinguished by Schimper as physiologically dry. As far as absorption goes we have already seen that the development of an osmotic pressure sufficient to withdraw water from a soil saturated with sea-water is characteristic of such plants. When this is so there can be no difficulty in obtaining the necessary supply. Schimper held, however, that reduction in transpiration in such plants would reduce the amount of toxic salt absorbed. Ruhland (1915) has shown that in the Statices and Armerias the root cells are highly impermeable to salts, while the glands with which the leaves are provided excrete the surplus. Furthermore, these plants seem to transpire as rapidly in saline as in ordinary soil. Yet Armeria from an inland station placed in saline solution, though it too excretes salt rapidly, soon shows signs of poisoning. This points to the presence of a constitutional difference between the coastal and the inland varieties—a difference unconnected with the water relations.

Delf (1911, 1912) has shown that, while the salt succulents are xerophytic in so far as they store water, and in many cases possess thickened cuticle, waxy bloom, and protected stomata, they do not withstand drought like the desert succulents, but flag and wither very rapidly. The question is still further complicated by the fact that succulence is not confined to the plants of the salt marsh, but is shown, often markedly, by those inhabiting dunes or rocks near the sea, *e.g.* by *Arenaria peploides*, *Cochlearia officinalis*, and *Plantago maritima*. The salt content of the soil of these stations must be relatively very low, though higher than inland. The succulents of saline and alkaline soils inland, where salinity is combined with a desert climate, are subject to entirely different conditions.

The desert succulent is a plant with a large water balance, reduced surface, and low transpiration, features clearly related to the conditions in which it grows ; but it is evident that the apparently xerophytic characters of the salt succulent cannot be so easily related to their environment.

Recently MacDougal, Richards, and Spoehr (1919) have offered a suggestion as to the causation of succulence. Their work is based on comparison of thin-leaved and fleshy-leaved specimens of *Castilleja latifolia*, of the same genetic origin, growing at Carmel, California, the latter on the foreshore in arid but not saline conditions, the former in the open forest. Their conclusion is that extreme reduction of water in the cell leads to a conversion of the hexose polysaccharides into pentosans, which have an enormous water-absorbing capacity. Aridity would thus lead directly to succulence by a change in the carbohydrate chemistry of the cell, and would automatically lead to an increase in water-retaining capacity. In the particular case of Castilleja this change to succulence would be a *direct* result of the reduction of the cell water below a certain point. This investigation opens up a whole vista of fresh possibilities in the investigation of succulence. It is quite possible that salt succulence may find its explanation on some such lines.

§ 31. Assimilating Roots

In some epiphytic orchids extreme reduction of the leaf is accompanied by the formation of a root system with abundant chlorophyll. In *Tæniophyllum Zollingeri*, a Javan species described by Goebel (1889), the roots are flattened, leaf-like, and pressed to the stems of the palms on which the epiphyte grows. The velamen is confined to the under surface of the root. Similar roots are formed in many species of Angræcum. The leaves are present only as minute scales. In the American *Aeranthus funalis*, the assimilating roots are cylindrical and hang partly in the air. It may be noted that the aerial roots of many orchids possessing well-developed leaves have a certain amount of chlorophyll.

Reference may here be made to the remarkable assimilating roots of the Podostemaceæ and Tristichaceæ, two families inhabiting torrential streams in the tropics. The root system is, in some cases, thalloid, closely applied to the water-worn rocks on which the plant grows, and is the only vegetative organ developed (see Fig. 23 and pp. 111, 294).

§ 32. The Significance of Xerophytism

In the account of the ways in which transpiration is restricted, we have met with plants showing xerophytic characters growing under a great variety of conditions, ranging from stations in which the water supply is ample or excessive and the transpiration on the whole low, as on the moors of Northern Europe, to an environment in which the most extreme aridity of soil is combined with untempered

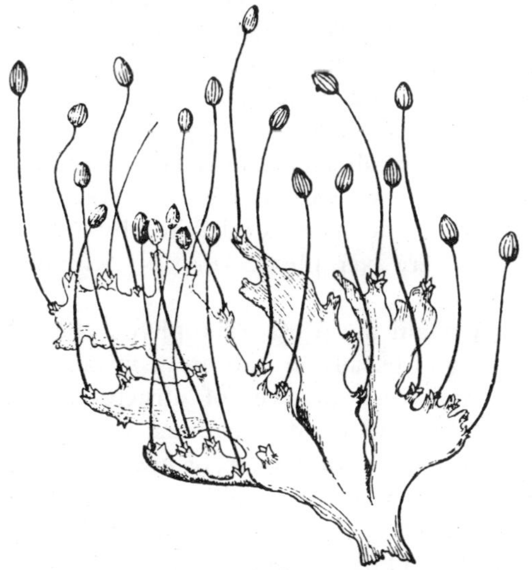

Fig. 23.—Podostemon. Nat. size. (After Warming.)

insolation, as in the typical desert. The term *xerophyte* is thus applied to the most varied types of plant growing under the most diverse conditions. A definition of the term is by no means easy to arrive at. As Delf (1915), in a discussion of the subject, in which a review of the work of previous authors will be found, points out, xerophytism is sometimes defined as a character of the plant and sometimes as a character of the habitat; the xerophyte is described as a plant with certain structural features, or as a plant growing in certain situations. Schimper combines the two ideas by first defining xerophytes as plants possessing means of aiding absorption and depressing transpiration, and then by stating that the xerophytic vegetation corresponds to physiological dryness. The conception of a xerophyte has undoubtedly had its uses in stimulating research ; its weakness lies in the fact that, arguing from a resemblance in the structural features of their vegetation to that of undoubtedly arid regions, Schimper tended to fit other habitats, such as the moor, to the Procrustean bed of " physiological dryness," without, as we have seen, sufficient experimental evidence. Xerophyte was first used to denote those plants inhabiting arid regions and showing markedly features such as might naturally be expected to diminish transpiration ; plants growing in much less markedly dry stations, or in stations not obviously dry at all, showed similar characters, and were included in the class ; various reasons were then advanced for believing that the stations inhabited by these " pseudo-xerophytes " were xerophytic—or, as Schimper put it, physiologically dry.

We must retain the word " xerophyte," because it is in universal use and because it indicates conveniently a group, very large and very diverse, of structural peculiarities. *Xerophyte* is a useful word to contrast with *mesophyte*, the *ordinary* land plant of our climate, or *hydrophyte*, the aquatic. But it does not indicate a natural biological group of plants, and should never be taken as a sufficient definition of a biological type. Only a beginning has been made with the analysis of the different xerophytic types and their habitats, as in the case of the succulents and the moorland plants. We may summarise the main factors to which xerophytism in the widest sense is related. They are : (1) High evaporating power in the atmosphere, either alone or combined with (2) difficulty in water absorption, which may be due to diverse causes. (3) Periodic low soil temperature. (4) Structural " defects," such as insufficient conducting power of the wood. (5) A direct reaction of the chemistry of the plant cell to water shortage. (6) Phylogenetic causes.

§ 33. Promotion of Transpiration

If we assume that transpiration is not simply a necessary evil, but has a useful *rôle* in the plant's economy—*e.g.* that of promoting the supply of mineral nutrients—we might expect to find in plants growing in situations where evaporation is very low, not only an absence of xerophytic characters but even the presence of features which might be interpreted as increasing transpiration. If such features are to be found, and if their effect can be definitely demonstrated, we shall have indirect but valuable evidence in favour of the importance of transpiration. Attempts have not been wanting to discover such features.

(*a*) The clearest case is the undoubted existence of a "transpiration current" in submerged plants. Aqueous vapour is not given off, but that there is an ascending stream of water in such plants as Elodea and Potamogeton has been shown by Snell (1908) and Thoday and Sykes (1909). This might be interpreted as a survival of an ancestral function. There is, in land plants, a regular increase in osmotic pressure from the root upwards, which may have something to do with the ascent of water in the stem. Such a gradient in a water plant might cause a flow from root to tip, and Hannig (1912) has shown that the gradient does exist. We might possibly believe that here we have a condition derived from some ancient land ancestor with no relation to the plant's present condition, like the functionless stomata on the lower sides of some floating leaves. But, as we have seen, it has been shown that these plants thrive better when rooted, and it seems highly probable that the reason is a better supply of salts, due to the transpiration current; but Brown (1913) suggests that the increased vigour is due to the anchoring of the plant near a rich source of carbon dioxide—the mud with its decaying organic matter—and Arber thinks that carbon dioxide may be carried up in the ascending current.

(*b*) Under conditions of reduced transpiration and favourable absorption—high temperature, well-watered soil, and high atmospheric humidity—many plants exude drops of liquid water—a phenomenon sometimes referred to by the inelegant term "guttation." The drops of water seen on the tip of every blade of grass after a close, damp night are not dew but drops excreted at the leaf tips. The process may be easily demonstrated by placing a pot of oat seedlings, 4 or 5 in. high, under a bell jar in a warm room, and watering thoroughly. Drops begin to appear on the leaf tips after a few minutes and, if removed, quickly reform. The excretion is active and abundant. It is due to a rapid supply from the root, forced up under root pressure : when evaporation is high transpiration can cope with the supply, when it is low the water is forced out as a liquid.

In many leaves, common examples being the fuchsia, the garden nasturtium and the balsams, special water-excreting stomata are formed at the tips of the serrations. These stomata may occur singly or in groups. They are commonly large and permanently open. Below them lies a loose parenchymatous tissue, with abundant intercellular spaces, called the *epitheme*, and below this lie the terminal tracheids of a vein, spread out in a brush. The whole organ is called a *passive hydathode*, passive because the water is forced through it from the tracheids. Epidermal hydathodes which, acting as glands, actively excrete water are also known.

Such hydathodes occur and function on leaves of water plants ; they have been found on the lower surface of floating leaves and on submerged leaves, and no doubt act in eliminating the water raised by the transpiration current. They are also frequent on the leaves of land plants, and in particular of plants of humid tropical regions. As Schimper writes, "Early in the morning, especially in the tropics, many plants, herbs as well as trees, are so covered with drops of water that not infrequently a drizzling rain seems to be descending from the forest canopy of leaves." According to F. Shreve (1914) hydathodes are infrequent in the Jamaican rain forest. The amount of water given off may be quite large ; the most celebrated case known is that

PLATE V

"SLEEP" MOVEMENTS OF LEAVES.

Above, *Oxalis sp.*, day position (left), and night position with the leaflets folded downwards (right); below, *Acacia cæsia*, day position (right), and night position with the pinnules folded upwards (left).

of *Colocasia antiquorum*, the taro, from the leaf tip of which nearly 200 drops may be spirted off per minute, while a single leaf may excrete 100 c.c. in a night.

We have here an evidently widespread mode of increasing the amount of water given off by the plant ; not an actual increase of *transpiration*, but a supplementary excretion. It is suggested that this is important because it increases the supply of salts, particularly at night when growth in the warm tropics may be rapid. It has also been suggested that harmful substances may be got rid of in liquid water. It has been further suggested that the excretion prevents an injection by water of the intercellular spaces of the leaf, which would of course hinder gas diffusion and consequently assimilation. This last suggestion is at complete variance with the other two, for it does not suppose the water stream to perform any useful function ; excretion merely gets rid of too great a supply of water. Decisive experimental evidence is lacking. At present, then, we cannot say whether excretion of water is of any use, or is simply a consequence of excessive water supply.

(*c*) A structural feature common among tropical plants of humid regions, such as the rain forest of the Cameroons and of West Java, is the elongation of the leaf tip into a narrow point. This point, or *drip-tip*, is supposed, by Stahl (1893), to facilitate the run-off of rain-water, so that the leaf dries more quickly after rain and consequently transpiration sets in again more quickly. As the diffusion of carbon dioxide through the stomata will also occur more quickly, this feature might equally well be regarded as promoting assimilation. According to F. Shreve, (1914*b*), however, the drip-tip has scarcely any effect in hastening the drying of the leaf. He determined the rate of drying for a number of leaves with elongated tips, for the normal leaf, and for the leaf with tip removed, and found practically no difference. This evidence seems decisive.

(*d*) Stahl (1893) also pointed out that many rain-forest leaves have satin-like texture due to papillose structure of the epidermal cells. He supposed this to lead to a rapid spreading out of rain-drops to a capillary film which would evaporate rapidly. A similar function has been assigned to the covering of soft hairs shown by such plants as our native *Stachys sylvatica*. Unwettable leaves—the surface of which is coated with wax—avoid wetting altogether. But that any of these features is of real importance in nature remains to be proved.

(*e*) A very striking character of pulvinate leaves is their assumption of a more or less vertical position at night. We have already seen that such leaves may become vertical in intense insolation. The same position, due in this case to an unequal *increase* of turgor, is assumed in the dark. Less frequently light and dark positions are assumed by non-pulvinate leaves, *e.g.* of Impatiens and Amarantus, as the result of differential growth rates in the petiole. The exact relation of these *sleep movements*, or *nyctinastic movements*, to external conditions is very complex. In a simple case like that of *Acacia cæsia* or *Acacia lophantha* the movement, here a folding up of the pinnules, follows within a few minutes of darkening, and the reverse opening movement follows illumination. But a reverse movement also takes place without illumination, whether as a delayed effect of the darkening, or as a reaction to the previous movement we do not know, though the latter is the more likely. In Acacia these movements may be repeated as often as twelve times in the 24 hours. The primary leaf of the scarlet runner, however, carries out its double movement only once in the 24 hours, and more frequent changes of light and dark scarcely influence it. In constant darkness it swings regularly in 24-hour periods for some days. Stoppel (1912, 1916) has shown that a seedling grown in uniform darkness exhibits this daily periodic movement unmistakably. She has attempted to show that under these conditions it is regulated by changes in atmospheric electricity, but her results have been challenged by Schweidler and Sperlich (1922), and by Cremer (1923), and the question of causation is undecided. Between these two extremes intermediate cases occur. In general the movement is closely related to the diurnal change in nature, but it may be carried on in uniform conditions, exhibiting a periodicity due either to the after effects of previous repeated movement, or, perhaps, to an inherent character of the plant. This should be compared with the account of flower movements.

These sleep movements have of course nothing to do with the repose of animals. What benefit the plant derives through its leaflets folding up, as in the clover, or down, as in the wood sorrel, is a question that has long exercised the ingenuity of naturalists. Darwin (1880) suggested that the vertical position, by diminishing radiation, saves the leaf from damage on cold nights. His evidence in support of this hypothesis was that clover leaves, kept open on cold autumn nights, were seriously damaged, while those in the natural sleep position were not. F. Darwin (1898) points out that, while this may not apply to plants of the tropics, where nyctinastic movements are commonest, sufficient cooling might take place even in these to prevent the ready translocation of starch, and so to interfere with assimilation on the following day. Stahl (1897) held that the chief effect of overcooling is the formation of dew on the leaf surface, and produced experimental evidence to show that dew formation seriously delays the beginning of transpiration the next morning. He supposed that the sleep position, by preventing dew formation, increases transpiration and through it the supply of salts. We may here note again that, if the avoidance of wetting by dew led to any increase in transpiration in the early hours of the morning, it would also increase the diffusion of carbon dioxide.

It is possible that dew formation acts directly on the stomata; and we may note the fact that the stomata of nyctinastic leaves are in many cases open throughout the night. Loftfield states that when a leaf is wetted the stomata usually open widely, and that when it dries they close. If this is the general rule, then the resumption of free gaseous diffusion in the morning would be further delayed. The dew might also block the stoma and even infiltrate the intercellular spaces of the leaf.

Recently Erban (1916) has investigated the distribution of stomata on leaves which show nyctinastic movements. She finds that in those cases where the leaflets fold downwards, so that the lower surfaces come together and are "protected," the stomata are wholly, or almost wholly, confined to the lower surfaces. Twenty species of Oxalidaceæ, *e.g. Oxalis hedysaroides* and *Biophytum sensitivum*, and Leguminosæ, *e.g. Robinia pseudacacia*, *Cytisus Laburnum*, all show this arrangement, and of these fourteen had *no* stomata on the upper surface. The result is the same whether the leaf is pinnate or trifoliate. Where the leaflets fold upwards, so that the upper surfaces come in contact, the stomata are most numerous on the upper surface, again the "protected" surface. Seven species were examined, and, though none had stomata confined to the upper surface, the distribution is quite unmistakable; on an average the upper surface bore twice as many stomata as the lower. In a third class of plants both sides of the leaflets are partially "protected" in the night position, owing to partial covering of one leaflet by another. This is seen, for instance, in the clover, where the two side leaflets fold together, and the end leaflet rises so as partly to cover them; in the sensitive plant a twist in the pulvini brings each pair of leaflets forward, as well as up, so that they overlap the basal parts of the pair in front. The conditions here are more complex, as is also the stomatal distribution. In Mimosa the stomata are most numerous on the upper surfaces, and on the lower halves of the under surfaces, while at the completely uncovered tips of the under surfaces they are few or absent. In some other plants examined no definite relation between protection and stomatal number was found. The most interesting case, that of *Marsilia quadrifolia*, though it is a fern, may be described. The four leaflets fold together so that two lie completely between the other two. On the two inner leaflets, with both sides "protected," the stomata are equal in number on the two surfaces The two outer leaflets have the under surfaces exposed, and the stomata are twice as numerous on the upper " protected " surface.

The general result is that there is an unmistakable tendency for the stomata to be more numerous on that side of the leaflet which is covered in the sleep position. It is of interest that this tendency is much more marked when the under surface, and less marked when the upper surface, is covered. As the normal distribution of stomata is greater on the under surface, in the latter case we have a partial reversal, in the former an exaggeration, of the normal distribution. There can be no doubt that Erban's results point to some sort of protection of the stomata, and it remains for critical studies of the transpiration and assimilation conditions in such leaves to discover the precise effect of this protection. When these leaves assume the vertical position as a result of strong insolation, we are safe in assuming that a reduction in transpiration will result. There is always the possibility that *day sleep* is the primary phenomenon, and that the much more striking and regular *night sleep* is secondary, the result simply of the mechanical properties and of the chemical processes in the pulvinus.

§ 34. Functions of Transpiration

Of the various features which have been described as tending to promote transpiration it cannot be said, in any case, that there is very good evidence of effective action. Nor can we draw from them much support for the theory that transpiration is of primary importance to the plant. At the beginning of this chapter we said that transpiration might perform two useful functions—it might increase the supply of salts, and it might reduce overheating. We may now reconsider these two possibilities.

Salt Supply.—In the behaviour of aquatic plants we have the only convincing evidence in support of the salt supply theory, but statistics of the ash contents of various leaves are suggestive. Czapek has gathered the results obtained by various investigators for ten different plants, and the figures show that the leaf has an ash content on the average double that of the stem when both are mature. In young plants, on the other hand, the ash content of the stem is slightly greater. We have here a distinct indication of a concentration of ash in the transpiring leaf. The xerophytic leaf of the conifers has a very much lower ash content than that of mesophytes in general. Taking conventional figures, we may put the one at 1·5 per cent. of dry weight, the other at 10 per cent. Moreover, the ash content of the pine needle mounts with the years of its age. For the young leaf of *Pinus austriaca* it is 1·6 per cent., at a year old 1·8 per cent., at 2 years 2·7 per cent., at 3 years 3·17 per cent., and at 4 years 4·55 per cent.; on the other hand, shade leaves are said to be richer in ash than sun leaves. Huber (1923) finds that the basal shoots of Sequoia transpire more vigorously than the apical, and that they have also a higher ash content. McLean (1919) found the ash content of the leaves of sun plants to be lower than that of the leaves of shade plants of the forests of Brazil. Such figures are suggestive, but even if they were unambiguous they would not be conclusive.

It might be thought that the relation of transpiration to mineral supply could be easily tested experimentally, but the difficulty of maintaining all conditions except transpiration rate equal over a long period is very great. Various investigations have been made, but without yielding a satisfactory answer. Hasselbring (1914) grew two sets of tobacco plants side by side, shading one set from the sun by gauze. The two sets attained exactly the same dry weight; the ash of the shade set was 11·2 per cent. compared with 9·6 per cent. in the sun set. The shade plants had absorbed 35 litres of water, and the sun plants 46 litres, or 30 per cent. more. The amount of minerals absorbed was therefore not proportional to transpiration, nor did the plants with lower transpiration suffer at all in this respect. But in this experiment other conditions besides evaporation were different. More important is the fact that transpiration (deduced from the amount of water absorbed) was very considerable even in the shade plants, and in all probability the supply of salts was sufficient in both sets and did not limit growth. The experiment really throws little light on the question at issue.

Muenscher (1922) grew barley in water culture and compared the behaviour of plants in different atmospheric humidities. With transpiration cut down to 50 per cent. of the maximum, the fresh weight attained was somewhat greater and the ash content 10 per cent. less ; calculated in terms of the dry weight the ash content was about equal. Much more ash was absorbed per 1000 c.c. of transpired water by the plant with lower transpiration, *e.g.* 0·795 compared with 0·42 grm. Mendiola (1922) found that diminished transpiration in the tobacco resulted in increased dry weight and decreased ash content. These three modern investigations agree in one point—that there is no proportionality between the amount of water transpired and the amount of ash absorbed. They disagree on the question as to whether more ash is absorbed under conditions of higher transpiration. The subject evidently requires further investigation. We may, however, note four points. (1) It would be surprising if there were a direct relation between transpiration and salt absorption ; the actual rate of entrance of salts into the plant can be affected to a slight degree only by the rate of entry of water. But if salts are swept up the wood vessels in the transpiration current diffusion from the parenchyma cells of the roots into the vessels must be accelerated. (2) The function of the transpiration stream, if it has one in this connection, must lie in the transport of the salts from the root to the shoot ; it is likely that quite a slow rate of transpiration would be sufficient to supply enough salts to the leaves and growing points to secure normal growth. (3) The experiments quoted all deal with what are really high transpiration rates. The effect of transpiration could only become evident if it were a *limiting factor*, and there is no evidence whatever that it is a limiting factor in these experiments. (4) What is required is a series of graded experiments which might demonstrate whether limiting effects on growth rate are produced at minimal transpiration values. Until we have such evidence discussion of this function is rather idle. It is, however, difficult to see by what means salts in sufficient quantity could possibly be transported to the leaves and growing points of, say, a beech tree in the absence of the transpiration stream. The evidence available is, however, definitely unfavourable to the view that lowering of transpiration, even by a very considerable amount, has an adverse effect on the supply of salts ; Stahl's interpretation of sleep movements and of various structural features of rain-forest leaves must therefore be abandoned.

Cooling.—The second suggested function of transpiration concerns the process directly and assumes that it is essential in lowering leaf temperature in the sun. We have abundant evidence that leaf temperature in the sun may rise very high. In Blackman and Matthæi's experiments (1905) a leaf of the cherry laurel, placed in direct sunlight at noon in July, showed a temperature of 39·8° C. when vertical, and of 44·6° C. when at right angles to the sun's rays. The air temperature in the sun was 30·5° C., so that there was a minimum excess of 9° C. and a maximum of 14° C. The temperatures reached were such as to have a rapid and progressive retarding effect on photosynthesis and respiration, and to produce intense transpiration. In a leaf enclosed in a glass chamber the temperature rose to 51° C., an excess of 20° C., and the leaf quickly showed visible signs of damage in the appearance and spread of brown spots of killed tissue. Blackman remarks, "This heating up of the leaf will, no doubt, partly be due to checking of transpiration in the enclosed space, but the effect comes on so quickly that it must chiefly be due to "the greenhouse effect," the imprisonment of the reflected dark heat-rays by the glass plate which is almost impervious to them." The temperature of a leaf in the sun must of course be lowered by transpiration, but such experiments do not tell us to what extent. Neither do comparisons of the temperatures attained by insolated leaves of succulents (supposed to transpire relatively little) and mesophytes. Askenasy (1875) found that *Sempervivum alpinum* attained a leaf temperature of 49·3° C. and *Sempervivum arenarium* of 48·7° C. in the sun, with air temperature 31° C. Under the same conditions the leaf temperature of the non-succulent *Aubretia deltoides* was 35° C. Stahl (1909) also registers high values for the temperature of succulents, *e.g.* for *Opuntia monacantha* 51·7° C. when exposed normal to the sun's rays, and 42·7° C. even in profile position.

How much of these very high temperatures is to be referred to reduced transpiration we cannot say ; a part at least of the excess must follow the more extensive absorption of heat rays by such thick leaves.

We can get at the effectiveness of the cooling effect of transpiration in another way. In an experiment of Brown and Escombe (1905) a sunflower leaf transpired at the rate of 3·96 grm. per 100 sq. cm. per hour in the sun. Taking the latent heat of steam as 590 calories (at 18° C.), this means that the energy used up in transpiration is 0·389 calorie per square centimetre per minute. Now, Brown and Wilson (1905) have determined the *thermal emissivity*—that is, the loss of heat by radiation, conduction, and convection—of the green leaf. For Helianthus the value, in a gentle breeze of 5 miles an hour, is 0·038 calorie per square centimetre per minute per 1° C. excess of leaf temperature above that of the surrounding air, or from the two surfaces of the leaf 0·076. In a leaf 10° C. in excess of the air temperature, the thermal emissivity would be 0·76 calorie. Transpiration at the rate taken would therefore account for 30 per cent. of the total loss of heat from the leaf. In perfectly still air the value for the thermal emissivity from both sides of the leaf is 0·3 cal. If we suppose the transpiration to be reduced to one-half of the previous value it will now account for 40 per cent. of the total loss of heat. These figures illustrate the very important effect which transpiration must have in cooling the leaf.

When we try to apply these results to xerophytic plants we find difficulties due to insufficient data. We may, however, consider the case of the succulents, which we have seen show very high temperatures. MacDougal's values for the transpiration of the cactuses are very irregular, but we may take two widely different examples. During May (arid month) an Opuntia weighing 140 grm. lost water at the rate of 0·337 grm. per day. During March an Echinocactus of 18 kg. lost 30 grm. per day. For the Opuntia this represents 200 grm.-calories, and for the Echinocactus 17,700. Even if we assume that this energy is all lost in the hours of daylight, and this is certainly not the case as has been shown by E. B. Shreve (1916), we find that the Opuntia loses in the course of the day about the same amount of energy as a piece of sunflower leaf weighing 0·025 grm. with an area of 0·7 sq. cm. The energy loss of the Echinocactus is equalled by 1·9 grm. of sunflower leaf measuring 65 sq. cm. The figures do not allow us to make a comparison for areas, but they are sufficient to show that the desert cactuses can be cooled by transpiration to an inappreciable extent only.

In halophytic succulents different conditions prevail, for, as Delf (1911) has shown, the transpiration may, for equal areas, be twice as vigorous as in an ordinary mesophyte ; here the lowering of temperature must be correspondingly important.

An investigation by Bergen (1904) gives some idea of the conditions in sclerophyllous trees. He measured the transpiration of sun and shade leaves of *Olea europæa*, *Quercus Ilex*, *Pistacia Lentiscus*, and *Rhamnus Alaternus*. These leaves do not differ in the same way as do the sun and shade leaves of deciduous trees. The sun leaves are smaller, usually paler, and, except in Pistacia, do not show more strongly developed pallisade ; their stomata are more numerous. With few exceptions the sun leaf transpired more rapidly than the shade leaf when both were insolated, or both shaded. The difference may be due to the different numbers of stomata. The sun leaf would seem to be more efficiently cooled.

The case of the Cactaceæ is at first sight a difficult one.

Just those plants which are most exposed to extreme insolation get the least good of the cooling effect of transpiration. We must, however, look at the economy of the plant as a whole, and, in any case, this is evidently not the only means by which plants meet high temperatures. The cactuses *do* survive temperatures which would be fatal to other plants. Stahl (1909) has shown this experimentally, and we are bound to suppose that their protoplasm is capable of accommodating itself to a greater range. There is nothing extraordinary, though much that is not explained, in this. The same is very probably true for many other xerophytic types, though we must remember that a plant which reduces its total transpiration by reducing the size of its leaves also reduces the heat and light absorbing area, and may at the same time transpire as vigorously, area for area, as a mesophyte. In fact, it is very likely that the lowering of temperature by transpiration is of greatest importance for the freely transpiring mesophyte.

The importance of this second *rôle* of transpiration we may take as established much more firmly than the *rôle* in salt supply. We should always remember, however, that, whether transpiration has useful functions or not, it *must* occur in advanced types of land vegetation.

§ 35. Inter-relation of Transpiration and Assimilation

We have insisted at the beginning of the chapter on the close relation between assimilation and transpiration ; and throughout our discussion it will have become increasingly evident that, in the first place, the opportunity for transpiration is enhanced by the necessities of the photosynthetic processes, and, in the second place, that assimilation must often be limited when transpiration is in any way reduced. The necessity of reducing water loss must frequently clash with the work of building up organic substances. It might be expected that some general relation should exist between transpiration and assimilation, summing up the reaction of the plant to its environment in these two functions. This relation has not yet been exhaustively studied, but some suggestive work has appeared.

We may refer to Livingston's paper (1905) on the "Relation of Transpiration to Growth in Wheat." He found that under different conditions the increase in leaf area, and so, practically, in the size of the plant, was more or less proportional to the total amount of water which had been transpired ; other American authors have found a similar proportionality. This does not mean that transpiration has any direct effect on growth ; it indicates rather that while transpiration is active, so is the diffusion of carbon dioxide, and that the amount of transpiration, as of assimilation, is a function of leaf area. No more striking illustration of another relation of transpiration to assimilation could be found than the result of Thoday (1910) we have already quoted. The assimilation of the sunflower leaf steadily decreases as transpiration reduces the turgor of the leaf, and ceases when the leaf is completely wilted.

We may refer to an investigation by Iljin (1916) on the relation of assimilation to transpiration in the plants of Russian steppes and neighbouring meadows and ravines. Assimilation was determined in an enclosed atmosphere rich in carbon dioxide, and transpiration from cut shoots dipping in water ; both are related to unit dry weight. The methods are open to objection, and the plants were not in natural conditions, yet certain general features are brought out. In its natural environment a mesophyte such as *Geranium pratense* or *Senecio doria* transpires at about the same rate, or rather less rapidly, than does a xerophyte such as *Phlomis pungens* or *Stipa capillata* in its natural habitat. Transferred to the steppe environment the transpiration of the mesophyte far exceeds that of the xerophyte. This is illustrated by Table XXXII.

TABLE XXXII

TRANSPIRATION OF MEADOW AND STEPPE PLANTS

Plant.		Transpiration cgrm. per grm. dry weight.	
		Meadow.	Steppe.
Mesophyte :	*Geranium pratense* ..	308	1216
	Senecio doria ..	313	617
Xerophyte :	*Centaurea sibirica* ..	—	335
	Stipa capillata ..	—	284

The assimilation is found to vary very greatly in different plants, when expressed in relation to dry weight as Iljin prefers ; the mesophytes assimilate more vigorously. If the assimilation is related to the amount of water transpired, then the xerophytes are shown to lose less water in assimilating a unit amount of carbon dioxide. This is sometimes the case when the two classes are compared each in its own environment ; it is much more marked when the mesophyte is allowed to assimilate in a xerophytic environment. These relations are shown in Table XXXIII.

TABLE XXXIII

ASSIMILATION RELATED TO TRANSPIRATION

Plant.	Habitat.	Rate of			
		Transpiration.	Assimilation.	Transpiration per cgrm. CO_2.	Determined in
Centaurea sibirica ..	steppe	335	5·22	64·2	steppe
Geranium pratense	meadow	308	10·16	30·0	meadow
Geranium pratense	meadow	1216	10·16	120·0	steppe
Stipa capillata ..	steppe	284	5·22	55·0	steppe
Senecio doria ..	meadow	313	3·99	78·0	meadow
Senecio doria ..	meadow	617	3·99	154·0	steppe

Iljin's conclusion is that the xerophytic plants of the steppe can survive in the arid conditions because they are able to assimilate more vigorously with a limited water supply, than are the mesophytic plants of meadows and damp ravines. He examined a considerable number of plants, and the general result is quite clear. It is, of course, difficult to adapt exact methods to work in natural stations. But the investigation points to an avenue of great possibilities, and discloses an interesting way in which these two great functions of the land plant co-operate in determining the suitability of a plant to its habitat.

CHAPTER III

SPECIAL MODES OF NUTRITION

§ 1. Parasites. § 2. Saprophytes. § 3. Mycotrophic Plants. § 4. Bacterial Symbiosis. § 5. Insectivorous Plants.

FROM the standpoint of nutrition the plant is fundamentally an independent organism. It synthesises organic compounds from inorganic carbon dioxide, water, nitrates and other salts as its raw materials and light as its energy supply. It absorbs the carbon dioxide from the air, the water and salts directly from the soil, and, although this substratum is in part a product of the activity of many organisms, it is none the less true that the plant growing on it is neither immediately nor necessarily dependent on these. A number of flowering plants, however, as well as the great and heterogeneous group of fungi, have departed from this characteristic mode of nutrition ; in one way or other they supplement or replace the supply of inorganic raw materials, by absorbing organic compounds. The species which depart widely from the ordinary mode of nutrition are only a small minority, but the variety of methods employed, the curious structural modifications shown, and the bizarre effects of unusual relations have always drawn to them a large share of attention. We may distinguish five biological groups, though these are sharply delimited neither from each other nor from normal autotrophic plants.

1. *Parasites*—plants organically united to and more or less dependent on other living plants, the hosts. Lower forms—fungi and bacteria—may be parasitic on animals.

2. *Saprophytes*—inhabitants of rich humus soils, on

the decaying organic matter of which they depend for a supply of carbon compounds. All saprophytes are also members of the next group.

3. *Mycotrophic plants*—living in symbiosis with fungi. The significance of this inter-relation for the higher partner is often obscure; in certain cases it is certainly to be found in the supply of nitrogenous compounds.

4. *Plants symbiotic with bacteria*—these usually benefit by an enhanced supply of nitrogenous compounds.

5. *Insectivorous plants*—these utilise animal food.

§ 1. Parasites

Systematic.—Regarding parasitic flowering plants from the systematic standpoint, we find that they belong to a very few families (*a*) The Loranthaceæ, the mistletoe family, includes about 21 genera and 850 species; only 2 genera with 5 species live as independent plants, the remainder are shrubby parasites on the branches of trees and shrubs. (*b*) The Santalaceæ, sandalwood family, includes about 250 species, most of which are parasitic on the roots of other plants. (*c*) The Myzodendraceæ, a small family with a single genus and about a dozen species parasitic on the branches of Chilian beeches (Nothofagus). (*d*) The Balanophoraceæ, with 14 genera and about 40 species, are all fleshy parasites on roots. (*e*) The Rafflesiaceæ, with 7 genera and about 24 species, and (*f*) the Hydnoraceæ, with 2 genera and about 8 species, are all parasites on roots or stems. (*g*) The score of specios of Cassytha are all twining parasites on the leaves and branches of tropical plants. Cassytha is an isolated parasitic genus of the Lauraceæ, bay-laurel family, which otherwise consists of independent plants. (*h*) The Orobanchaceæ, broom-rape family, has 8 genera and about 120 species, all of which are root parasites. (*i*) Of the tribe Rhinanthoideæ of the Scrophulariaceæ, foxglove family, about 11 genera with some 350 species are root parasites. (*j*) The Lennoaceæ is a small family of sub-tropical North America with 3 genera, the 4 species of which are large, fleshy root parasites. (*k*) Cuscuta, with about 100 species, all twining parasites, is an isolated parasitic genus belonging to the Convolvulaceæ. (*l*) The 2 species of Krameria are parasites on the roots of shrubs of the Arizona deserts. The genus is of uncertain systematic position, but may be referred to the Leguminosæ.

Partial Parasites.—All parasites do not show the same degree of dependence on their hosts. Complete parasites possess no chlorophyll and must draw a supply of elaborated organic food from the host. They show a great range of reduction of their vegetative organs. As *partial parasites* we distinguish those which do possess chlorophyll, and which probably absorb only water and salts from the tissues of the plant on which they live. Such partial parasites as the mistletoe can none the less develop only in connection with a suitable host, while others, like the yellow rattle, can come to maturity as independent plants. We have, therefore, all grades from facultative parasites, which may grow independently, and which differ little, if at all, from normal plants, to such advanced parasites as Rafflesia, in which complete dependence on the host is accompanied by loss of chlorophyll, and extreme reduction of the vegetative organs. The Rhinanthoideæ are particularly instructive. The tribe includes independent genera, facultative parasites such as the yellow rattle, complete parasites such as the tooth-worts, and genera showing intermediate stages. We may begin a more detailed survey with this group.

Rhinanthoideæ.—Among the little specialised partial parasites are such common British plants as the cow-wheats (Melampyrum), the yellow rattles (Rhinanthus), the eye-brights (Euphrasia and Bartsia), and the red rattle (Pedicularis). These plants all possess green foliage though it may be rather scant, the plant giving an impression of scragginess ; they also have a normal though rather poorly developed root system with root hairs. They do not appear very different from other common plants among which they grow. The seeds germinate easily and, if sown in soil kept free of other plants, they produce free-living seedlings. Rhinanthus and Euphrasia can complete their development without a host, but *Melampyrum sylvaticum* and *M. pratense* do not develop fully unless they can fix on the roots of neighbouring shrubs or herbs. In nature the close meadow and pasture vegetation in which these plants grow, extraordinarily favours their parasitic tendency. The soil is full of roots and the parasites show little or no specialisation ; they can prey on any neighbouring plant.

They become attached by side roots, which remain quite short and swell into tubercles. If such a root touches the root of a grass, for instance, it partly envelops it, and then sends into it an absorptive process in which tracheids arise connecting the wood of the host with that of the parasite. A single parasite attaches itself to many roots of one or several hosts, belonging to one or more species. Under such circumstances the parasite thrives more vigorously than by itself.

The green leaves of these plants are functional. It has been shown that they form starch in light when isolated from possible hosts, so that the assimilation must be quite vigorous. Kostychew (1922) has shown that the intensity of assimilation of Rhinanthus, Melampyrum, Pedicularis, Euphrasia, and Bartsia is about the same as in the autotrophic *Veronica longifolia*, and *Linaria vulgaris*. The root system is not luxuriant, and, though certainly capable of absorbing water and salts, it cannot cover the normal requirements of the shoot. The same investigator found that a cut shoot of one of these partial parasites absorbed water from twice (Euphrasia) to ten times (Melampyrum) as rapidly as could the same shoot through its own roots. In the case of Linaria or Veronica, absorption through the root system was as rapid as through a cut surface. There is an evident disproportion between the parasite's water requirements and the powers of the root to satisfy them, and in the increase of the water and salt supply we see the chief advantage derived from the parasitic attachment. In Euphrasia, where independent existence is possible, the disproportion between the supplying power of the root and the shoot's requirements is least. Structurally these plants show practically none of the reduction which is so marked a feature of more advanced parasites. We will return to this group later.

We may here note the parasitism of *Krameria canescens*, a desert shrub described by Cannon (1910). It is a partial root parasite on such small trees as Parkinsonia, and its behaviour is much like that of Rhinanthus, though differences in the details of the suckers exist.

Santalaceæ.—At the same level of parasitism are the semi-parasitic members of the Santalaceæ. The family is widespread in the tropics and sub-tropics, chiefly in dry regions ; a number of genera occur in the Mediterranean region, and one, Thesium, has several species in Central Europe, and one in England. Some are small trees, like the Sandalwoods (Santalum) ; most are shrubs, like the Mediterranean *Osyris alba*, or herbs like Thesium. It has been shown that *Santalum album* can thrive independently, though it is normally a root parasite. A good many santalaceous plants are normally independent ; most, however, require the assistance of a host. There is no high degree of specialisation ; several different species may serve as hosts.

Germination occurs normally, and suckers are developed first on the side roots ; they resemble those of the Rhinanthoideæ, partially surrounding the host root and then sending in an absorbing process in which strands of tissues make contact with the wood of the host (Fig. 24). The leaves show a tendency to reduction in size, perhaps not more than those of other xerophytes ; the plants seem able to assimilate normally. It is again likely that the chief demand on the host is for water and salts. Two genera, Phacellaria and Henslowia, occur as bushes parasitic on the *branches* of trees in eastern Asia. Their mode of attachment and relations are not known.

Loranthaceæ.—In the Loranthaceæ this is the common habit. This great family is predominantly tropical ; it spreads out into the sub-tropics of all the continents, but in the temperate zones it is poorly represented. A few species occur in Europe, and one of these, *Viscum album*, the mistletoe, reaches England. We may take this most familiar of all parasites as a representative of the family.

The white pseudo-berries contain each one seed which is deposited on the branches of trees, either after passing the alimentary canal of a bird like the missel-thrush, or, more usually, from being rubbed off against the bark as the bird endeavours to free its beak from the viscid pulp of the fruit wall. This slime, used as bird lime, fixes the seed to the branch till rising temperature and longer illumination permit germination to take place, about the beginning of May in central Europe. The seed germinates on any substratum, even on a sheet of glass, and it can be made to germinate in winter if the temperature be high enough, and if artificial illumination be employed, for it germinates only in light (Heinricher, 1916). The base of the hypocotyl—no radicle is formed—leaves the seed coat. For a short time, as Heinricher (1916) has shown, it is negatively geotropic, but it soon loses its power of reacting to gravity and becomes negatively phototropic, so that its free tip curves round and grows towards the branch on which it lies; the tip becomes pressed to the surface and flattens out into an adhesive disc. From the middle of this a peg-like sucker grows out

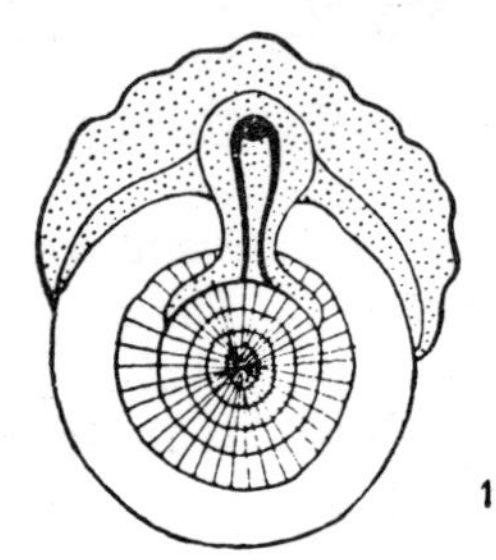

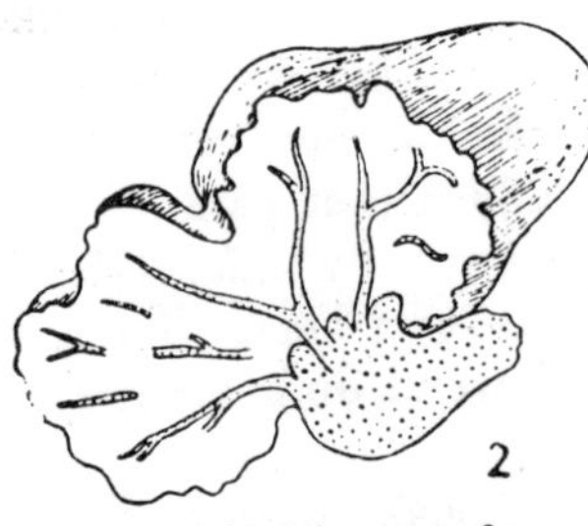

FIG. 24.—Attachment of parasites: 1, cross-section of host root with clasping root of *Osyris alba*; the splayed-out end of the sucker has been enclosed by secondary thickening; 2, section of tuber of *Balanophora globosa* penetrated by tissue from the host root. 1, slightly magnified; 2 = ⅔ nat. size. (After Solms-Laubach.)

and penetrates the cortex of the host. The penetration seems to be due to mechanical force and to take place most easily through weak spots such as lenticels and cracks. When it reaches the wood growth of the peg stops. At this stage it consists of large-celled, pitted parenchyma, but later a strand of wood vessels develops and connects with the wood of the host.

Only at the beginning of the second year do the first two leaves of the young plant appear, and for the next year or two growth is slow; later it becomes more rapid, and the familiar bush is formed with its forking branches and paired leathery leaves. Meantime there arise from the neck of the first sucker exogenous branches, which grow within the cortex of the host plant. Their tips are formed of spreading brushes of slimy cells. In the main they run *along* the branches; sometimes they grow crosswise for a time and then bend anew in the longitudinal direction. They never encircle and strangle the branches of the host. From these branches secondary suckers descend to the wood. At intervals adventitious buds are formed which burst through the cortex and give rise to fresh bunches of mistletoe, and at the same points new branches of the absorbing system arise. The utilisation of the host plant is thus extended, and a single seed may give rise to many bunches of mistletoe; cutting off the visible bunches does not free the host from the parasite (Fig. 25).

As the host branch continues to form secondary wood the sucker would, in the course of a few years, be enveloped and destroyed, were it not that it possesses a meristematic zone at the level of the host's cambium, through the activity of which it lengthens by just the amount that the host increases in thickness. Increase in thickness also occurs at this point, so that a longitudinal section through a branch bearing an old mistletoe shows the wood penetrated to different depths by numerous formidable pegs. At the point of attachment of the bush the host and its parasite usually have a club-like appearance.

We have not referred the absorbing system of the mistletoe to any morphological category, for the reason that its morphology is very obscure. At first sight it looks like an adventitious root system modified in the special circumstances in which it grows. But neither the suckers nor their branches in the cortex have any root-like characters apart from their absorbing function. Anatomically they are not root-like; they are produced exogenously, they have no root caps. The actual suckers of the mistletoe, as of the parasites already described and of those about to be considered, are frequently called *haustoria*, a term with a very

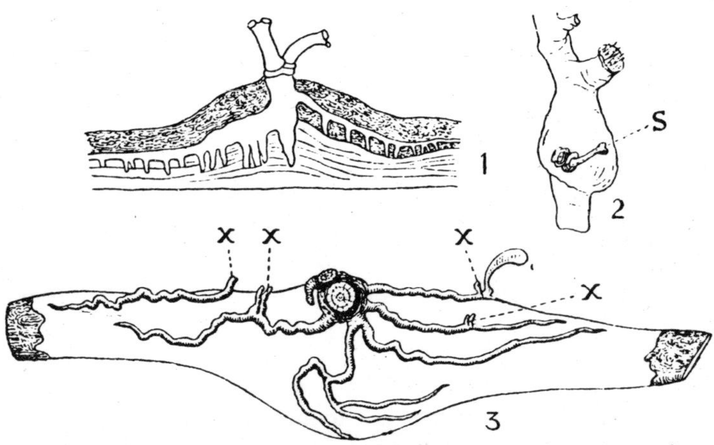

FIG. 25.—Mistletoe: 1, long section through host and parasite; 2, seedling mistletoe at S; 3, branch of host with bark removed to show course of absorbing system of mistletoe from which adventitious buds arise at X. All nat. size. (After Unger.)

wide and vague application, which is perhaps better left to describe certain definite organs of some parasitic fungi. In the Rhinanthoideæ the suckers are, definitely, modified side roots, and this is probably the case in the Santalaceæ; but in Viscum the whole absorptive system may best be regarded as a group of special organs which has arisen in relation to the peculiar circumstances of the plant.

The mistletoe possesses abundant green foliage and has even chlorophyll in the absorbing system. Its connection with the host is primarily through the wood; it, too, is a partial parasite, assimilating carbon dioxide, probably normally, and drawing from the host only water and salts. Its parasitism is, however, a good deal more advanced than that of Thesium or of Euphrasia. Its seed may germinate on a dead surface, but in these circumstances the seedling survives a few months only. It has a highly specialised absorbing system which gives it no possibility of independent life in the soil or as an epiphyte. Further, it is much more exigent in the matter of hosts. Mistletoe is to be found on a very large number of conifers and deciduous trees; but, as Heinricher and Tubeuf (1923) have shown, this appearance of wide choice is deceptive. There are, in fact, three distinct races of *Viscum album* which are morphologically identical (or practically so), but which require different hosts. The first, which grows on the pine, cannot be transferred—by seed or otherwise—to deciduous trees, nor to any conifer except one or two closely related species of Pinus and to the larch. The second occurs on the silver fir, and can be transferred only to some other species of Abies. The third occurs on such broad-leaved trees as the apple, pear, poplar, oak, and is, in its turn, divided into sub-races with more or less sharply marked preferences. The oak is perhaps less often attacked than any other tree, and this rarity may have enhanced the magical properties of the oak mistletoe in the eyes of the Celtic races. The reasons why a particular tree should alone serve as a host to a particular race of mistletoe are not yet known. We here wish to note this specialisation as marking an increased dependence on the host plant.

It might be thought that in its relations to its host the mistletoe would resemble the relation of a rose grafted on a briar. The rose assimilates carbon dioxide and draws its salts and water from the stock. So intimate is the union that the stock, deprived of foliage, is supplied by the scion with organic food and kept by it in healthy life. The parasite gives no such return. Molisch (1921) has demonstrated this by an interesting experiment. Small apple-trees had several mistletoe seeds sown on them. When these had developed into vigorous plants the apple foliage was entirely pruned off. For a year the mistletoe throve, and then the apple stocks died off and with them the mistletoe. The parasite supplies nothing to its host, and can draw supplies from it only when it is living.

In other members of the family methods of attachment very different from that described are to be found. In *Loranthus europæus*, the golden-green mistletoe which grows on evergreen oaks in the Mediterranean countries, the branches which arise from the primary sucker grow, not in the cortex, but in the cambium and into the wood of the host, thus making direct connection. When the resistance of the older wood becomes too great to allow of further penetration the tips turn outwards, and the repetition of the process gives rise to a curious step-like appearance of the absorbing organs of the parasite in longitudinal section. These branches keep pace for a time with the growth in thickness of the host wood, but may ultimately become more or less completely embedded. In some American genera, *e.g.* Phoradendron, the mistletoe of the United States, and also in some species of Loranthus, after the penetration of the primary sucker the tissues of the adhesive disc grow out marginally ; stimulated by the contact of the parasite the host tissues grow out also, forming a disc or cup of some size. In the most extreme cases the wood of the host forms a convoluted, lobed cup over 6 in. in diameter, completely surrounding the disc of the parasite. These remarkable objects are known in Mexico as *Rosa de Palo* (Fig. 26).

In the majority of tropical species true adventitious roots are produced either from the stem above the original adhesive disc or, in species with long twining stems, at any point on these. The behaviour of these adventitious roots is very varied. In some species they grow irregularly, forming an interlacing network with each other and with host branches ; in others they grow along the host branches ; in yet others they behave like tendrils twining securely round any solid object. In all cases they give rise to adhesive discs from which suckers penetrate shoots of the host or even other branches of the parasite itself. On reaching the wood these suckers splay out to form an absorbing disc in contact with the wood, and this in its turn sends absorptive filaments into the medullary rays. Finally, some twining species of Struthanthus produce suckers directly from the stem.

The securing of the seed to the host by a viscid layer of the fruit wall is universal in the family. Keeble (1895) states that the fruits of Singalese species of Viscum and Loranthus are greedily sought by small birds, which extract the pulp and wipe the seeds off their bills on to branches: "On the single telegraph line there are every year hundreds

FIG. 26.—Rosa de Palo; the cup formed by the host at the point of attachment of a Phoradendron. ⅓ nat. size.

of seedlings of *Loranthus loniceroides*, all in early stages of germination. It can hardly be supposed that the seeds arrive in this anomalous position as a consequence of being voided, but rather that the birds free their beaks of them by striking or rubbing against the wire." In some Loranthaceæ the fruit is explosive; the swelling of a mucilaginous layer expands the outer wall, the fruit breaks off, and the seed is then expelled. Peirce (1905) found the seeds of *Arceuthobium occidentale*, parasitic on the Monterey Pine, to be flicked to a distance of 15 to 25 ft. Explosive fruits were described by Johnstone (1888) for *A. oxycedri*. According to McLuckie (1923) the fruit of

Loranthus celastroides merely falls on to lower branches, the seed being squeezed out : wider dispersal depends on birds. Peirce also considers the explosive fruit inefficient in securing fixation. It should be borne in mind that the "fruit" of the Loranthaceæ is partly composed of the hollowed axis of the flower.

Throughout the family the nutritive relations of host and parasite are, on the whole, the same as for Viscum. Only the West Australian *Nuytsia floribunda* and the small American and Australian genus Gaiadendron grow independently as trees. The rest of the family is parasitic. So far as is known the majority are not very highly specialised as regards their hosts, but in this respect only *Viscum album* has been thoroughly studied, and the results make caution necessary in drawing general conclusions. Weir (1918) has shown that each species of the American genus Razoumowskia has a limited range of host conifers. The leaves are usually well developed, but in some species of Viscum, Phoradendron, and other genera they are reduced ; in these, however, the stems take on the work of assimilation and show ridges, or flattening such as is common in other xerophytic plants. Interesting is *Viscum Crassulæ*, parasitic on the succulent Crassulas and Euphorbias of arid regions of South Africa ; like its hosts it is a succulent, with short internodes and very thick orbicular leaves. In *Arceuthobium minutissimum*, a Himalayan species growing on *Pinus excelsa*, only the flowers come above the bark of the host ; its tissues are otherwise internal, and presumably it is completely parasitic.

Of interest are the transpiration relations of such plants. Our mistletoe, which is evergreen, grows on both evergreen and deciduous trees ; *Loranthus europæus* is deciduous. The leaves of the mistletoe are leathery and strongly cutinised, and we have seen that xerophytic features are shown by other species. Kamerling (1914) has shown for several tropical species, *e.g. Loranthus pentandrus* on *Mangifera indica* in Java and *Loranthus dichrous* on *Psidium guajava* in Brazil, that the transpiration of the parasite is more rapid than that of the host, whether equal areas or weights of leaves are compared. He concludes that the chief danger to the host is the withholding of water from leaves lying above the point of attachment of the parasite, through the competition for the supply by the latter. He states that leaves in such a position do frequently wither. McLuckie (1923) notes that the hosts of *Loranthus celastroides* in New South Wales may be killed by the withdrawal of water. Future work must show if the relation is general.

Complete Parasites.—The mistletoe may be taken as the most advanced of the semi-parasites, and we may now return to the Rhinanthoideæ and trace out another line along which parasitism has evolved. In the genus Tozzia, with one species in the Alps and another in the Carpathians, the habit is somewhat that of a large eyebright or a yellow rattle, but with distinctly fleshy shoots and pale green leaves obviously poor in chlorophyll. The life-history is curious. For two or three years after the germination of the seed the plant exists as a subterranean rhizome with scale leaves, wholly parasitic on the roots of the hosts; then the leafy aerial shoot is sent above ground, and the plant leads a subaerial and semi-parasitic existence for a few weeks before flowering.

Finally four genera of the group are complete parasites. Of these the best known is Lathræa with two (European) species, *L. Squamaria*, the toothwort, which occurs in Britain, and *L. clandestina*. The toothwort has a thick flowering axis above the ground; it is purplish in colour, possesses only small scale leaves, and ends in a raceme with numerous flowers. Neither axis nor scales possess chlorophyll; the plant is completely parasitic. The inflorescence springs from a thick, branched, subterranean rhizome bearing fleshy scale leaves in four rows. The structure of these scales is peculiar. The upper part is reflexed and united to the base, forming a hollow with a narrow opening; this enclosed space communicates with branching canals extending into the leaf tissues. The inner surface is covered with epidermal glands which excrete water. Various functions have been assigned to these remarkable leaves.

It seems certain that they are food-storage organs. Goebel regards the excretion of water by the glands as a substitute for transpiration. The amount of water given off and finding its way to the soil through the opening of the scale may be large. Chemin (1920) has determined it at 3 to 4 per cent. of the weight of the plant in twenty-four hours. He regards the excretion as getting rid of superfluous phosphates, sulphates, and ammonia—a somewhat improbable explanation. It has also been suggested that small soil organisms—protozoans, diatoms, worms—find their way into the leaf hollows and are digested and absorbed by the plant, but no proof of this exists. It may be noted that the subterranean leaves of Tozzia show features like those of Lathræa in a simpler form. The margins are bent back, and in the hollow thus produced water-excreting glands are found.

According to Heinricher (1910) the seed of Lathræa cannot germinate without contact with the host roots, but the process is not adequately known. The two cotyledons are unfolded and a true root with side roots develops. On the root arise numerous disc-like suckers, whether in response to the stimulus of a host root, to the stimulus of contact with solid particles, or spontaneously, has not been definitely settled. The suckers are proliferations of the cortical tissue of the root. In contact with a host root they spread out over it or almost enclose it, becoming fixed by numerous unicellular filaments like root-hairs. An absorbing process, dissolving its way through the cortical tissues, penetrates to the wood. Connection between the wood of the host and of the parasite is made by the development in the sucker of a strand of tracheids ; other prosenchymatous cells make contact with the parenchyma and the bast.

The Lathræas have a certain amount of latitude as regards host plants ; normally they occur in shady places where they parasitise the roots of woody plants, *L. Squamaria* affecting in particular the hazel.

Broom-rapes.—The Lathræas lead to the Orobanchaceæ, in which family they are sometimes placed. The family is best represented in the warm temperate regions of the old world; about half a dozen species of Orobanche, the broom-rape, occur in Britain. *Orobanche rubra* is parasitic on roots of the thyme, *Orobanche major* on leguminous shrubs, *Orobanche caryophyllacea* on bedstraws and brambles. Some species are thus restricted in the choice of a host, others are less exigent.

The inflorescence axis appears above the soil; in *Orobanche major* it is 1 to 2 feet high and stoutish, in other species it is smaller. It bears a few small scales and ends in a crowded spike. Below the surface it terminates in a swollen base with many scales. The plants possess no chlorophyll and have dull colouring in tints of brown, yellow, and purple. Brilliant colours are developed in other members of the family, the Caucasian Phelipæas having bright scarlet flowers.

The seeds of the Orobanchaceæ are minute, with an undifferentiated embryo embedded in an endosperm. Scattered by the wind, they are washed into the soil by rain, and germinate only in contact with living roots of host plants. A chemical effect must be involved, but of what nature we do not know. A filamentous embryo 1 mm. long is produced, in which only relative position distinguishes a root end from a shoot end; the latter remains in the seed coat, the former comes in contact with the host root and sends a sucker down to the wood. Vascular tissue then develops in the seedling, and an extremely intimate connection is made with the host, wood with wood, bast with bast, cortex with cortex, and epiderm with epiderm. The upper part of the seedling now grows into a tiny tuber with a lumpy surface, and the apical portion usually withers away. From this tuber arise adventitious outgrowths which extend to other host roots and produce secondary suckers. They arise exogenously, and, if a root cap is present, it is much reduced. They may, perhaps, be looked on as adventitious roots of an aberrant nature. The flowering stem arises as a rule singly and adventitiously from the tubers. Flowering stems may also arise from the adventitious roots. The plant may be annual or may take several years to produce its flowering shoot. After flowering it usually dies.

If we look back over the series we have traced, we see that the partially parasitic Rhinanthus shows little departure from the normal. Tozzia shows reduction of the aerial shoot which is short-lived and has diminished chlorophyll; there is a well-developed subterranean rhizome. In Lathræa the vegetative part of the plant is entirely subter-

FIG. 27.—*Cytinus hypocistis*, two inflorescences rising from the root of a Cistus shrub. Nat. size.

ranean though well developed, only the flowering axis comes above the ground, and chlorophyll has completely disappeared. Finally, in the broom-rapes the vegetative body is reduced to the little tuber and its peculiar roots, with purely absorptive functions. Only the flowering shoot attains any size.

Balanophora, etc.—The Balanophoraceæ are root-parasites on shrubs and trees. This family is nearly confined to tropical rain-forests, a few species occurring in savannah and bush vegetation, and one, *Cynomorium coccineum*, the

Maltese Sponge, growing on halophytic shrubs in the Mediterranean region. The vegetative body consists of large branched tubers attached to the roots of the host. From it arise massive, club-shaped or globular, often branched inflorescences with scale leaves, and many flowers, frequently brilliant in colouring.

Even in young stages there is no differentiation of root and shoot. The tuber may be in contact with several host roots, which it envelops. No suckers are produced. The host root, the cortex of which is resorbed, sends branching extensions of vascular tissue into the parasitic tuber, and even into the inflorescences. We have here a case in which the union of parasite and host is mutual and extremely intimate; the tuber is, in a way, a joint organ (Fig. 24, 2).

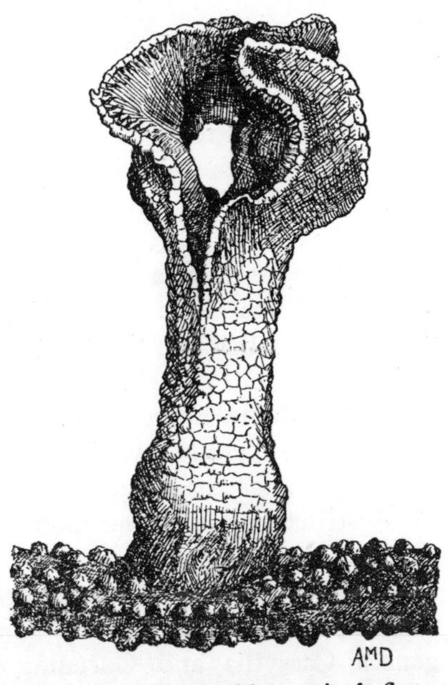

FIG. 28.—*Hydnora africana*, single flower rising from the thick rhizome-like vegetative body. Nat. size.

The Hydnoraceæ and Rafflesiaceæ are also characteristically tropical families. One species, *Cytinus hypocistis* (Rafflesiaceæ) is parasitic on roots of Cistus shrubs in the Mediterranean and southern Atlantic coastal regions of France (Fig. 27). The Hydnoraceæ possess branching rhizomes which make connection with the roots of the host apparently after the fashion of the broom-rapes. From these arise several solitary leathery or fleshy flowers (Fig. 28).

The flower of the Rafflesiaceæ is also usually solitary.

That of the Javan *Rafflesia Arnoldi*, blood-red in colour, stinking like carrion, and measuring a yard across, is the largest known flower. Cytinus has short fleshy scapes with several flowers, the brilliant yellow of which, contrasting with the equally brilliant scarlet of the scaly bracts, makes them lovely objects as they appear above the bare brown soil below the Cistus bushes of the garigue. The flowers or inflorescences of the Rafflesiaceæ spring straight from the cortex of the host plants. When they break through the cortex of a stem, *e.g.* Pilostyles on vines, they present a very remarkable sight, the trunk of the host appearing to blossom. The vegetative body exists entirely inside the host. In most cases it is reduced to an irregular weft of branching filaments, which wander through the parenchymatous tissues of the host. In Cytinus and some others undifferentiated strands of tissue are formed. The inflorescence starts as a parenchymatous swelling inside which a vegetative point is ultimately laid down; this gives rise to the bud which breaks through the tissues of parasite and host to the exterior.

In the Rafflesiaceæ we have the ultimate stage of reduction of the vegetative body, which is at a level of organisation no higher than that of a fungus. Of the flowering plant structure the flower alone remains, and we may note that the seeds are minute, the embryo is undifferentiated, and the development of the gametophyte is abnormal.

Cassytha and Cuscuta.—The two isolated parasitic genera Cassytha and Cuscuta, differ widely from other parasites. They possess extensive branching stems which twine round the stems and leaves of the hosts. The leaves are reduced to minute scales; chlorophyll is often present, but usually in such small amount that the plant appears quite yellow. Cassytha, according to Mirande (1905), has abundant chlorophyll. It is capable of slow growth for as much as eight months without a host, but develops rapidly only when it has attached itself to another plant. Some dodders, *e.g.* the tropical *Cuscuta reflexa*, are markedly green. Photosynthesis may in these be sufficiently vigorous to produce oxygen in light, but not to meet the constructive requirements of the parasite. Cassytha is tropical: Ernst (1908) describes it as growing in festoons on the strand grasses of Krakatau. Cuscuta is widespread from the tropics to the temperate zones, several species occurring in Britain. *Cuscuta Epithymum* is common on whin, thyme, and heather, *Cuscuta Trifolii* grows on clover, and *Cuscuta Epilinum* on flax.

The seed of Cuscuta germinates on the soil, usually, so far as European species are concerned, late in the spring, when other vegetation has sprouted, and young shoots of host plants are therefore available. The late germination may be determined by high temperature or light requirements. The root leaves the seed coat first and pushes slightly into the soil, from which it absorbs water. Cotyledons are absent or rudimentary. As the young shoot, a fine yellow thread, grows, it appears to creep forward on the soil. The tip is raised above the surface and circumnutates. If it fails immediately to meet a host it may grow forward for some time at the expense of the basal parts, which wither away, but it ultimately dies. Spisar (1910) found the maximum term of life without a host to be seven weeks in *Cuscuta Gronovii*; in smaller species it is less. If it meets a living plant it commences to twine round it, behaving in turn like a twining plant and like a tendril; that is, it alternates a series of loose elongated coils with a series of close tight ones. Peirce (1894) agrees with older investigators that the seedling dodder can only twine round living plants, though the mature parasite can twine round any support, living or dead. More recently, Mirande (1900) and Spisar (1910) have found that the seedlings of the small species, *e.g. C. Epithymum*, can twine round a dead object if it is moist, while seedlings of the large species, *e.g. C. Gronovii*, can twine indifferently on wet or dry, dead or living, supports. On the surface of the tight coils, in contact with the host, epidermal adhesive discs, attached by papillæ which grow between the cells of the host, are formed in response to the contact stimulus. The papillæ also function as absorbing organs. From these discs suckers are sent down into the host tissue, which is dissolved away by secreted enzymes. The suckers spread out inside the host ; vascular tissue is formed and connects up with the wood and bast ; prosenchymatous filaments penetrate the cortex and even the pith, and tap living cells.

The nature of these suckers is not clear. Some botanists regard them as modified adventitious roots, and in some species, *e.g. Cuscuta europæa*, they are produced endogenously.

After connection has been made with a host growth is very vigorous. The stem passes from shoot to shoot, branching freely ; the parasite may completely exhaust and smother the host plant, and may do extensive damage to crops of clover and flax. The little bunches of flowers are produced abundantly.

It is interesting to note that in Cuscuta and Cassytha reduction runs on lines different from those common in other advanced parasites. The foliage indeed is gone, but the stem is well developed ; this is of course to be related to their unique mode of attachment to the host, and brings with it the advantage of rapid spread over available host plants. Alone among the parasites, Cuscuta can be grown independently of host plants in a sugar solution. In such cultures *Cuscuta monogyna* flourishes, and produces flowers and fruit (Molliard, 1908).

General Considerations.—Reviewing this account of parasitic flowering plants, we see that, while we are well informed as to the structural relations between them and their hosts, little of the physiological relation is known. We do not even know exactly whether green parasites receive only water and inorganic salts. As to the forms in which organic compounds are absorbed by complete parasites we know nothing. Nor do we know much of the factors which bind a parasite to its particular hosts. It may be surmised that some plants are unsuitable as hosts because they offer difficulty to penetration, but this explains little. In Cuscuta enzymes play a part in securing penetration ; to what extent they are active in other cases is unknown.

Research on water relations has just begun. Here is a wide, and it should be added a difficult, field to be explored.

An enticing problem, more of a speculative nature, is the origin of the parasitic habit. The distribution of parasitic groups in the system of flowering plants shows that the habit has certainly arisen more than once in evolutionary history, but that it has not arisen very frequently. Cassytha, Cuscuta, and Krameria are parasitic genera in otherwise autotrophic families, which are not related to any of the other groups of parasites. In the Scrophulariaceæ the habit appears in various genera of a single tribe; the closely related Orobanchaceæ are completely parasitic. The small family of Lennoceæ is not related to any other parasitic family. The Santalaceæ and Myzodendraceæ are related, as are the Loranthaceæ and Balanophoraceæ, and the Hydnoraceæ and Rafflesiaceæ. These families stand near each other in the natural system. It may be noted that there are no parasitic monocotyledons nor gymnosperms. It looks as if in certain plant groups there has been, and probably still is, some tendency of an unexplained nature which has made parasitic development possible. This tendency is present in the Rhinanthoideæ, and in the very closely related Orobanchaceæ. It shows itself in the group of families Loranthaceæ, Balanophoraceæ, Hydnoraceæ, and Rafflesiaceæ, and in the Santalaceæ and Myzodendraceæ. It crops out quite unexpectedly in the isolated genera Krameria, Cassytha, and Cuscuta. It may be that the nature of the tendency in the various cases has been quite different. It is certain that in every case it has been the acquisition of some *positive* quality or qualities which has made possible the connection with the host. Parasitism has almost certainly not been a consequence of diminished capacity for independent existence, nor of reduction of vegetative structure; these have followed. Parasitism, in other words, is primarily a positive capacity, and only secondarily does it mean degeneration.

A factor in the evolution of parasitism has been opportunity. Parasitism in Cuscuta has been favoured by the opportunities for attachment in twining plants so characteristic of the bindweed family. The opportunity of the Rhinanthoideæ lies simply in the abundance of roots of other plants in the soil of meadows and pastures, and the frequent contact with them which must be made. Given the start of root parasitism, it is not difficult to picture the evolution of the more advanced and specialised types, culminating in such forms as Rafflesia. The problem of the origin of the mistletoes is more difficult. One might think of them as originating from epiphytes. But in all the multitude of the true epiphytes of the tropics there seems to be no indication of any sort of parasitism The distribution of the Loranthaceæ to tree branches is now secured by the slimy berry, but whether this appeared before or after the assumption of the parasitic habit we do not know. It is quite conceivable that the Loranthaceæ, too, started as root parasites and became branch parasites later, for two genera of the Santalaceæ are branch parasites though root parasitism is dominant in the family; it is also quite possible that they were originally lianas, for some tropical species, as we have seen, have tendril-like roots and rambling stems. We have, in fact, no means of deciding how the parasitic Loranthaceæ took to their present mode of life.

Some efforts have been made to determine experimentally what conditions will permit of parasitic existence. It is quite possible to grow many ordinary plants for a time in wounds of others. Peirce (1904) planted germinating peas in slits of a bean stem; the root system grew down the hollow internodes and drew a supply of water from the walls of the cavities. The peas reached maturity and bore seeds. Molliard (1913) grew cress in wounds in the hypocotyl of the French bean, in a saturated atmosphere, and found that sucker-like side roots were formed.

MacDougal and Cannon (1910) made an extended study of the growth of several plants used as cuttings in wounds of various cactuses. In several cases the cutting survived for upwards of two years and showed considerable growth. Agave produced so vigorous a root system as to destroy much of the tissues of the "host" cactus. Opuntia remained alive for a long time, absorbing water through its parenchymatous cells, but forming no roots. But Peirce's pea plants were stunted. Slow growth and poor root formation were characteristic of MacDougal's plants. MacDougal (1911) lays stress on the necessity of an osmotic pressure superior in the parasite to that of the host. Senn and Hägler (quoted from Tubeuf, 1923) have shown that Viscum, Thesium, Euphrasia, Orobanche, and Pedicularis have in fact osmotic pressures from 0·025 to 0·25 atmos. higher than those of their hosts. Harris and Lawrence (1916) found that the Loranthaceæ of Jamaica had in almost all cases osmotic pressures higher than those of the hosts. It is of interest to note that Harris (1918) found the osmotic pressure of Jamaican *epiphytes* to be always very much lower than that of the trees on which they grow.

It is tempting, too, to institute a comparison between the behaviour of a plant grafted on another and the relation of parasite to host. The stock is of course normally treated so that it bears no foliage, but a scion may quite well be grown on a leafy stock, and so reproduce the external features of a mistletoe and an apple-tree. As we have seen, however, the inter-relations are quite different.

It is extremely doubtful whether the behaviour of the scion grafted on a stock, or even such experiments as those of MacDougal, throw any light on parasitism or its origin. Grafts are successful only within strictly limited bounds of relationship; the graft is really a regeneration or wound-healing phenomenon, which results in the formation of what is, from the standpoint of nutritive physiology, a single organism—the scion supplying organic compounds, the stock water and salts. Parasitism is normally a relation between two plants which do not stand close to each other phylogenetically, and which may easily stand at the opposite poles of the plant system—*e.g.* a Viscum on a gymnosperm. It is essentially an *attack* of one organism on another, an active incursion into its system; and the union, however intimately connected the two organisms may be, is essentially dualistic and antagonistic.

As regards MacDougal's and Peirce's experiments, we may say that they offer no indication of how even the most elementary form of parasitic connection might be brought about; none of their plants, even though planted in wounds, give any sign of forming an organic connection with the "hosts." They are, in fact, simply plants growing in an abnormal substratum and showing some sign of the fact in hunger stunting. To interpret the stunting as in any way analogous to the reduction seen in parasites seems unnecessary. Such experiments are certainly interesting, but they are only the first beginnings of experimental investigation of this problem. Molliard's cress approaches parasitism rather more closely.

§ 2. Saprophytes

Saprophytic flowering plants are much less numerous than parasites. Complete saprophytes as a rule lack chlorophyll and show reduction in vegetative structure. Even so typical a saprophyte as the bird's-nest orchis, however, possesses a little chlorophyll masked by a brown pigment; assimilation is probably insignificant in such cases.

The saprophytes live in soils rich in humus, and most characteristically in the mould of woods. Their subterranean system consists of rhizomes or roots. In some, *e.g.* the coral-root orchis, roots are absent, their function being performed by the much-branched rhizomes. All saprophytes are symbiotic with mycorhizal fungi which evidently play a part in their nutrition; the saprophytic orchid *Wullschlægelia aphylla* alone is said to be an exception to this rule (Ramsbottom, 1922). What little is known of this in detail will be referred to in connection with the general problem of mycorhiza. As we shall see, a great many green plants which have no appearance of leading a saprophytic existence also possess mycorhiza and grow in humus soils. Between these and typical saprophytes intergrades exist, so that it is likely that there are many cases of partial saprophytism which have not been recognised. Or we may look at the relation in another way. It is not certain that any saprophytes really draw organic food directly from the soil. The fungus may in all cases act as an intermediary. The word "saprophyte" would thus be a misnomer, and these plants would properly be regarded as the end of a series, exhibiting the extreme results of the mycorhizal habit, having become parasitic on their fungi. It is, however, convenient at present to distinguish some humus plants which have reduced leaves and chlorophyll as a separate class.

Systematic.—While there are no parasitic monocotyledons the majority of saprophytes belong to that group. They occur in the families Orchidaceæ, Burmanniaceæ, and Triuridaceæ. Among the dicotyledons the families Pirolaceæ (sometimes included in the Ericaceæ), Gentianaceæ, and Polygalaceæ, include saprophytic species. Among the gymnosperms neither parasites nor saprophytes are known.

(*a*) Polygalaceæ: the 2 species of Epirrhizanthes of the Indo-Malayan region are saprophytes.

(*b*) Gentianaceæ: of the 60 genera of the family six, which occur in four different tribes, with in all about 30 species, are saprophytes. The North American Bartonia and Obolaria have sufficient chlorophyll to give them a distinct green tinge, and may be regarded as partial saprophytes. The rest, with a distribution that includes tropical America and Africa, the West Indies, Ceylon, and the Himalayas, are devoid of chlorophyll. In all cases the leaves are represented only by scales, often of minute size.

(*c*) Pirolaceæ: 8 of the 10 genera, with 10 species, are saprophytes, without chlorophyll and with only scale leaves. Typically they inhabit the leaf mould of woods. Thus, *Monotropa Hypopithys*, the bird's-nest, occurs in beech and fir woods in Britain; *Monotropa uniflora* is the Indian pipe of North American woods. The flowering shoots arise from buds produced by the roots. Parasitic connections with roots of other plants have been described in some cases, but this has not been confirmed by all investigators. In addition to the complete saprophytes one American species of Pyrola, *Pyrola aphylla*, with little chlorophyll, is a partial saprophyte.

(*d*) Triuridaceæ: this small family, with 3 genera and about 40 species, is exclusively tropical. All are small yellowish or reddish plants with scale leaves.

FIG. 29.—*Thismia Aseroe*. Nat. size. (After Groom.)

(*e*) Burmanniaceæ: another small tropical family, with its chief centre in Borneo and New Guinea, and with representatives in America. Most of the genera are completely saprophytic; Burmannia has autotrophic and saprophytic species. One or two species are doubtfully parasitic. The family is remarkable for the extraordinary forms of its flowers (Fig. 29). Some species possess rhizomes which may be coral-like; others produce shoots from the roots.

(*f*) Orchidaceæ. This great cosmopolitan family has more than 400 genera and 7000 species; about 50 species, belonging to a dozen genera, are saprophytes, some only partial. It is likely that the family includes many other species of partially saprophytic habits: practically all harbour mycorhizal fungi. Three saprophytic species occur in Britain ; *Corallorhiza innata*, the coral-root, and the much rarer *Epipogon aphyllum* are devoid of roots, possessing coralloid rhizomes ; *Neottia Nidus-avis*, the bird's-nest orchis, possesses roots. These plants are brown or yellowish in colour, though Neottia possesses some chlorophyll, and have leaves reduced to scales. Epipogon and Neottia occur in shady woods in rich humus soil ; Corallorhiza is peculiar in inhabiting sandy copses.

General Remarks.—As has been said, all saprophytes are mycotrophic ; this habit is, as we shall see, extremely widespread, and, in conjunction with growth in rich humus soils, would seem to offer admirable opportunities for a dependent existence. Yet complete saprophytes are few in number.

A good deal of work has been done in investigating the possibility of normal plants absorbing organic substances from the soil. It is certain that the normal root system may absorb soluble organic substances like sugars. Robbins (1922) has recently shown that amputated root tips of the maize may grow vigorously and branch on organic culture media. Complete plants supplied with sugar show increased growth and also increased chlorophyll content. Brannon (1923) has grown various plants in the dark on glucose and other sugars and for periods of many weeks. Increase in dry weight took place, and peas even produced flower buds. This tells distinctly against any idea that chance absorption of organic substance might have led to loss of chlorophyll and further reduction. The fact is established that humus compounds are not available for higher plants, and this makes it all the more likely that the saprophytes draw on the soil only through their symbiotic fungi. They must have been derived from green mycotrophic plants. They have originated, as have the parasites, in widely different regions of the system of flowering plants. Of the causes which have led to diminution of chlorophyll and further reduction we know nothing, except that in ordinary plants mutations with little or no chlorophyll occasionally occur, *e.g.* in *Lychnis dioica.* In such mutations in plants with advanced mycotrophic habits it is possible that the saprophytes have had their origin.

§ 3. Mycotrophic Plants

The term *mycorhiza* was first used by Frank (1885) to designate the association of a *fungus* with the *root* of a higher plant. It is commonly regarded as a kind of *symbiosis*, by which is meant an intimate partnership of two organisms in which both partners benefit. But the details of mycorhiza are so varied, and we know so little of their physiological relations that it is not safe to assume mutual benefit or even one-sided benefit in all cases. The term is somewhat naturally used to cover the association in a plant like Corallorhiza which has no roots but only a root-like rhizome, and it is extended to plants in which the fungus inhabits shoot structures which have no resemblance to a root at all, as in the curious Japanese orchid, Gastrodia. Despite the violence it does to etymology this usage is established, though the term *mycotrophic* is more exact.

Mycorhiza associations were divided by Frank into two classes, *ectotrophic* and *endotrophic*. In the latter, typically seen in the orchids, the fungus inhabits the root cortex; in the former it occurs as a mantle outside the root, and only penetrates *between* the epidermal cells, as in the pine, the beech, and many other forest trees. This distinction, striking enough in extreme cases, cannot always be applied. Intermediate types are known, and, even in typical cases of the one or the other extreme, the fungus is not so sharply localised as the terms imply.

Mycorhiza is an extremely widespread phenomenon. It is exhibited by all the orchids, by the autotrophic as well as by the saprophytic, and by all the Ericaceæ, in endotrophic form; the majority of our forest trees possess ectotrophic mycorhiza. These are the best known cases, but it has been shown by Janse (1897), Stahl (1900) and Gallaud (1905) that most families include plants which more or less frequently have their roots associated with fungi. Only in the Cruciferæ and Cyperaceæ did Stahl fail to find any mycorhizal species. He found that mycorhiza was most frequent in soils poor in nitrogen and mineral salts, and founded the theory that the chief benefit derived by the flowering plant from its fungus was an increased supply of salts.

Ectotrophic mycorhiza is, as we have said, seen in its most characteristic form in forest trees. It is particularly well developed in the pine, the larch, and other conifers, in the beech, oak, hazel, and other cupuliferous trees. Young side roots are infected by fungi growing in the rich humus soil. The roots have their growth in length curtailed and the infected tips are club-like. Infected pine roots fork repeatedly and thicken into a coral-like mass. The mature mycorhiza forms a felted or hairy mantle round the root, from which numerous hyphæ penetrate between the walls of the epidermal cells. Root-hair formation is prevented (Fig. 30). It has been recently shown by Melin (1921), for the pine and the spruce, that infection takes place through the root-hairs or epidermal layer, and that at first the fungus exists inside the cortical cells, in which it is ultimately digested ; later it passes between the cells of the epiderm and forms the typical mantle. In trees growing in wet bogs the fungus exists exclusively in the cortical cells. Melin isolated three distinct fungi which can form

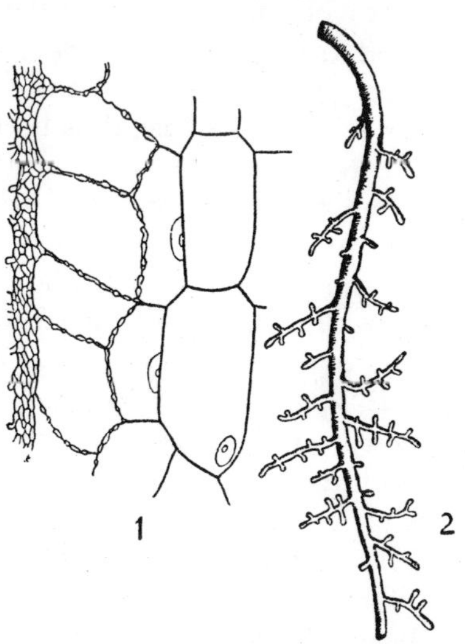

FIG. 30.—Ectotrophic mycorhiza ; 1, section of root of the hornbeam, the fungus forming a mantle outside and penetrating between the epidermal cells ; 2, root system of seedling fir. 1, × 320 (after Frank). 2, Nat. size.

mycorhiza with the pine, and showed that they belong to the Hymenomycetes and to the genus Boletus. The endotrophic mycorhiza found in bogs is due to a fourth form.

It has long seemed probable that forest tree mycorhizas are formed by the large toadstools characteristic of woods, some of which are constantly associated with a particular species of tree. This has been finally proved by Melin who has shown, by infection of sterile seedlings with pure cultures of the fungi, that the mycorhiza of the larch is derived from *Boletus elegans*, a toadstool that is common in larch woods. One of the fungi of the pine is *Boletus luteus*. Mycorhiza has thus been synthesised.

Peyronel (1921) has demonstrated mycorhizal connection between the roots of forest trees and a number of large humus fungi belonging to the Tuberales and Basidiomycetes. Thus the mycorhiza of the larch may be formed by *Boletus elegans*, *B. laricinus*, *B. cavipes*; of the aspen by *Boletus rufus*; of the beech by *Cortinarius proteus*, *Boletus dysenterica* and *B. cyanescens*, *Hypochnus cyanescens*, and *Scleroderma vulgare*. It may be taken that older work in which the mycorhizal fungus was referred to species of Penicillium, Mucor, etc., was vitiated by a faulty technique which easily admits to cultures the omnipresent spores of these fungi.

The ectotrophic mycorhiza of forest trees does not seem to be essential to their existence. At least in some soils the trees can exist without the fungus. It is not yet certain whether under favourable conditions sterile trees grow as well as infected. Nor is there a certain answer to the question of the function performed by the mycorhiza when normally developed. From the nature of the case it must replace the root-hairs in the transference of water and salts to the root; but it does not follow that it supplies these more efficiently than would the root-hairs themselves. Melin states that the mycorhiza of pine and spruce is poorly developed in mild humus and well developed in raw humus, and that it is necessary to the success of the tree on drained peat. He thinks that in one way or another the fungus transfers nitrogen from the soil to the plant. Though the fungus does not fix free nitrogen in isolation, it is possible that it may do so when associated with the root. It is very likely that in general the mycorhizal fungus has an important action in making available for the plant the otherwise unavailable inorganic nutrient substances present in the humus. W. B. McDougal (1914) looks on ectotrophic mycorhizas of broad-leaved trees as a chance association in which the fungus is a parasite. The maples he regards as possessing an endotropic mycorhiza which is symbiotic.

Endotrophic mycorhiza : Calluna.—The best investigated case is that of the Ericaceæ and particularly of *Calluna vulgaris*, the heather, which has been cleared up by the work of Rayner (1913–1922). The fungus is not confined to the root, though there lies the region of infection and of its principal development. It extends in an attenuated form through the stem and leaves, and into the ovary. From the intercellular spaces of the leaves hyphæ extend into the air. From the ovarial wall it stretches over to the minute seeds, and, when these are shed, they carry with them fragments of mycelium in and on the seed coat.

On the germination of the seed, as the radicle begins to elongate, and even before it has left the seed coat, it is infected by the fungus hyphæ through its external cells, no root-hairs being formed. The fungus passes rapidly from cell to cell dissolving a passage through the cell walls, and is presently found like a coiled skein of thread in almost every cell of the root cortex and epiderm. It spreads to the branch roots and to the shoot. Later the fungus pushes hyphæ between the epidermal cells, and forms a fine network on the outside of the root. Indeed, although its principal development is intracellular, and we may conveniently class it with the endotrophic forms, it has been described as ectotrophic and is a good example of an intermediate type.

By suitable means the Calluna seeds may be sterilised and uninfected seedlings raised, but these never develop properly. The cotyledons and a few leaves unfold ; the root system is represented only by a few minute stumps at the base of the hypocotyl, when, at a corresponding stage in an infected seedling an extensive branched system of fine transparent roots has been formed. The fungus-free seedlings never get beyond this stage (Fig. 31).

Here, then, we have an association much closer than that exhibited by the mycorhiza of forest trees ; for in these, at least in certain conditions, good development can take place in absence of the fungus, while in the case of Calluna —and, it is probable, in many other Ericaceæ—normal development, and in particular root formation, can take place only if the fungus is present. Furthermore, regular

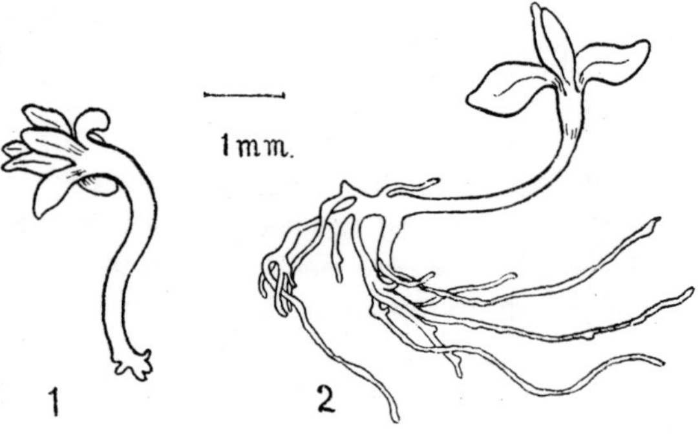

FIG. 31.—Seedlings of heather (*Calluna vulgaris*) ; 1, in sterile culture ; 2, infected with mycorhizal fungus ; both five months after sowing. (After Rayner.)

infection by the fungus is ensured by the presence of its hyphæ on the seed. The manner in which development is influenced by the fungus is at present quite obscure. We now know, however, that the fungus definitely benefits the general growth of the plant in a particular way. It has been shown by Ternetz (1907) that fungi isolated from the roots of the Ericaceæ are capable of assimilating free atmospheric nitrogen. The identity of Ternetz's fungi—species of the genus Phoma—with the mycorhizal fungus, has been confirmed by Rayner (1922), who also showed that normally infected seedlings could flourish in a medium quite free from nitrogen compounds, and this the ordinary green plant cannot do. The presence of fungal hyphæ outside the plant, both in the air and in the soil, is probably of importance in relation to the supply of nitrogen. The heather thus receives nitrogen compounds from the fungus, and the fungus in its turn must receive organic food from the plant. Its main development is intracellular and its only source of carbon compounds is the organic matter of the cell. It does not seem to act harmfully, no degeneration being visible in the cells of normally infected plants. But the balance between plant and fungus is a very delicate one, for, if the plant is enfeebled in any way, the fungus takes the upper hand and overmasters its host. In normal conditions the fungus never fructifies, but in a weakened plant the fructifications of the fungus indicate its dominance. This altered relation is beautifully seen in seedlings grown in calcareous soils. Calluna is a well-known example of a plant that lives habitually in acid humus, and when grown in chalk or lime soils is stunted and chlorotic, as the result of some difference in the soil chemistry or physics the precise nature of which is at present not understood. In seedlings grown in such soils, or even in watery extracts from them, the fungus of the mycorhiza is seen to become dominant ; so that even an environmental change which may occur in nature, is sufficient to upset the balance normally maintained between the two symbionts. An account of the Calluna symbiosis by Christoph (1921) differs from Rayner's in many points ; he finds that Calluna can thrive without the fungus, which he regards as a parasite. Rayner (1922*b*) has shown that Christoph's conclusions are not well founded.

We have, then, a very complete though not exhaustive account of the mycorhizal relation of Calluna. We do not know the mechanism of the effect of the fungus on root formation. It is not certain that the assimilation of nitrogen is the only way in which the fungus serves its host. It is quite possible that it affects the supply of salts and of water. We may remark on the prevalence of mycorhiza in the heaths and their allies, typical inhabitants of peaty soils with a low salt content. As regards the fungus we do not know the details of its food relations, nor do we know how the fructification is suppressed.

The Orchids.—More attention has been paid to the mycorhiza of the orchids than to that of any other group. In several cases they have been exhaustively studied from the morphological standpoint, and their biological relations have been the subject of a great deal of work during the last twenty years. Our knowledge of this side is largely due to the French botanist Noel Bernard (1909), and to the amplification of his work by Burgeff (1909). A useful account of recent work is given by Rayner (1916*a*). The general life-history of the association is fully described by Magnus (1900) for Neottia. The fungus is sharply limited to the three or four external cortical layers of the root; it is never found in the inner cortex, nor in the central cylinder, and only sparingly in the epidermal layer (2 to 3 cells thick). Very few hyphæ find their way outwards into the soil. In the rhizome as many as six cortical layers may be infected, and the fungus even reaches a short distance into the flowering axis. Infection of new roots takes place, at a very early stage, from the rhizome. The mycorhizal cells are further sharply separated into two distinct classes—an outer and an inner layer of "digestive cells," and between these a layer of "host cells."

In the host cells a coil of rather thick-walled hyphæ clothes the inner surface of the walls, and from this thinner hyphæ traverse the protoplasm and vacuoles, and probably are absorptive in function. Throughout the life of the root these cells present the same aspect. In the digestive cells the fungus at first forms coils of thin-walled hyphæ. Soon these die and are evidently digested by the plant cell; the remains of the wall substance collapse and, together with some plasma, are separated as an indigestible "clump," which is surrounded by cellulose material—a sort of internal excretion of waste matter. Starch appears at the time of infection in small grains, which soon disappear and are reformed after digestion. The digestive process is accompanied by characteristic changes in the nucleus of the host cell seemingly of the nature of an interrupted division.

This highly specialised differentiation into digestive and host cells is found in many, perhaps all, other orchids, but the arrangement in definite layers is not usually so sharp ; nor is the quite definite localisation of the fungus and its regular occurrence in all the roots general.

Magnus describes, too, the case of *Orchis maculata* which is, he says, characteristic of the European green geophytic orchids. In these old well-developed roots are frequently found with no trace of fungus, while in other roots the fungus occurs only locally. In the two external cortical layers the cells are host cells, and further in digestive cells predominate. Reinfection of digestive cells may take place. In Listera sometimes the whole of the infected cells carry out digestion. In Orchis the two-layered epiderm is sparingly infected, and from it numerous branching hyphæ run out through the root-hairs into the soil. In some species, *e.g. Platanthera chlorantha* and *Orchis mascula*, the rhizome is free from fungus and the young roots are infected from the soil through their root-hairs when they are 3 cm. long. In the rhizome of Corallorhiza the fungus occupies the external layers of the cortex as a host region, and the middle layers are digestive ; numerous hyphæ pass into the soil.

Germination of Orchid Seeds.—The germination of the orchid seeds is closely related to the presence of the fungus. In the investigation of this relation, the first of the kind known, Bernard was the pioneer. The orchid seed, *e.g.*, of Phalænopsis, is minute, about a fourth of a millimetre long, covered by a loose coat, and consisting of an undifferentiated embryo of a few hundred cells. Its germination is peculiar. It swells into a little spherical body, a tiny tuber, from the lower end of which absorbing hairs grow out, while the upper end becomes green. (The tuberous orchids of our meadows, however, remain colourless underground, leading a saprophytic existence for several years.) Only after four to five months' growth does the first root appear at the lower end. At this stage the little plant is only 3 to 4 mm. long. Then a bud with one or two leaf rudiments appears at the upper end (Fig. 32).

This stage is never reached unless infection by the symbiotic fungus takes place from the soil. Bernard followed the mode of infection with sterilised seeds grown in association with a fungus isolated from mature orchid roots. In one case, *Bletilla hyacinthina*, germination proceeded to the leaf stage without the fungus, but no roots were formed. In another, a hybrid of a Cattleya with a Lælia, germination proceeded for three months, to a point at which the tuber possessed a little green ring above and a few hairs below ; degeneration then set in unless infection took place. In a third case, a Cypripedium hybrid, no germination occurred unless the embryo became infected at the start. Burgeff obtained similar results. Recently an American investigator, Knudson (1922), has obtained germination of a hybrid Lælia-Cattleya on substrata containing sugar, without the presence of the fungus ; confirmation of this result will be awaited with interest. Bernard's results showed that there was a varying degree of necessity for the presence of the fungus, and wider examination may show still greater differences of behaviour.

Bernard believed that there was a high degree of specificity in the fungus. In experiments with the seeds of a Phalænopsis he found that normal germination occurred with the fungus isolated from Phalænopsis roots ; with a Cattleya fungus the seeds were infected but the fungus killed the seeds ; while with an Odontoglossum fungus infection took place and germination started, but the fungus was soon totally digested by the plant, and no further development took place. Other species were less specialised. Burgeff tested the germination of the seeds of a Lælia-Cattleya hybrid with fungi isolated from seventeen different orchids, many of them European ; in four cases germination was normal, in others it proceeded to various stages, but the fungi were either too strong, invading the embryo too vigorously, or too weak, undergoing digestion by it.

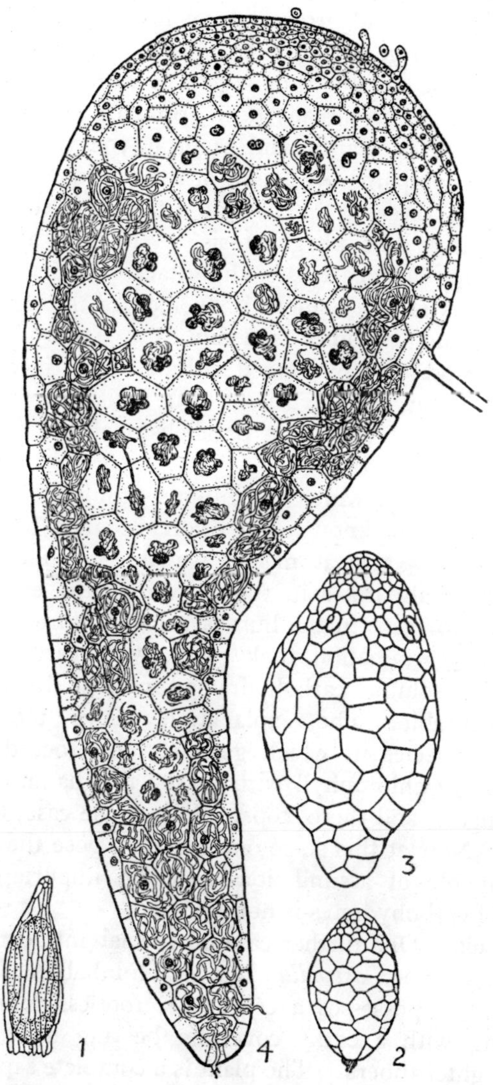

FIG. 32.—Germination of orchid (Phalænopsis); 1, seed, with seed coat; 2, embryo after some days without fungus; 3, after three months without fungus; 4, long. section of embryo, fifty days old, infected by fungus, showing host and digestive cells. All × 75. (After Bernard.)

Bernard's discovery has already had important practical applications. Orchid growers have long found it difficult or impossible to germinate the seeds of many kinds. The provision of the suitable fungus gets over the difficulty, and the new method has already been applied by some growers on a large scale (Costantin and Magrou, 1922).

Bernard compares the action of the fungus on the plant to a state of disease, " a benign disease " ; the reaction of the plant is like phagocytosis. With the appropriate fungus a nice balance exists, the attack is strictly limited. If the fungus is too virulent it overcomes the defences of the plant and becomes speedily and fatally parasitic. Bernard found that prolonged culture on artificial media made his fungi more virulent, but this has not been confirmed by Burgeff. The parallelism of the relation to that in Calluna is striking.

The manner in which the fungus promotes the orchid germination is not known, but reference must be made to Knudson's view that it makes available for the seedling insoluble carbohydrates in the culture media employed— or naturally present in the humus. His cultures, in which fungus-free germination took place, were carried out in media with soluble carbohydrates, fructose being found most favourable. The explanation seems too simple. There are species in which germination proceeds to the production of chlorophyll without the fungus and without soluble sugars, and there stops. This is the case, too, with Calluna. Now at this stage one must suppose the seedling to be capable of assimilation, and the importance of a transfer of carbohydrates is not obvious.

Gastrodia.—One further case of special interest may be described, that of *Gastrodia elata*, which inhabits oak woods in Japan. It possesses a colourless, rootless tuber about 6 in. long with a corky covering, bearing a number of small daughter tubers. The plant is a complete saprophyte. The large tubers are infected by a species of a fungus, Armillaria, strands of which spread over its surface and pass into the soil, where they often bear their typical fructifications. Special branches penetrate the tuber and spread through the outer cortex ; in the outer zone the appearance of the cells is that of typical host cells, in the next zone the fungus absorbs the cytoplasm, in the innermost zone the cells digest the fungus. Tubers can live separately and attain their full size only if they are infected, and only in this state can they flower. An account of this case, investigated by Kusano (1911), is given by Rayner (1916).

Relation to Fungus.—The systematic position of the orchid fungi, except in the case of Gastrodia, is uncertain. Many can be isolated and grown in pure culture, but in some cases, *e.g.* Neottia and Corallorhiza, so intimate is their association with the plant that isolation has not yet been possible. The orchid cannot grow without the fungus. What the physiological relation between the two is in the mature plant we do not know. There is evidence that the fungus cannot fix free nitrogen, and thus differs from that of the Ericaceæ. We are left with the assumption that it helps in converting insoluble organic compounds in the soil into forms available for the plant, or that it is active in the transfer of water and salts. The latter alternative would apply to the green orchids, and the former to the saprophytes. Both possibilities may be realised in a single species. Apart altogether from our lack of knowledge of the way in which transformation of organic compounds may occur, the case of some of the saprophytes is particularly difficult to understand. As we have seen, in Neottia the connections of the fungus with the soil are extremely sparing. In such a case it seems almost necessary to believe that absorption is carried on directly by the roots and that the action of the fungus is purely supplementary to the metabolism within the cell. Corallorhiza grows in sandy and Ophrys in chalky soil, where humus is scanty. The fungus is presumed to benefit by obtaining food from the plant ; but here again the case of the saprophyte is difficult to understand if we are to assume that, in the first place, the fungus supplies the plant with all its food, and in the second withdraws some from it. The saprophytes are sometimes spoken of as being parasitic on their endophytic fungi, a description which scarcely helps us to an understanding of the relation. Only in the Ericaceæ have we any real grasp of the situation—and not even here of the causation of root formation. An explanation in other cases must wait on more and very difficult research.

Other Cases.—Endophytic mycorhiza has been described for many other plants, *e.g.* by Stahl (1900), Janse (1897), and Gallaud (1905). The fungi are hyphomycetous, without

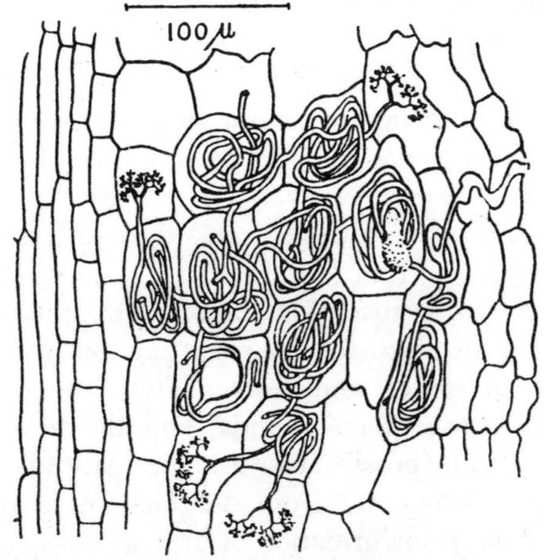

FIG. 33.—Mycorhiza of black bryony (*Tamus communis*); long. section through root. (After Gallaud.)

cross walls, and infect the roots through the epiderm. They spread both inter- and intra-cellularly, and typically show hyphæ ending in fine branches the ends of which are converted by the host plasma into lumps or "sporangioles" which are digested (Fig. 33). Of their importance for the plant little is known. In most cases it is probable that the fungus is not necessary to the plant. We must remember that the conditions for the infection of a root by fungi are extraordinarily favourable in humus soils, where a rich fungus flora exists. There is doubtless a constant struggle between potentially parasitic fungi and the more or less resistant roots of higher plants. In such conditions the closer relations of typical mycorhiza have been evolved.

We may here mention Bernard's theory (1911) that the formation of tubers is, in general, dependent on the presence of fungi, and that the tuberous habit is a direct consequence of fungal infection. The development of this theory was cut short by Bernard's early death. But it has been recently revived by Magrou (1921). He extends it from tuberisation to the formation of perennial subterranean organs in general. The case of the potato presents difficulties because our cultivated plant is not infected, though the related wild species, *e.g. Solanum maglia*, are. It is suggested that the cultivated plant, constantly grown in rich soil, has become independent of its hypothetical fungus. Magrou compares related perennial and annual species, and shows, for instance, that *Mercurialis annua* has no mycorhiza, while *Mercurialis perennis* has. He compares the development of infected and uninfected plants of *Orobus tuberosus*, and shows that only the former produce tubers. The extension of such experimental work will be awaited with interest, though the too daring phylogenetic theories that have sprung from it—that the flowering plants have been derived from liverworts with fungus symbionts—must be viewed with much scepticism.

We may close this account of mycorhiza by reference to the isolated and peculiar case of *Lolium temulentum*, the darnel. This grass is normally infected by a fungus in its aerial parts. The fungus spreads through the intercellular spaces of the stem and leaf base, and passes into the ovary, where the nucellus, and later the plumule of the embryo are infected. Hannig (1907) isolated fungus-free races and showed that the infected plant had a slight power of assimilating atmospheric nitrogen. He also obtained the interesting result that the well-known poisonous effect of the darnel is due to the infecting fungus and not to the darnel itself.

§ 4. Bacterial Symbiosis

Leguminosæ.—It has been known from antiquity that the growth of vetches and lupins enriches the soil. Pliny, in the seventeenth book of his "Natural History," wrote: "Every one agrees that nothing is better for manuring the fields than green lupins ploughed or dug into the ground before the pods are formed; and that there is nothing better for trees or for the vine than to bury, at the foot, handfuls of this plant." In the eighteenth book he says of the lupin: "It thrives in dry sandy places and requires no cultivation. . . . We have observed elsewhere that it enriches the fields and vineyards where it is sown. Far from requiring manure for its cultivation it itself takes the place of excellent dung. . . . Vetches, too, enrich the soil." Theophrastus wrote of the bean: "Beans are in other ways not a burdensome crop to the ground, they even seem to manure it . . . wherefore the people of Macedonia and Thessaly turn over the ground when it is in flower."

The use of these and of other leguminous crops as green manure has long been an agricultural practice; but only in the 'eighties of last century was the reason for this improvement of the soil satisfactorily cleared up and connected with a well-known case of bacterial infection of the nodules or tubercles on the roots of the Leguminosæ. This was due in the first place to the work of Hellriegel and Willfarth (1889), who showed (1) that in sand lacking nitrogen salts, lupins and peas could make satisfactory growth, while cereals could not; (2) that this satisfactory growth did not take place in sterilised sand, but set in if the sand was watered with an extract of arable soil; and (3) that this was linked with the formation on the roots of the well-known bacterial tubercles which could evidently be formed when the sterilised soil was infected with the necessary organism from arable land. Where nodules were formed a definite and considerable gain in combined nitrogen was registered. The bacterium was isolated—a matter of difficulty, as it does not grow under the conditions suitable for most bacteria—by Beijerinck (1888), and the details of infection studied by Prazmowski (1890). Since then much work has been done on the nature of the symbiosis, the conditions of nitrogen fixation, and the practical importance of the association. Accounts will be found in Jost and Russell, where the literature is cited.

The bacterium, now generally referred to as *Bacillus radicicola*, is found in soil in which leguminous plants have grown, finding its way there from disintegrating roots. It infects the young root through the root hairs. In the hair it multiplies enormously and travels inwards in its millions embedded in a slimy cord. The intervening cell walls are dissolved and the thread of slime passes into the cortical cells and stimulates these to abnormal growth. The resulting hypertrophy of the root cortex forms the familiar root tubercle or nodule, which may be found on the roots of all leguminous plants. The nodules vary in size and form with the species and the size of the root; they may be the size of a pin-head, or they may be as large as a hazel nut, as on the stout tap-roots of the lupins (Fig. 34). They consist of large-celled parenchymatous tissue, and are traversed by forked vascular strands, arising from the central cylinder (cp. Spratt, 1919). In the cortical cells the bacteria leave the infection thread, and it is after this that the stimulation to division of the root cells occurs. The parenchyma cells in the mature tubercle are so densely packed with the bacteria that they have a granular appearance under low powers of the microscope. In the central cells of the tubercle the bacteria lose their characteristic form; they swell up or form short forked bodies. These are degeneration or involution forms, and have been named bacteroids.

That the leguminous plant in conjunction with the bacteria could fix free nitrogen was shown in the cultures of Hellriegel and Willfarth, by comparison of the nitrogen balance of plants grown in sterilised and infected sand in which no nitrogenous compounds were available. In a typical case a lupin grown in sterile sand attained a dry weight of 0·933 grm. and showed a nitrogen *loss* (as

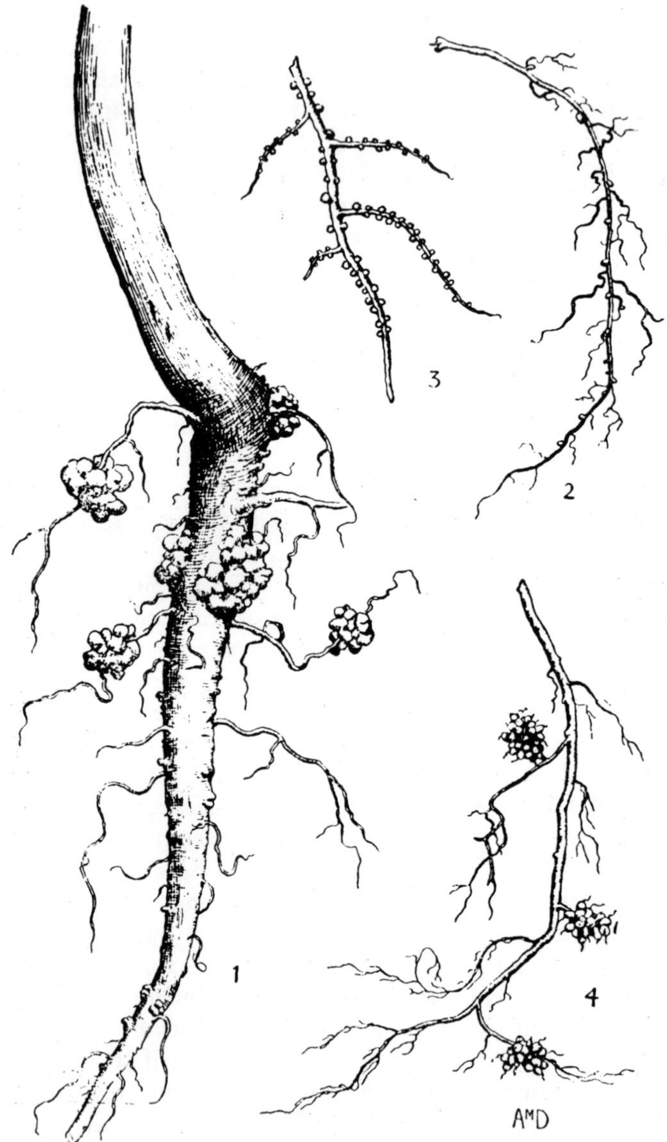

FIG. 34.—Bacterial nodules on roots. 1, *Lupinus albus*; 2, *Trifolium repens*; 3, *Podocarpus chilina*; 4, *Myrica Gale*. (3 after Spratt.)

compared with the content of the seed) of 0·008 grm.; in sand infected with the micro-organism the plant attained a dry weight of 40·574 grm., and showed a nitrogen *gain* of 1·049 grm. The proof that the bacterium in isolation could fix nitrogen was difficult. The conditions of existence in pure culture and in symbiosis with the plant are, of course, very different. Golding (1905) has shown that the products of the metabolism of the bacillus, which accumulate in artificial culture media, have an adverse effect on the fixation of nitrogen; when these are removed by suitable means, fixation can be demonstrated.

The mutual benefit of the two organisms is thus quite clear; the higher plant receives a supply of nitrogenous compounds, the passage of which from the nodule to the rest of the plant has been demonstrated, and the bacterium receives the balance of its food requirements, chiefly carbon compounds. The relations between the two is not so close as in endotrophic mycorhiza, for either organism can exist very well apart from the other—the leguminous plant if it is grown in soils with normal nitrogenous manuring, the bacterium on suitable culture media. Yet normally they are not independent. For it is evident that in nature only in the plant does the bacterium obtain those exact and peculiar conditions which enable it to synthesise nitrogen compounds from free nitrogen; and only in conjunction with the bacteria are the leguminous plants able to luxuriate in such barren soils as they commonly affect—the rest-harrow and bird's-foot trefoil on sand, the whin on sandy heaths, the petty whin on peaty heaths, and so on.

Practical Importance.—The importance of the symbiosis for plant life in general must be very great. Fixation of nitrogen takes place in ordinary fertile soils by bacteria in the soil itself. But in poor soils such fixation may be much diminished, and the action of denitrifying bacteria may in some cases adversely affect the nitrogen balance. In poor soils leguminous plants frequently thrive, and their decaying roots and nodules enrich the soil to the benefit of other vegetation. Even on good soils an abundance of leguminous plants in the vegetation probably increases the nitrogen supply to a greater extent than do the soil bacteria. Russell gives an illustration of the effect of clover on a Rothamsted soil. A plot was divided into two portions on one of which clover was grown, on the other barley. In the clover harvest there was removed 151·3 lb. of nitrogen per acre compared with 37·3 lb. per acre in the barley crop. In the soil there was left 0·1566 lb. of nitrogen per cent. in the clover plot, and 0·1416 in the barley. In the following year barley was grown in both plots; in the barley harvest from the former clover plot there was 69·4 lb. of nitrogen per acre, while in the control there was only 39·1. This shows the great importance which leguminous crops have in actual practice. The most valuable results have been obtained in the farming of poor soils, and especially in the use of lupins as a preliminary crop in the reclamation of barren moor and heathland. On such soils innoculation with the suitable organism has been practised with success.

Biological Races.—Although the symbiotic organisms of all the legumes are included in the one species, *Bacillus radicicola*, it is certain that this consists of a number of biological races, each capable of infecting only certain groups of species. Hellriegel and Willfarth had shown that the organism of the pea could not infect the lupin, and *vice versâ*. Later work has confirmed and extended this observation. Thus Klimmer and Krüger (1914) tested by serobiological methods the bacteria isolated from eighteen different leguminous plants and found that they belonged to nine different races; distinct races, for example, inhabited Lupinus, *Vicia sativa*, *Vicia Faba*, and Melilotus, while the last harboured the same race as Medicago and Trigonella. Various attempts have been made to induce the leguminous bacteria to enter into symbiosis with non-leguminous plants, such as cereals, but so far without success. Could such a symbiosis be produced it would have far-reaching practical results.

Other Plants with Root Tubercles.—This type of symbiosis is not confined to the Leguminosæ; it is found in a limited number of other families, though not always in all their members (cp. Kellerman, 1910). These are the Cycadaceæ, Podocarpaceæ, Eleagnaceæ (Eleagnus, the oleaster, Hippohaës, the sea buckthorn, and Shepherdia), Rhamnaceæ (*Ceanothus americanus*, and *C. velutinus*), Myricaceæ (*Myrica Gale*, the bog myrtle or gale, and *M. asplenifolia*), Betulaceæ (Alnus, the alder), and perhaps Casuarina. The nature of the organism in some of these cases is doubtful. Peklo (1910) assigns the symbionts in Myrica and Alnus to the peculiar bacterial genus Actinomyces. Shibata (1902) regards the nodules of Podocarpus as mycorhiza. Moeller (1890) describes the symbiont of Alnus as a non-septate fungus. Nodules on Casuarina have been described by Miehe (1918) as a case of mycorhiza. Bottomley (1912 *a* and *b*, 1915) and Spratt (1912*a* and *b*, 1915) have, however, isolated the organism of Alnus, Myrica, and Eleagnus, Ceanothus, Podocarpus and Cycas. They state that in all cases it is a form of *Bacillus radicicola*. All these plants are capable of thriving in nitrogen-free media if infected with the appropriate organism, which fixes free nitrogen.

In the non-leguminous plants the bacillus infects the root through the root hairs, passes into the cortex and there infects a young lateral root as it passes outwards through the tissues (Fig. 35). This lateral root then grows out and becomes hypertrophied, producing the mass of parenchyma which is inhabited by the symbiont. Except in the Podocarpaceæ the lateral root forks repeatedly, and the tubercle assumes a more or less coralloid form (cp. Fig. 34). These tubercles are perennial (a condition confined in the Leguminosæ to the tribe Mimoseæ), and, as growth continues year after year, they may assume large dimensions. In the alder they may be as big as a cricket ball.

Especially interesting are the two gymnospermous families the Podocarpaceæ and Cycadaceæ, in all the genera of which so far examined tubercles have been found. The tubercles of the Podocarpaceæ are simple, and the fine roots beset with them have the appearance of a loosely threaded string of beads (cp. Fig. 34). The conditions in the Cycadaceæ are very complex. In most genera a secondary alteration in structure produces a large internal space in the tubercle in which lives the blue-green alga Anabæna. The significance of its presence is not known. In addition there is present *Bacillus radicicola* and also *Azotobacter chroococcum*, a nitrogen-fixing bacterium found otherwise only in the soil. We have here the remarkable case of a fourfold symbiosis of Cycad, Anabæna, Azotobacter, and Bacillus.

Leaf Nodules.—A still more intimate bacterial symbiosis is found in the leaves of certain tropical Myrsinaceæ,

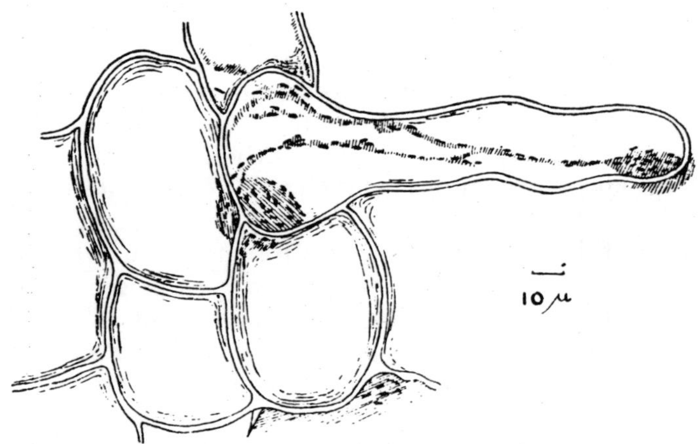

FIG. 35.—Infection of root hairs of Phyllocladus by *Bacillus radicicola*. × 480. (After Spratt.)

e.g. Ardisia crispa, and some other species of this genus, Ambylanthus and Ambylanthopsis, and among Rubiaceæ, *e.g.* Pavetta, Psychotria, Grumilea, Ixora. The symbiosis in the Rubiaceæ was first discovered by Zimmermann (1902), and later investigations are due to Faber (1912, 1914). The investigation of Ardisia has been carried out by Miehe (1911, 1914, 1919). (Cp. also Orr, 1923, on Dioscorea.)

In Pavetta the bacterium, *Mycobacterium Rubiacearum*, causes the formation of small knots or galls on the leaves. It also lives free in the stipular cavity of the embryonic leaves, at the vegetative point. As each leaf unfolds, it is infected through the stomata. From the growing point the ovary and the seeds are infected, the latter through the micropyle. In the seeds the bacterium lives between the embryo and the endosperm, and infects the growing point anew as the seed germinates. It was found possible to sterilise the seeds by careful heating, and on the sterile plants which were obtained no galls appeared. Such plants grown in nitrogen-free soils gave very poor

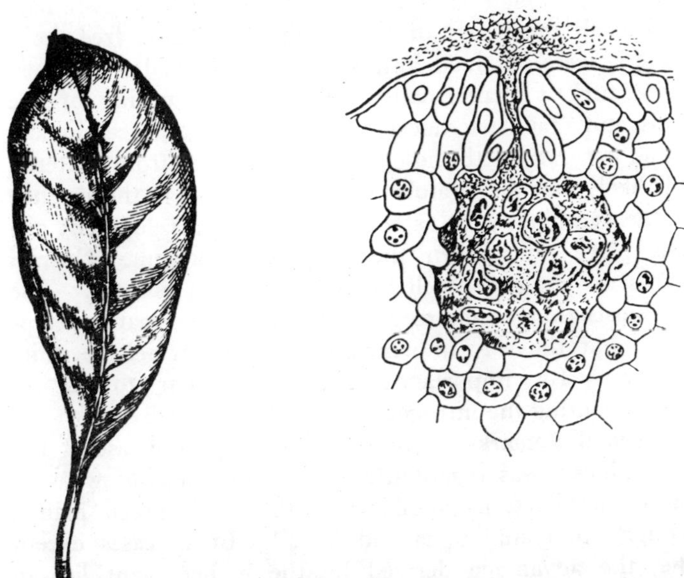

FIG. 36.—Bacterial nodules on leaves. 1, *Pavetta Zimmermanniana*, showing nodules along mid-rib. ⅔ nat. size. 2, Infection of leaf of Psychotria through stoma, and dissolution of leaf cells. × 620. (After Faber.)

growth compared with infected plants. Faber isolated the symbiont, showed that it could fix free nitrogen, and that sterile plants infected with the cultures produced normal galls and normal growth (Fig. 36).

The picture presented by *Ardisia crispa* is less clear. The bacterium again occurs at the vegetative point, and again invades the leaves, the ovary, and the seeds, thus passing on to the next generation. Infection of the leaves takes place through the hydathodes on the leaf margin and is confined to the epitheme, a hypertrophy of which produces the knots on the leaf margin. After the entrance of the bacteria the hydathode is closed by ingrowth of cells, but this may occur in the absence of infection.

Two bacteria were isolated from the seeds, and one of them, named *Bacillus foliicola*, is supposed to be the symbiont, though this is not certain. It may assimilate very small quantities of free nitrogen. Plants free from bacteria were obtained from seed sterilised by heat. They show curious malformation—a tuberous swelling of the buds, and a failure to develop leaves. As plants occasionally found naturally with these characters are free from bacteria, it is likely that the teratological forms are due to the absence of the symbiont; or rather that normal development can take place only when the bacillus is present. Normal plants do not grow well in soils free from nitrates. The symbiosis thus seems to be necessary for normal development, but the way in which the bacterium acts is quite obscure. Its importance does not seem to lie in supplementing the nitrogen ration of the plant.

General Remarks.—Not many years ago the case of the Leguminosæ was regarded as almost unique, but we now know that bacterial symbiosis with higher green plants, though not common, is widespread. In all cases except one, the advantage derived by the higher plant lies in increased nitrogen supply. One cannot but believe that many more cases will be discovered. The interesting leaf symbiosis of Ardisia and Pavetta has further widened our view by showing that symbiosis is not confined to the root system. In general the symbiosis is not obligate; the higher plant gets on very well without the bacterium if it is suitably nourished otherwise. Ardisia seems an exception to this rule too. Ardisia and Pavetta present a further remarkable development in that the symbiosis is congenital. We may relate this with the fact that chance infection of the shoot system in the air is, of course, much less likely to occur than infection of the root system in the soil. This is, in fact, practically certain to take place in soils where former generations of the species have grown. We may recall here that the *Lolium temulentum* symbiosis is also congenital.

§ 5. Insectivorous Plants

" Scarcely another region of botany has in recent times had so much attention drawn to it in wider circles as the so-called *Insectivorous Plants ;* and this is chiefly due to Darwin's extensive work, which has given rise to many accounts." So writes Goebel at the opening of his own memoir on Insectivores in his " Pfanzenbiologische Schilderungen," a work which must share the honour with Darwin's book on " Insectivorous Plants " as providing the solid ground of our knowledge of this quaint group.

Systematic.—Insectivorous plants belong to five different families. (*a*) The Droseraceæ include 5 genera and about 100 species, of which some 90 belong to the genus Drosera, the sundews. The most familiar of the insectivores, Drosera, and the most famous, *Dionæa muscipula*, the Venus fly-trap, both belong to this family. (*b*) The Nepenthaceæ, pitcher plants, include a single genus, Nepenthes, with about 50 species. (*c*) The *Sarraceniaceæ*, American pitcher plants, include 3 genera and about 10 species. (*d*) The *Cephalotaceæ* are represented by a single species inhabiting Australia. These four families stand near each other in the natural system ; they all belong to the Archichlamydeæ. (*e*) Among the Metachlamydeæ there is the single insectivorous family, the *Lentibulariaceæ*, with 5 genera and about 300 species, the great majority of which belong to the genus *Utricularia*, comprising the bladderworts, while the genus *Pinguicula*, comprising the butterworts, comes second in point of numbers. These two genera include familiar British species. The insectivores thus form two small groups far apart in the phylogenetic system. The habit has arisen twice at least in evolution. It is probable that its origins are much more numerous than this, for, in spite of the small number of species, the ways in which the insect prey is captured and utilised are very diverse.

(*a*) **Droseraceæ.**—Drosera, the sundew, is a cosmopolitan genus and has most numerous representatives in the Cape and Australia. It is represented in Britain by two species and a hybrid. The sundews are small plants of wet peaty moors. Their leaves form a rosette from which rise in summer the flowering stems with cymes of white flowers. The root system is poorly developed. The leaf is stalked and terminates, in the commoner British species, in a rounded blade of a reddish hue. This blade is beset with numerous "tentacles." The tentacle is an emergence of peculiar structure. The stalk is traversed by a strand of tracheids which, in the club-like head of the organ, ends in a massive group. The tracheid group of the head is surrounded by three layers of cells, the outer secretory, filled with red sap, the inner a bundle sheath. The tentacles at the margin of the leaf have much longer stalks than those of the central region, and the glandular head is asymmetrical. The lower surface of the leaf is devoid of tentacles. The head is normally covered by a glutinous and sticky secretion, and this entangles small insects alighting on the leaf—it is not known whether or not they are definitely attracted thither. The glands which the insect touches immediately begin to pour out a much more abundant secretion. If marginal tentacles are affected they begin, in less than a minute, to bend over at their bases towards the centre. They bring the insect in contact with the central tentacles, which, in their turn, are stimulated to secrete. These tentacles do not move, but they communicate a stimulus to the unaffected marginal tentacles which now close in. In the course of a few minutes all are bent over to the centre of the leaf, and their approach is intensified by an incurving of the leaf margin. The insect is entirely covered and submerged in the secretion.

The movement of the tentacles is determined by two different stimuli—by contact and by the chemical action of nitrogenous compounds. These stimuli act either together or separately. The contact of indifferent solid bodies of extremely small weight, such as splinters of glass, induces a movement. Drops of water, even if they strike the tentacles with considerable force, have no effect. The stimulus is perceived by the head only of the tentacle, and the reaction, due to a differential growth rate, takes place at the base of the stalk ; a conduction of the excitation occurs. Very small quantities of such nitrogenous compounds as ammonium sulphate or white of egg produce the same result ; less than one-thousandth of a milligram of ammonium sulphate is effective. If the central tentacles are stimulated they do not respond by movement, but they do perceive the stimulus and transmit it to the marginal tentacles, which then bend inwards. The stimulus is not transmitted from one marginal tentacle to the others.

The increased secretion, which is accompanied by a change in the vacuoles of the cells, is also a result of the stimulation, contact here being less powerful than chemical action. The nature of the secretion, too, is altered ; it contains an enzyme like pepsin and an acid. The enzyme is capable of digesting proteins in an acid medium, breaking them down, as has been shown by White (1910), to peptone, which is then resorbed by the leaf cells.

After the closing in of the tentacles upon the insect the process of digestion goes on and may take several days, depending on the size of the booty. When it is complete the tentacles slowly fold back, and the fully opened leaf is capable of capturing and digesting fresh prey.

The process is similar in all the species of Drosera. Some of the sub-tropical species are much larger than our native sundews and have very different leaf forms. In *D. capensis* the leaves are very long and narrow ; in *D. binata* they are long and forked. In long-leaved types the movement of the leaf blade may be extensive ; it may double up or twist round the captured insect. In *Drosophyllum lusitanicum*, a small Portuguese shrub, we meet a less active condition. The leaves are long and narrow, and a single one may catch dozens or even hundreds of little flies.

Goebel mentions that it is used as " fly-paper " in Portugal. The stimulation of the glands induces increased secretion but no movement ; the capture is purely passive. Roridula, a Cape genus, and Byblis with two Australian species, are also passive. Doubts have been expressed as to the capacity of the former to digest an animal diet.

A different type of mechanism is exhibited by *Dionæa muscipula* and *Aldrovanda vesiculosa*. The former occurs in bogs in the Southern Atlantic States of America. Its leaves are arranged in a rosette. Each possesses a winged stalk and a blade, the margins of which bear long fine teeth. On the upper surface of each half of the blade are three multicellular bristles with jointed bases. If one of these bristles is touched the two halves of the leaf snap together in less than a second, the marginal teeth interlock, and the insect or small worm which has released the mechanism is securely trapped. The movement is due to turgor changes, and is fixed by growth (Brown, 1916) : it is accompanied by electrical phenomena similar to those which occur in stimulated animal muscles (Burdon-Sanderson, 1882, 1889). The six bristles alone, and no other parts of the leaf, are sensitive to the shock stimulus. Brown and Sharp (1910) have shown that, at ordinary temperatures, two successive touches within about 20 secs. are required to produce a reaction. Movement follows the second touch immediately. At high temperatures (35° C.) a single touch suffices. The numerous sessile glands which cover the surface respond to chemical stimulus by pouring out a digestive secretion. After resorption of the products of digestion the leaf opens, by growth, and is capable of renewed action. This movement is, with that exhibited by the leaves of *Mimosa pudica*, the most striking in the plant kingdom.

The European *Aldrovanda vesiculosa* is a submerged rootless aquatic, which sends only its flowers above the surface of the water. Its leaves resemble those of Dionæa but are much smaller—about a third of an inch long. Sensitive hairs occur on the upper surface, and stimulation of these causes the leaf halves to close together. Glands are also present, but it is doubtful whether they excrete a digestive fluid. A description of this remarkable plant is given by Arber (Fig. 37).

(*b*) **The Nepenthaceæ** are old-world tropical plants most abundant in the Malay Archipelago. They are frequently cultivated in hot-houses and many hybrids have been raised. They are usually small shrubs, epiphytes, climbers, or ramblers ; some attain a length of 90 feet. The leaves may be very large—over a yard long. The broad sheathing blade runs into a tendril. Usually this tendril, after making a turn round some suitable support, bends vertically down and then curves up and terminates in the characteristic pitcher, as large as a quart pot in some species, in others not bigger than a thimble. The extreme tip of the leaf forms a lid arching over the pitcher mouth. The pitchers are often brilliantly tinted in reds and purples (Fig. 38).

FIG. 37.—*Aldrovanda vesiculosa*, shoot with two whorls of leaves. × 2. (After Caspary.)

The pitcher is thus a modified part of the leaf tip. It is not formed by all leaves. Those near the spike-like inflorescence frequently end in the tendril. Goebel (1889) modifies an older observation of Sachs, that pitchers are formed only if the tendril has twined round a support, by adding that this is true only in the older plants of certain species. Thus *Nepenthes ampullaria* forms no pitchers on functioning tendrils. Seedlings and cuttings regularly form pitchers before climbing sets in. The tendril of Nepenthes must be looked on as serving not only to support the plant as a whole, but in particular as supporting the considerable weight of the large pitchers half filled with liquid.

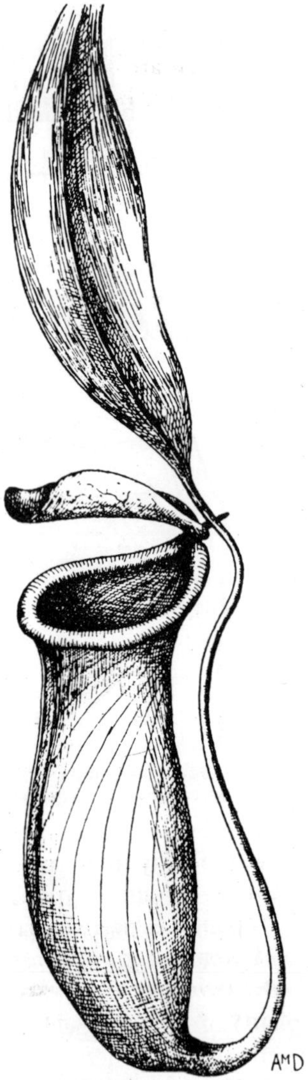

The edge of the pitcher is strengthened by a strong rounded and ribbed rim in which vascular tissue predominates. The mouth is thus kept full open. The structure of the inner wall is complex. It is usually divided into two sharply differentiated zones, the upper covered by an unwettable wax coating, the lower extending halfway up the pitcher, or, in some cases, nearly to the rim, glandular and glistening. The glands are multicellular, half sunk in epidermal cavities, and secrete the watery fluid which is found in the pitcher. Glands are also found on the under side of the lid and about the rim, and these are said to secrete nectar which attracts insects.

FIG. 38.—Pitcher of Nepenthes. ½ nat. size.

The non-glandular upper zone of the pitcher consists of smooth cells, among which are some projecting over the surface with the free part pointing downwards. These are modified and abortive stomatal guard cells. This zone serves as the trap for any insect which has strayed on to it in pursuit of nectar. It can find no foothold, and slips down to the glandular zone or into the fluid. The mode of action of this "slipping" zone is remarkable, and has recently been experimentally studied by Knoll (1914*a*). He used wingless ants, which possess curved claws with which they walk on rough surfaces, and adhesive cushions with which they can climb even on a perfectly smooth glass plate. The claws are useless on the smooth surface; even the projecting cells, as they point downwards, afford no grip. With the adhesive papillæ the ants can walk properly on a smooth wax surface, but the wax on the Nepenthes cells forms a coating of minute scales; these adhere to the papillæ and slip off the surface. The papillæ covered with slippery wax scales are useless, unless the ant has an opportunity of cleaning them, which, on the vertical wall of the pitcher, it has not. The trap is deadly.

As has been mentioned, pitchers are formed in some species on leaves which have not functioned as tendrils. This tendency reaches its climax in those cases where they are produced on the surface of the ground or almost buried in the soil, on leaves which are otherwise scarcely developed. Such pitchers are efficient traps for small animals creeping about the soil surface—worms, leeches, centipedes, and so on.

The liquid in the pitcher, before it has received any prey, is neutral in reaction and contains no digestive enzymes. The presence of an insect, or of fibrin or albumen artificially introduced, stimulates the glands to an energetic secretion of acid—perhaps formic acid—and of proteolytic enzymes. Digestion of fragments of fibrin or albumen takes place within an hour. It seems likely that the enzyme is peptic, breaking down the protein to peptone which is quickly absorbed; Vines (1905) found that peptones were broken down to amino acids by the action of erepsin (cp. Hepburn, 1919). The liquid appears to have antiseptic properties, perhaps in consequence of its acidity, which prevent the development of the bacteria of decay.

Of great interest is the fact that in these pitchers, the function of which is the digestion of organic food, there is constantly to be found a community of small plants and animals. Oye (1921) investigated the flora and fauna of terrestrial pitchers and found representatives of the following groups: Myxophyceæ, Desmidiaceæ, Diatomaceæ, Rhizopoda, Nematoda, Acarina, Poduridæ, Diptera, and larvæ of Diptera and Lepidoptera. These organisms must be protected in some way, probably by anti-enzymes, against the action of the digestive enzymes (Hepburn and Jones, 1919).

(*c*) **The Sarraceniaceæ** are marsh plants of tropical and sub-tropical America. Their leaves are arranged in a rosette, the mature leaves having the form of pitchers, standing upright, inclined outwards, or lying almost horizontal. In Nepenthes the pitcher is part of the leaf; in Sarracenia the whole leaf except the short stalk is modified. The pitcher is asymmetrical; along the inner edge—that towards the centre of the rosette—is developed a more or less strongly marked wing, which increases the assimilating surface. The outer side of the rim is prolonged, in Sarracenia, into an arched lid which partially protects the mouth. In Darlingtonia the upper part of the pitcher is completely arched over the mouth which opens downwards, flanked at each side by a wide wing. In the third genus, Heliamphora, the mouth is widely splayed open, and the lid is rudimentary.

The edge of the pitcher, in *Sarracenia psittacina* for example, is furnished with a stiff rim rolled outwards. About the rim, the lid, and even the outer surface of the pitcher nectar glands secrete drops of sweetish fluid. In *Sarracenia variolaris* and *Sarracenia rubra* so strong is the secretion that the rim appears as if smeared with syrup. Insects seeking nectar are thus led towards the inside of the pitcher. Below this region the inner surface of the pitcher is provided with a slipping zone of quite different structure from that of Nepenthes. The smooth cuticularised epidermal cells have their lower edges projecting over the cells next below; the appearance is that of a tiled roof. Then comes a zone which bears long bristles, also pointing downwards. Nectar glands occur here and there, and these, combined with the difficult footing, secure the easy descent of the insect into the fluid below. " It is," writes Goebel, " a unique spectacle to watch with what certainty and speed ants, for example, which seek the nectar glands of *Sarracenia flava*, vanish into the pitcher. If they once slip in they never reappear."

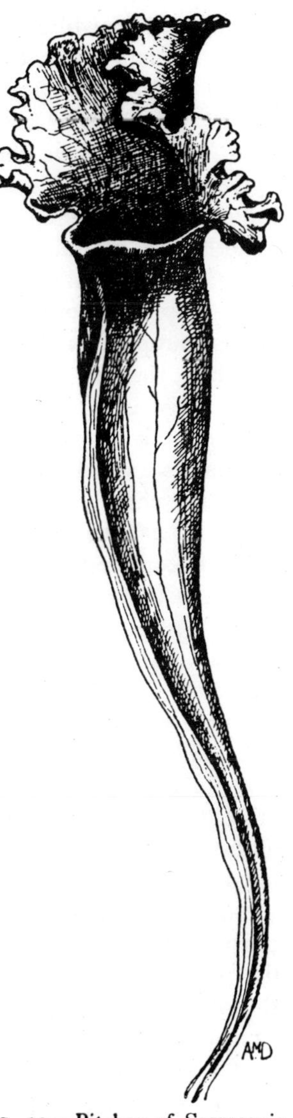

FIG. 39.—Pitcher of Sarracenia. ½ nat. size.

The bottom of the pitcher is covered by a smooth epiderm devoid of glands and hairs. Different species show differences in the relative space occupied by the various zones.

Young unopened pitchers of Sarracenia may contain a little fluid. It is doubtful whether it is secreted by the nectar glands or by the epiderm. In the opened pitcher some fluid is usually present, but the pitcher may be quite dry. As the mouth of the pitcher is usually imperfectly protected, it is likely that the liquid is at least partly rainwater. The best protected pitchers are those which are long and more or less erect, *e.g.* those of *Sarracenia Drummondi* and *Sarracenia flava*, and of Darlingtonia. They contain little liquid; were they filled they would be mechanically incapable of remaining erect.

The way in which the insect food is made available for absorption is a matter that has not as yet been satisfactorily cleared up. It has generally been supposed that no digestive enzymes are secreted. Recently, however, Hepburn (1920) claims to have demonstrated the presence of a proteolytic enzyme in Sarracenia, and its absence from Darlingtonia. The liquid in the pitcher is not antiseptic; bacteria are abundant, and, under the action of these, decay of animal matter takes place. The products of this decay may be absorbed by the plant. To what extent this process supplements or is supplemented by the enzyme of the plant, remains to be seen. Sarracenia seems to be on a lower stage of organisation than Nepenthes; the pitcher is imperfectly supported and contains little liquid, the insect trap is perhaps less efficient, the digestive arrangements are incomplete.

(*d*) **Cephalotus follicularis,** the Australian pitcher plant, which is confined to the region of King George's Sound in West Australia, has been little investigated. The leaves are in a rosette and arranged in two tiers. The upper are normal, flat and broadly elliptical. The lower are the pitchers—short and fat with a very prominent ribbed rim, a double wing down the middle, and a single wing at each side. The rounded mouth is covered by an arched lid. The lid is borne on a short stalk. A "slipping" zone is present, and there are glands as to the functions of which nothing certain is known. The pitchers are usually half filled with fluid containing numerous small insects. How these are utilised we do not know. The brilliant red-purple colouring of the outer side of the pitcher and of the lid may be noted: "As the pitchers stand in a whorl under the foliage leaves . . . the Cephalotus plant must be as striking as if it bore brilliantly coloured flowers."

(*e*) Of the **Lentibulariaceæ** the simplest relations are shown by Pinguicula. The British *Pinguicula vulgaris*, the butterwort, is a common moorland plant; two other native species are rare. The genus is distributed chiefly in the north temperate zones. Pinguicula has a rather poor root system. Its leaves, arranged in a rosette lying flat on the soil, are broadly elliptical with margins upturned, and glisten with a sticky secretion from numerous glandular papillæ, some of which are stalked, others almost sessile. The flower stalks, each bearing a single fine violet flower, spring from the middle of the rosette.

Insects are caught in the viscid secretion; whether they are attracted by it is not known. The glands are then stimulated to increased secretion and the margins of the leaves roll in markedly. The secretion of the stimulated leaves is acid in reaction, and contains a peptic enzyme. Digestion goes on vigorously.

Much more remarkable are the conditions in the remaining genera. The many species of Utricularia are chiefly tropical, and in the tropics they show the greatest diversity of habitat, occurring as water, land, and epiphytic plants. The few species that occur in the temperate zones are all aquatics. Such are our native *Utricularia vulgaris*, *U. intermedia*, and *U. minor*, the first and the last being not uncommon plants of moorland pools. The flowering axis rises above the water but appears rarely in this country; the rest of the plant is submerged. Roots are absent throughout the genus. The vegetative body seems to consist of a number of shoots bearing numerous much-divided leaves. Many of the ultimate segments of these are replaced by the bladders. The bladder is a small (about a tenth of an inch long) oval or pear-shaped structure, attached at one side by a short stalk to the leaf. In *Utricularia intermedia* and others the bladders are borne on special shoots, while other shoots bear only much-divided leaves. At the pointed end of the bladder is an opening to the interior. This is closed by a valve-like lid, joined to the margin above and partly down the sides, with its free lower edge resting firmly on the thickened rim of the opening. From the outer surface of this valve arise a few long, branched hairs; on the surface of the bladder beside and above the opening tufts of stout hair-like appendages are frequently borne (Fig. 40).

The bladder serves for the capture of animals—small

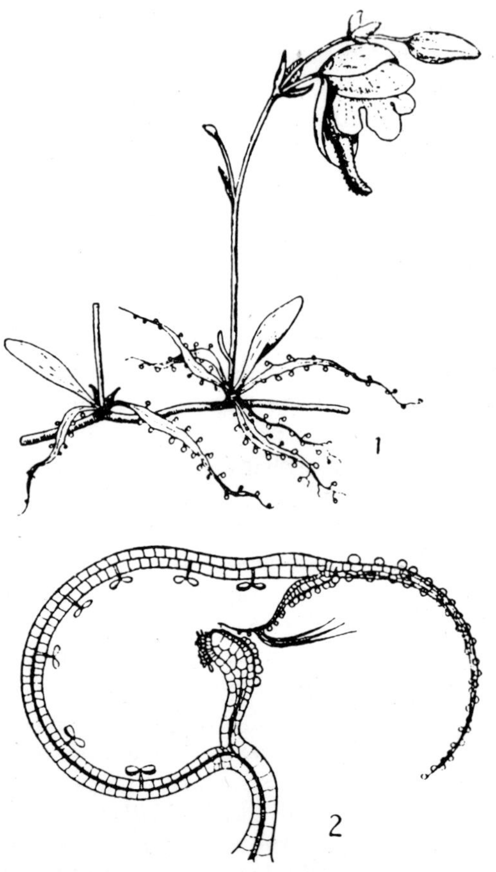

FIG. 40.—Bladderworts: 1, *Utricularia Jamesoniana*, an epiphytic species with two types of foliage leaf, the narrower bearing bladders. Nat. size. (After Oliver.) 2, Section through bladder of *U. reniformis*. × 120. (After Luetzelburg.)

water crustaceans, such as Cypris and Daphnia, as well as rotifers, and infusoria. It was long thought that the mechanism was passive, as in an old-fashioned mouse trap, the visitor, entangled in the bunch of hairs, pushing open the valve, swimming in, the valve falling to behind it, not to be opened by pushing from the inside. But Czaja (1922) has recently confirmed an observation by Brocher (1911) that the bladder mechanism is active. The walls are impermeable and water can pass through them only very slowly if at all. On the inner wall are situated numerous four-armed glandular hairs, and these withdraw water from the interior of the bladder. As water is withdrawn the walls are pulled in; they show a "dimple" on each side instead of being distended. A considerable tension is set up by the mechanical tendency to expand. If now an animal touches one of the hairs on the outside, the valve is levered slightly open; the tension is relieved, the walls expand, water rushes in, and the animal is drawn in with the current; the valve closes and escape is impossible. A pricked bladder does not function as no tension is set up, water passing freely in from the outside. Merl (1922) regards the opening of the valve as due to a stimulus; but Czaja's explanation is more satisfactory.

No digestive enzyme has been found in Utricularia, perhaps because of the small size of the bladder. It is generally assumed that the captives swim about inside till they die, that they then undergo bacterial decay, and that the products are absorbed, probably by the glandular hairs.

We have said that the vegetative portion of Utricularia *seems* to consist of shoots bearing divided leaves, but this interpretation of the morphology is probably incorrect and certainly insufficient, as is shown by comparison with the tropical land species. From the base of the flowering axis of *Utricularia Hookeri*, a West Australian species, there spring three different types of organs—simple linear foliage leaves, ½ cm. long, bladders on short stalks, and long rhizoids which penetrate the soil. All three are homologous; in their relation to the parent axis, and in their developmental origin they are identical. The bladders and the rhizoids must be regarded as modified leaves, and this gives an idea of the range of form which may be shown by leaves in the genus. In *U. Jamesoniana*, a species creeping on the trunks of trees in Ecuador, there are elliptical foliage leaves and linear foliage leaves, the latter alone bearing on the margins stalked bladders (Fig. 40). In *U. affinis*, no foliage leaves are produced from the base of the inflorescence, but only " rhizoid " leaves and homologous " runners." The latter bear bladders and spatulate foliage leaves. These runners are probably not shoots but again modified leaves. In other species the leaves, borne by such runners, may give rise to runners, or to other leaves. It is thus probable that, what in our native species appear to be axes bearing leaves, are leaves only. These complicated relationships are discussed by Arber and by Goebel. Utricularia is in fact a plant in which the ordinary mode of organisation of the dicotyledons is largely abandoned ; the distinctions between leaf and axis can no longer be logically carried through. A different interpretation of the morphology is given by Compton (1909).

Biovularia is a monotypic West Indian genus, in habit like a small water Utricularia : Polypompholyx, with three tropical species, resembles small land Utricularias. Genlisea, however, a genus with ten species, mostly Brazilian, has a remarkable and unique habit. It is a land plant with a dense rosette of small spatulate leaves, from which arise the inflorescences. It has no roots, but is anchored in the soil by long white rhizoids, each with a forked tip. Each of the two branches of the fork is twisted into a close corkscrew (Fig. 41). The rhizoid, which is homologous with the leaf, has a slender basal stalk a few centimetres long : it then swells out into a small bladder, about 1 mm. in internal diameter, the hollow of which is prolonged into a narrow canal, not more than $\frac{1}{5}$ mm. in diameter, which runs through the distal portion of the rhizoid and opens in a narrow slit below the forked end. In the young rhizoid the branches of the fork lie closely together so that penetration of the soil is possible ; later they spread out. The whole arrangement is an extraordinary transformation of the Utricularia bladder, suited to the capture of minute soil organisms, with the remains of which Goebel found the bladders filled. These may enter through the passage formed by the terminal twisted branches, or directly through the slit which terminates the canal. In either case they cannot retreat, for the passages and canal are furnished with rings of hairs, all pointed towards the bladder. Whether the prey is digested or the plant merely absorbs the products of bacterial decay is not known: glands occur in the bladder.

FIG. 41.—Bladder leaf of Genlisea. × 2½. (After Goebel.)

General Considerations.—The capture of animals has been conclusively demonstrated in all these insectivorous plants; in many cases, as we have seen, active digestion occurs. F. Darwin (1880) showed for Drosera that plants fed with insects, or with egg albumen, or with pieces of meat thrive better than those deprived of a flesh diet. Büsgen raised sundew plants from seed, and found that those provided with an animal diet produced more flowers, more and larger capsules, and more numerous seeds than controls. It is generally believed that the chief advantage lies in an increase of nitrogenous food supply. Stahl lays stress on the addition of other salts. It is likely that this, too, is important, for the root systems are typically scanty, but exact experiments on this point are wanting. There is the further possibility of an absorption of organic carbon compounds. But Kostychew (1923) has shown that, in an atmosphere rich in carbon dioxide, Drosera and Pinguicula assimilate as vigorously as plants like the coltsfoot.

As to the origin of the habit not much can be said with

certainty. It has arisen, as we have seen, at least twice in the course of evolution. The variety of mechanism displayed suggests a still greater diversity of origin.

There are two main types of organisation. The pitcher is exhibited by the various pitcher plants, and of a somewhat similar nature is the bladder of Utricularia. Now pitcher formation on leaves is not uncommon as a sport in many plants, *e.g.* the cabbage. It is likely that the permanent acquisition of such bladders or pitchers was the first step, and that greater specialisation followed. The second type is the viscid gland type of the Droseraceæ and Pinguicula. Hairs with a sticky secretion are common in many families, *e.g.* the Saxifragaceæ with which the Droseraceæ are probably allied. Insects are freely caught by such hairs, and this may have been the foundation of the insectivorous habit along this line of organisation.

It is interesting to note that both types are represented in each of the two widely separated groups of insectivores.

In these modes of abnormal nutrition (except in completely parasitic and saprophytic forms which are wholly dependent on other organisms), the exact relations are in many cases obscure, but in very many the supplementing of the supply of nitrogen seems to be the chief advantage secured by the higher plant. This is so in many mycorhizal plants, in the plants with bacterial symbiosis, and in the insectivores. These are all predominantly inhabitants of soils naturally poor in nitrogen compounds ; the vegetation of peaty soils includes examples of all three groups. Nitrogen is the element the supply of which is most precarious, and this varied display of special means to secure it emphasises the difficulty and the importance of obtaining a sufficient supply.

CHAPTER IV

MECHANICAL PROBLEMS : PROTECTION

I. Mechanical Problems

§ 1. Mechanical Tissues. § 2. Mechanical Features of the Root System. § 3. Mechanical Features of the Stem. § 4. Mechanical Features of Leaves. § 5. Aquatic Plants. § 6. Climbing Plants.

THE support of the great canopy of foliage with its scaffolding of branches in an ordinary broad-leaved tree can be effected only by a first-class mechanical system. The efficacy of the system is exhibited by the way in which such trees resist violent storms of wind, even in early autumn, when the leaves still offer great resistance to violent gales. When the tree does fall it is usually because it has been uprooted, not because the sub-aerial mechanical system has given way. The excellence of the materials of which the system is constructed is testified to by the uses to which wooden beams are put by man. In lesser plants the same necessities of support exist in varying degrees ; it is in herbaceous plants that the advantageous *disposition* of mechanical tissues is seen most strikingly. The mechanical tasks of different organs are, of course, different, and the necessities vary with habit and habitat. Our knowledge of the architecture of the plant from this point of view is chiefly due to Schwendener (1874) and Haberlandt.

§ 1. Mechanical Tissues

Rigidity in stem and root may be due to three different causes : (*a*) to turgidity of living parenchymatous cells ; (*b*) to the presence of non-lignified mechanical tissue termed collenchyma; (*c*) to the presence of woody or lignified tissue, as in the various elements of the wood, but more especially in the wood fibres, and in the sclerenchyma cells of the ground tissue, cortex, and bast.

Turgor.—The action of turgor and the presence of collenchyma is of special importance in growing regions. The effect of turgor is demonstrated by the behaviour of any young stem or leaf on wilting—the flaccidity is due to collapse of the cells, following on loss of water; the rigidity in the fresh state is consequently the effect of turgor pressure, the inflation of the cells by water, and depends on the osmotic pressure of the solutes of the cell sap. Another demonstration may be given by soaking a young stem, the scape of the dandelion is excellent material, in a strong salt solution; plasmolysis occurs, and the rigid, brittle stem becomes quite flaccid. At the same time its length diminishes by 2 to 5 per cent., showing that the young cell membranes are elastically expanded by turgor pressure in the fresh condition.

Turgor pressure produces rigidity in exactly the same way in which aerostatic pressure produces rigidity in a sausage balloon. Its effect is heightened by the fact that the stem or leaf is built up of millions of small cells. It is also frequently increased by the occurrence of tissue tensions. Again we may take the example of the dandelion scape. If a strip is cut it rolls up into a coil; if this is plasmolysed it straightens out and becomes limp. The inner tissues are more extensible than the outer, which act as a resistance, against which the inner tissues try to expand, so that in the intact stem the rigidity is thus increased. The same feature, in less marked degree, may be seen in most young stems which, if split longitudinally, bend away from the centre.

Collenchyma.—The turgor effect is produced in any living cell plentifully supplied with water, and exists also in the cells of the collenchyma. This tissue consists of parenchymatous or prosenchymatous cells, the walls of which are usually much thickened by the deposition of layers of cellulose. The thickening is most marked at the corners of the cells, and this gives the collenchyma a highly characteristic appearance, the darker cell contents standing out from the glistening white lozenges of cellulose thickening. Here turgor effects are supplemented by the mechanical strength of the thick walls. As the walls are, however, composed of cellulose, and parts of the walls are thin, they remain capable of extension, and the tissue can keep pace with the elongation or expansion of a growing region.

Woody Tissues.—We may distinguish between four types of lignified cell; they are not sharply marked off, but are united by intermediate forms, and show varieties of type and mode of origin, many of which have received special names (De Bary, 1884; Jeffrey, 1917). They have this in common, that the walls are impregnated with the pentosans and aromatic compounds which convert the cellulose into wood, and are also thickened. The completion of the lignification process coincides with the end of growth; the death and disappearance of the protoplasmic contents follow. In the mature state these cells have great mechanical strength and but little extensibility.

(*a*) Wood vessels or tracheæ, which are absent from most gymnosperms, are fused rows of wide elongated cells, the walls of which are thickened in characteristic patterns; the thickened parts increase mechanical strength, the thin bands or pits permit the ready passage of water. The vessels have primarily the function of conducting water, and the thickening of their walls may be regarded more as securing them from collapse under the pressure of neighbouring tissues, especially in their young state, and as withstanding the tension in the rising column of water, than as increasing the general rigidity of the axis. (*b*) Tracheids are single elongated cells, usually narrower than vessels, and with more heavily thickened walls. They form the bulk of the wood of conifers, and are present in most angiosperms. They conduct water, they probably act as water stores, and they also serve as mechanical elements. The tracheids are probably the primitive elements of the wood from which the other types have been evolved, on the one hand vessels, the perfect conducting elements, and on the other fibres, the most efficient mechanical cells. (*c*) Fibres are very narrow elongated cells with sharp, tapering, sometimes split, points, which are, as it were, spliced into each other by sliding growth in the course of development. Their walls are so much thickened that only a narrow lumen is left, and the pits are minute. They are purely mechanical in function. Of the same nature are the bast fibres associated with the phloem, as in the lime, and the sclerenchyma fibres which occur in strands and belts in the cortex and ground tissue, and are specially prominent and important in monocotyledonous stems and leaves. (*d*) The short, thick sclerotic cells, or stone cells, have much thickened lignified walls with small pits; they occur often isolated in leaves, and form the stony tissues of fruit and seed walls. A well-known example of stone cells is afforded by the gritty flesh of the pear.

The woody fibres are extremely strong. Haberlandt quotes figures for the sclerenchyma of various plants; we may give one typical example, that of *Hyacinthus orientalis*, the fibres of which support a weight of 12 kilos. per square millimetre of cross section without losing their elasticity, and of 16 kilos. at the breaking point. The corresponding figures for wrought iron are 13 kilos. and 40·9 kilos.

§ 2. Mechanical Features of the Root System

The land plant is fixed in the soil by its roots; as the stem moves in the wind the roots on one side are brought under a strain. In unsymmetrical plants a unilateral strain may be more or less permanent. In any case a longitudinal strain is the chief force which roots must resist. For a given material the resistance offered to such a strain is directly proportional to the area of the cross section; the *arrangement* of the material does not matter. In fact we find in roots a great uniformity of arrangement. In the young root the wood bundles are grouped near the centre of the organ; vessels may completely occupy the centre, or there may be a certain amount of pith. The secondary thickening, if it takes place, increases the thickness of this central core; sclerenchyma may be added in the cortex.

Fixation to the soil particles takes place by the root hairs, but this may not have much actual importance in securing the plant in the soil. The innumerable branching roots and rootlets, following a sinuous course in every direction, offer an enormous frictional resistance to withdrawal. It is, of course, to the laterals that stability is chiefly due; even a very strong vertical tap would be unsuited to maintain a tree erect. The multitude of laterals also ensures a distribution of the strain. In herbaceous plants and shrubs the root system is perfectly efficient; only with the immense weight and resistance to wind offered by the foliage of a tree does the root system fail to meet all possible demands, and rupture may occur in a storm. A wind-felled spruce, however, lifts the soil with it; the surface distribution of the laterals, and not their strength, is here faulty. This is the case only in shallow root systems.

A special mode of securing stability is shown by some plants which form " buttress " or " prop " roots. In the maize roots grow out from the lower nodes above ground and arch away from the stem before entering the soil. The mangroves offer varied examples. In *Bruguiera gymnorhiza*, which also possesses knee-shaped pneumatophores, prop roots, like thick blades, slant away from the base of the tree into the mud. In *Rhizophora mucronata*, according to Schimper, " a regular scaffolding of bow-shaped stilt roots supporting the stem represents a complete system of anchors, which is strengthened by new roots growing down from the branches to support the growth of the crown." The mangroves require such special supports to withstand the wash of tide and wave. Among our own trees the elm sometimes shows quite characteristic buttress roots formed by excessive growth in thickness on the upper edge.

Contractile Roots.—Mention may here be made of

another mechanical function fulfilled by roots and hypocotyls. In rosette plants with a vertical root stock such as the primrose, or in plants like the crocus, in which new corms are formed yearly at a higher level than the old, the new growth would soon project above the ground if it were not pulled down. This is done by special "contractile" roots (Fig. 42). They are usually rather thick, and, after becoming firmly fixed, they undergo an active contraction by the alteration in the shape of definite tissues. The epiderm does not contract, and comes to have a curious wrinkled appearance. This is well seen in the roots of young crocus corms in early summer. In some seedlings, too, a contraction of the primary root and of the hypocotyl brings the little plant well into the soil; this is particularly the case in plants in which the hypocotyl takes part in the formation of storage organs, which are thus pulled down to a definite level.

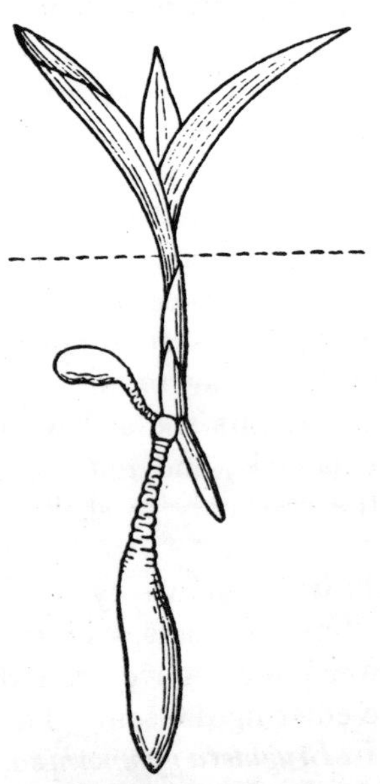

FIG. 42.—Contractile roots of *Chloræa membranacea*: the ground level is indicated by the dotted line; the base of the first root has been pulled far down by the action of the second. Nat. size. (After Rimbach.)

§ 3. MECHANICAL FEATURES OF THE STEM

The requirements of the stem are very different. Did the weight of the foliage and branches bear downwards quite symmetrically the stem would be subjected to longitudinal compression like a pillar. The stem is, however, normally subject to continuous movement by the wind, and, even in still weather, the bearing of the crown is never exactly central. The result is that the stem is principally subject to bending stresses, though at the same time it acts as a pillar. In trees, where the wood occupies nearly the whole bulk of the axis, no special arrangement of the mechanical tissues exists, unless we count as such the series of concentric cylinders formed by the closer textured and mechanically stronger summer wood, separated by the alternating cylinders of spring wood. We may note, however, that the base of the trunk is often markedly thicker than the top, and that this helps to resist longitudinal compression as well as bending stress.

In herbaceous plants, the mechanical tissues occupy only a fraction of the total bulk of the stem, and it becomes of interest to see how they are arranged with regard to the function they must perform. If a bar of material is bent the substance on the convex side is elongated, that on the concave side compressed ; inwards these changes become less and less towards the centre, where they vanish. Stress and strain, therefore, are felt most at the edges, and much less towards the centre, and on the flanks. So it comes about that in constructing a beam which has to resist bending it is important to distribute the material so that it will lie mostly on the two edges and least in the middle. Resistance to bending depends not only on the area of the cross section, but on the distribution of the material. The common shape of a steel beam is a T or I in cross-section, the two flanges united by sufficient material to join them rigidly, often in the form of a lattice work. Or the construction may take the form of a hollow tube where all the material is on the periphery and is distributed so that a stress in any direction can be met.

The distribution of mechanical tissue in a stem follows the same plan. Schwendener and Haberlandt have described a great many types, especially among monocotyledons, where the absence of secondary thickening makes the arrangement of sclerenchyma important. Two examples may, however, suffice here (Fig. 43). In the square stems of the Labiatæ, *e.g.* in *Lamium album*, there runs down each angle a thick strand of collenchyma, and these four strands, connected by the ground tissue of the stem, may be regarded as two crossed girders joined by a lattice work. If the stress acts against a face it is opposed by two girders, if against a corner by one. In the older stem this arrangement is reinforced by secondary woody tissue which forms an internal cylinder, or tube, of fibrous elements. The complete mechanical system is thus composed of two crossed girders reinforced by an internal tube.

In the stem of *Molinia cœrulea*, the purple heath grass, a broad band of sclerenchyma runs completely round the

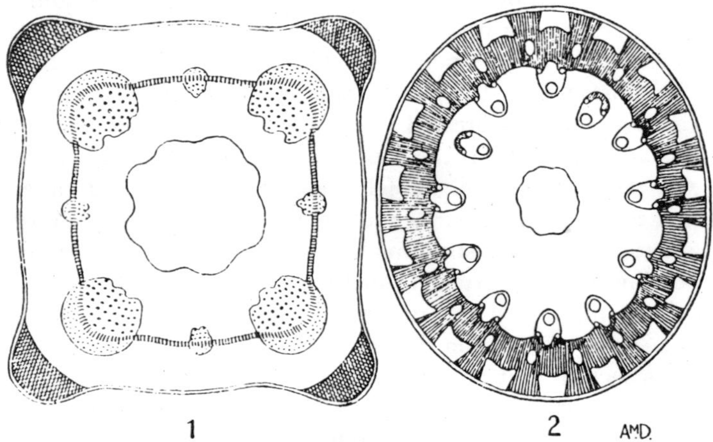

FIG. 43.—Arrangement of mechanical tissues: 1, cross-section of stem of Lamium; 2, cross-section of stem of Molinia; the sclerenchyma is shaded, the collenchyma is cross-hatched.

stem, outside the irregular circle of vascular bundles. From the external surface of this band there spring a number of ridges of sclerenchyma, which reach to the epiderm broadening outwards. On the inside a somewhat similar series is formed by the vascular bundles, the sheaths of many of which are fused with the sclerenchyma band. Here we have a hollow cylinder reinforced outside, and, to a less extent, inside, by T-shaped girders, the whole forming a very strong type of construction.

Apart from the arrangement of mechanical tissue the

hollow stem, so common in herbaceous plants, is itself an example of the hollow column. We see this in such plants as the Umbelliferæ and the grasses. Such a hollow tube has this weakness, that, if bending takes place over a considerable length, it may buckle ; this is largely overcome if the tube is partitioned, and such partitioning, more or less complete, usually occurs at the nodes. In some aquatics diaphragms occur at more frequent intervals (Snow, 1914). In many rushes the hollow is filled with a spongy pith of little strength, but serving as a latticed matrix for the tube ; in *Juncus articulatus* and others the pith is confined to numerous partitions. It is interesting to note that buckling of grass stems is infrequent, unless in violent gales, except in the cultivated cereals, lodging of which by heavy rain or wind is a source of much loss. Here we are dealing with artificial races bred by selection for heavy cropping, though, of course, the quality and resistance of the straw is also considered. They are plants in which the balance of load to support has been deliberately altered in favour of the former, and this explains their weakness as compared with natural races. Shading accentuates the weakness.

The bending strain is especially great in the case of horizontal branches bearing, towards their extremities, a load of leaves, and here we find special modifications. At its junction with the stem the branch springs upwards, and, in older branches, the development of wood is increased on the lower side. The branch is borne on a buttressed arch. The secondary thickening of the branch is asymmetrical. In general the development of wood is stronger on the upper side in broad-leaved trees, and on the lower side in conifers (" red " wood). The elements of the lower wood in the conifers are more strongly thickened. In some broad-leaved trees, such as the oak, the upper wood is distinguished by a greater proportion of mechanical elements ; in others, like the beech, the structure of the wood is the same above and below. In all cases the eccentricity means increased resistance to bending, by withstanding strain better in the broad-leaved trees, and stress in the conifers.

The cause of the different types of eccentricity is not yet completely known. Ewart and Masson-Jones (1906) found the formation of red wood in the conifers to be caused by gravity. Recent work by Engler (1918) seems to show that in the conifers the compression of the cambial cells is determining, while in the broad-leaved trees the gravitational stimulus is effective.

§ 4. Mechanical Features of Leaves

Leaves are particularly subject to bending. Ordinary broad leaves are in most cases sufficiently supported by their turgor, helped out by the network of interlacing veins. In long sword-shaped and linear leaves, such as those of the grasses, the tendency to double over is much greater. The leaf may be attached round nearly the whole circumference of the stem, as in the sedges, or it may be actually sheathing as in the grasses ; in both cases the tubular form of the basal portion ensures that this region shall be erect. Frequently *resupination* of the free portion of the leaf-blade is to be seen, that is, twisting on its axis once or more often ; this is a modification of, or approach to, the tubular type of structure. The sclerenchymatous tissue which makes the leaves of so many grasses tough is habitually arranged in girder fashion, the strands running from one surface of the leaf to the other, or bands on the upper and lower surfaces being joined by the vascular bundles and their sheaths.

The broad expanse of the leaf renders it peculiarly susceptible to tearing. This danger may be lessened if the leaf is more or less divided or is compound, one might say naturally torn, so that the resistance to air currents is diminished. The great leaves of the palms with their feathery divisions afford a good example of this. The epidermal cells of the leaf margin are normally more thickly walled than those of the two surfaces. Along the margin, and in the bays of lobes, where tearing is most liable to occur, strands of sclerenchyma, or marginal veins, may give extra strength, as, for example, in the leaf of the holly.

Similarly the slender petiole, when it is present, allows the leaf to move readily in the wind and so reduces resistance. Perhaps the quivering leaf of the aspen with its delicately balanced, laterally compressed petiole represents a higher degree of efficiency in this direction.

§ 5. Aquatic Plants

In water plants the mechanical system is much modified. In still water the whole weight of the shoot is borne by the medium in which it floats; the plant does not sink to the bottom, chiefly because the presence of air bubbles in its tissues makes its specific gravity the same, or nearly the same, as that of the water. The root system is, as we have seen, often relatively feeble.

Plants with floating leaves, and those which live in running water, show special features. If the leaf is to float it must be non-wettable, and this is secured by a strong development of cuticle which gives the leaves of pondweeds, water lilies, and the amphibious persicary, their characteristic polished appearance. In addition to this, the margins of the leaf may be turned up, a feature very noticeable in *Victoria regia*, the leaf of which is large enough and sufficiently buoyant to support the weight of a baby. Movement of the surface water involves strains which are met by the leathery texture of the floating leaf, sometimes reinforced by sclerotic cells, as in the water lily. The long lax petioles allow the leaf to follow changes in the water level.

Leaves of plants growing in running water are frequently much dissected; the fine segments are extremely pliable and turn readily with the current. Such is the case in *Ranunculus fluitans* and *Myriophyllum spicatum*. Here we see again the tearing action of the current, in this case of water, lessened by lowered resistance. The dissected type of leaf is, however, also common in submerged leaves of plants of still water, and it may be related to other requirements, such as the ready absorption of carbon dioxide and of salts through a larger surface, or to the absorption of light.

Submerged dissected, and floating leaves may occur on the same individual, as in some of the water crowfoots. The very long, linear leaf is another common submerged type. It, too, offers little resistance to currents. The pondweeds show this form in some species which may have floating leaves as well. The arrow-head, the water plantain, the flowering rush have submerged linear leaves and subaerial leaves of various shapes.

Submerged flowering plants in temperate climates are not found in violently rushing water ; they are confined to streams flowing gently over sandy or muddy bottoms in which the plants root. Those subject to most violent water movement are the few peculiar flowering plants of the sea, represented, on British coasts, by two species of Zostera, the grass-wracks. This and other genera, such as Cymodocea and Posidonia, are very common in the Mediterranean. They grow on flat, muddy or sandy shores, and may be left uncovered by the retreat of the tide. They root in the sand, and the root system is well developed, though not remarkably so. The leaves are linear, and, in *Zostera marina*, may reach a yard in length. These plants do not seem to be thoroughly fitted to meet the great force of tide and wave in a trying environment, for they are readily torn up by storms and cast on the shore in enormous masses.

In torrential streams of the tropics of America, and, to a lesser extent, in Africa and Asia, there occur two remarkable families, the Podostemaceæ and Tristichaceæ. They are not found in quiet streams. The plants are attached to the rocks and are firmly fixed even on smooth water-worn stones. In simple cases a branched, dorsiventral root system creeps over the rock face ; on the lower side exogenous attachment organs (*haptera*) are formed, which are closely applied to the inequalities of the stone, and may exude a sticky adhesive material, *e.g.* in Tristicha. This mode of attachment is analogous to that of the tendrils of certain vines such as the Virginia creeper. In other cases the root system develops as a thallus, *e.g.* in Dicræa and

Hydrobryum (see Fig. 22). It drifts or is firmly fixed to the substratum by fine hairs, and is at the same time the assimilating organ of the plant. These plants are perfectly suited for life in the most violent water currents. It may be noted that in such water the air supply is very good, and, in relation to this, the plants of these two families, unlike most aquatics, possess practically no internal air spaces. These remarkable plants have been investigated by Warming (1881) and Willis (1902); an account will also be found in Arber's book.

The mechanical tissue of the stem of aquatics is very much reduced. The vascular bundles show only a few vessels, and there is no sclerenchyma. The most prominent anatomical feature is the abundance of air spaces. The difference between typical land and water stems is well seen by comparing the land and water forms of a single species such as those of *Polygonum amphibium*.

§ 6. Climbing Plants

Not all land plants support unaided the weight of their foliage. The ramblers, the root climbers, the twiners, and the tendril climbers form a very large and varied biological class.

Ramblers.—By ramblers we mean such weak-stemmed plants as the bramble or goose-grass, which scramble over the surrounding vegetation. They have no special means of climbing, though we may regard the thorns and prickles which they often possess as enabling them to maintain a position once gained. These may be single cells or large epidermal and cortical emergences; they are frequently reversed or hooked, and so give a good purchase against the downward pull of the shoot.

Root Climbers.—We have an excellent example of a root climber in our native ivy. The adventitious roots are produced in great numbers on the unilluminated side of the stem. They are ageotropic and negatively phototropic. They penetrate cracks and crevices in the wall, or in the bark of trees, but are only moderately well fixed. The security of the ivy is due rather to the great number of roots than to their individual efficiency; and, indeed, ivy growing on a wall is not really secure, for it is very readily torn off. Growing round a tree it is much better off, and this is the more natural position. The anchoring roots are not much use in water absorption, being soon cut off by cork formation.

Root climbers are not infrequent in tropical rain forests. Schenck (1892) describes examples from some twenty different families in Brazil. Many are epiphytic, and in all the genera which include root climbers epiphytic species are also found. The possibilities for the evolution of epiphytic types from root climbers are obvious. Only rarely are adventitious roots produced as a supplementary mode of attachment by plants climbing by other means.

Twining Plants.—A much higher type of organisation is shown by the twining plants in which the apex of the stem carries out a regular circling movement or *circumnutation*, which enables it to twine round a suitable support. This curious movement is induced and regulated by the action of gravity on the inclined tip of the stem. On the clinostat a twining stem, such as that of the scarlet runner, ceases its regular circling and shows only irregular swaying movements; if the plant is inverted the tip bends up and circles, in the opposite direction relative to the plant, but in the same direction relative to the axis of the earth. The movement is due to the more rapid growth of one flank of the stem; as the circling movement is necessarily accompanied by a torsion this flank is constantly changing. If a particular longitudinal strip of tissue lies below at a given instant, it gradually twists round till it lies at the side; the increased growth rate now sets in and carries on the circling movement. As this proceeds the strip we are following goes on twisting till it lies above, then on the opposite side, and finally once more below. A complete twist on the axis accompanies each complete revolution of the tip. The stimulus to more rapid growth occurs when a given strip lies below. Gradman

(1921) has shown that the whole movement can be referred to a succession of negative geotopic reactions in an organ which responds vigorously with a strong over-curvature.

As the stem apex twists and circles it twines round any support which is nearly vertical—the precise angle of inclination which prevents successful twining differs in different plants—and which is not too thick. The rather loose coils at first formed are tightened up by subsequent straightening of the stem, a process in which gravity seems to play the chief part, though contact stimulus may have something to do with it. The grip of the twiner on its support may be further improved by the maturing of prickles or hooks, as in the hop. Only when twining is complete do the leaves of any region of the stem expand fully. As a rule the direction of twining is constant and specific; usually it is counter-clockwise, as in the scarlet runner, less often clockwise, as in the hop.

Tendrils.—Tendril climbers show the highest degree of specialisation in making use of external support; in these one organ or another is modified as a tendril, the slender, whip-like part which grasps the support. Tendrils are characterised by extreme sensibility to contact with a solid body. This may be admirably seen in the bryony. If the tendril is once or twice lightly stroked with a match on the lower side it may be seen to bend downwards at the point of contact after the lapse of a minute or two. The details of this reaction have been exhaustively studied, *e.g.* by Darwin (1875*b*) and Fitting (1903); a general account is given by Jost. In few cases does stimulus to any side of the tendril lead to bending in the direction of the stimulus; usually the reaction follows stimulation of the lower side only. Curiously enough this is not due to lack of sensibility of the other sides, for a tendril if stimulated equally above and below does not react at all. Thus both sides are sensitive, though only in the lower is stimulation followed by the response of movement. Stimulation of the upper side is perceived, indeed, but the reaction is only an inhibition and not an independent movement.

The sensitiveness of the tendril is very great. In favourable circumstances the contact of a cotton thread, moved by light air currents, induces a reaction. The contact of water drops, however violent, is ineffective, and this has an obvious biological advantage, in that rain does not cause a movement which could be of no use to the plant. It is probable that an uneven deformation of the protoplasm of the epidermal cells is the immediate effect of the contact. Haberlandt (1906) has described special thin regions in the walls of the epidermal cells of tendrils, sensitive pits, and thin-walled sensitive papillæ, as suited to the perception of contact stimulus.

The reaction is carried out by a considerable increase in the growth rate of the side opposite that stimulated, combined with a decrease on the stimulated side. About half an hour after stimulation this condition is reversed and the tendril straightens out.

The response to contact stimulus is termed *haptotropism.* It is most prominent in the tendrils, but is also seen in the tentacle of sundews, and in the stems of the dodders. A recent survey by Stark (1915, 1917) has shown, however, that haptotropism is very common in all sorts of plants and plant organs which do not encircle supports, though in such cases it must be looked for. Of sixty-three species of non-climbers investigated one-third were found to be sensitive to contact stimulus in the petiole, the shoot axis, or the peduncle. Of twining plants, over 50 per cent. show sensitiveness to contact stimulus, and in the case of tendril climbers the sensitiveness is not usually confined to the tendril, but is present, to a lesser extent, in other organs, a condition already recognised by Darwin. Haptotropism is also shown by root tips in the soil. It is thus a very general property of growing parts ; in tendrils the sensitiveness and the extent of the reaction have become intensified and specialised in relation to a definite and important function.

The chance of contact with a suitable support in nature is increased by the circumnutation of the apical portion of the tendril. Contact is not necessarily, or even usually, transitory ; as the tendril bends round the support new regions make contact, and the movement proceeds till the whole of the tip is used up. Reverse movements also take place, and these, in conjunction with continually renewed stimulus, result in the winding on of a part of the tendril below the point of original contact, and in a closer grip of the whole. When the encircling is complete a fresh reaction sets in, due primarily to the mechanical strain of the plant's mass. The mechanical tissue of the moving, growing tendril is collenchyma ; after the support is grasped sclerenchyma develops and the resistance to the pull of the plant is much heightened. In many tropical plants the tendril undergoes enormous secondary thickening. Frequently the free portion of the tendril, between the support and the plant, coils into a corkscrew. As it is impossible to make a simple coil between two fixed ends the direction of the twist is reversed at least once (A Fig. 44). The corkscrew acts as a spring ; in gusts of wind it pulls out somewhat, and closes up again, thus diminishing greatly the risk of tearing. Our native Bryonia and the tropical Lagenaria are good examples of this. Fixed by hundreds of tentacles to a fence or hedge, the bryony is extraordinarily secure. We may recall Darwin's description of a bryony hedge in a storm : " It is this elasticity which prevents both branched and simple tendrils being torn away from their supports during stormy weather. I have more than once gone during a gale to watch bryony growing in an exposed hedge with its tendrils attached to the surrounding bushes ; and as the thick and thin branches were tossed to and fro by the wind the tendrils, had they not been excessively elastic, would instantly have been torn off and the plant thrown prostrate. But as it was, the bryony safely rode out the gale, like a ship with two anchors down, and with a long range of cable ahead to serve as a spring as she surges to the storm."

Nature of Tendrils.—Tendrils are derived from the most various organs. In the simplest case the organ, unaltered except as regards its reactions, acts directly as a tendril. Examples are the leaf blades of *Corydalis claviculata*, the leaf-stalks of *Tropæolum sps.*, the stalks of the leaflets of *Clematis sps.* Next to these we may place the liliaceous Gloriosa, in which the leaf tip is prolonged into a tendril. In Nepenthes the same thing is seen, though,

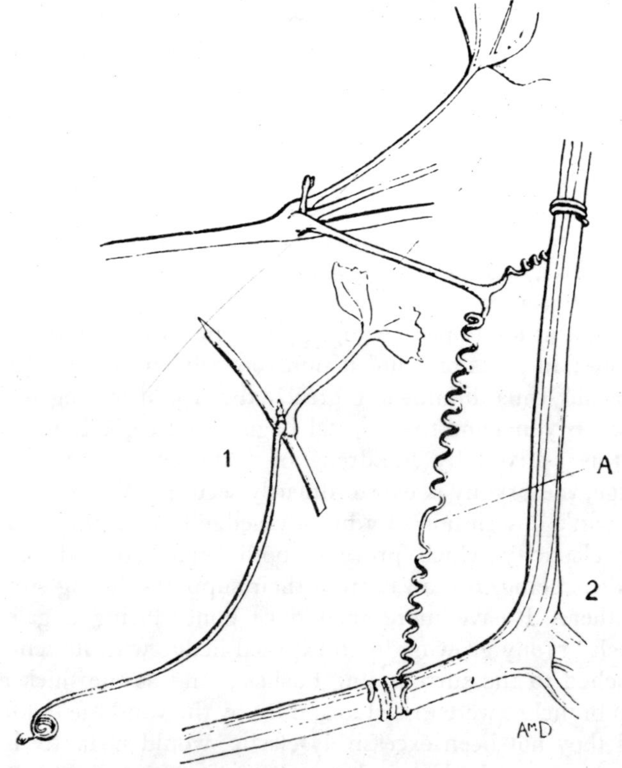

FIG. 44.—Shoot tendrils: 1, Adenia; 2, Lagenaria. Nat. size.

most frequently, the presence of the pitcher complicates matters. In the sweet pea and many vetches the branched tendrils represent several of the apical leaflets and their stalks. In *Lathyrus Aphaca* the whole leaf forms the tendril, the leaf function being taken over by the enlarged stipules. The tendrils of Smilax occupy the position of stipules, though the absence of stipules in monocotyledons makes their homology with these organs doubtful; Goebel regards them as outgrowths of the leaf base, of a unique character (Fig. 45). In *Capparis adunca*, the simple tendrils, springing from the leaf axils, are to be regarded as flower stalks; in Adenia the tendrils, occupying a similar position, are branched, and

FIG. 45.—Leaf tendrils: 1, Gloriosa; 2, Tropæolum; 3, Smilax; 4, Lathyrus. All ½ nat. size.

are modified inflorescences (Fig. 44). In the Cucurbitaceæ (Lagenaria, Cucurbita, Bryonia, etc.) and the Vitaceæ (Vitis and Ampelopsis) the nature of the tendrils is more obscure. In the vines the tendril arises opposite to a leaf, in the bryony at the side of a leaf. Goebel holds that in the vines the tendril is a modified shoot; in the Cucurbitaceæ it is of leaf nature when simple, and when branched it is a shoot bearing leaves (Fig. 44). In some plants, *e.g. Ampelopsis Veitchii*, *Glaziovia bauhinoides*, the tips of the tendrils instead of twining round the support apply themselves to solid surfaces and then form little sucker-like organs, which make an extremely close union ; the stimulus of contact produces

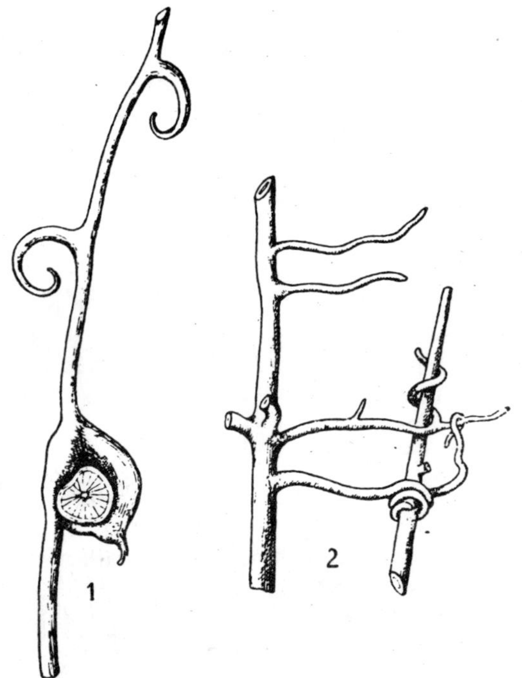

FIG. 46.—1, Hook tendrils (sympodial branches) of *Ancistrocladus Vahlii* ; one grasping a support has become thickened ; 2, Root tendrils of *Dissochæta sp*. Nat. size. (After Treub.)

a very peculiar result in this case. In some cases shoot tendrils have the form of thorns, and may grip a support, though they do not twine round it ; they show subsequent great thickening ; *Strychnos nux-vomica* is an example.

Root tendrils are rare, and are not very sensitive. Examples are the aroid *Philodendron melanochrysum* and the orchid *Vanilla planifolia* (Fig. 46).

The Lianas.—The term *liana* was used originally to denote the twining plants of the tropical rain forests. Schimper extended it to cover all plants, which climb by any means, belonging to any part of the world; it could not be denied to our native honeysuckle or hop. Yet this "plant guild," as Schimper aptly terms such a biological group, reaches its most magnificent state in the tropics, just as does the guild of epiphytes.

Haberlandt (1893) writes, "The wealth of species of lianas in the tropical woods is astonishingly great. While in central Europe only a few wood climbers occur, such as the ivy, the honeysuckle, the wall vine, and the number of herbaceous forms can scarcely amount to much over a hundred, the number in the tropics has been reckoned at 2000 and more, of which most show woody stems. Climbing plants occur in the most different divisions of the plant kingdom. Certain families are marked out by peculiar richness in climbing species, thus the Sapindaceæ, Malpighiaceæ, Menispermaceæ, Bignoniaceæ, and Leguminosæ."

Wallace, in his "Tropical Nature" (1878), gives a vivid picture: "Next to the trees themselves the most conspicuous and remarkable feature of the tropical forests is the profusion of woody creepers and climbers that everywhere meet the eye. They twist around the slender stems, they droop down pendent from the branches, they stretch tightly from tree to tree, they hang looped in huge festoons from bough to bough, they twist in great serpentine coils or lie entangled in masses on the ground. Some are slender, smooth, and root-like; others are rugged or knotted; often they twine in veritable cables; some are flat like ribbons, others are curiously waved and indented. . . . They pass overhead from tree to tree, they stretch in tight cordage like the rigging of a ship from the top of one tree to the base of another, and the upper regions of the forest often seem full of them. . . . In the shade of the forest they rarely or never flower, and seldom even produce foliage, but when they have reached the summit of the trees that support them they expand under the genial influence of light and air, and often cover their foster parent with blossom not its own."

The lianas form the subject of an exhaustive study by Schenck (1892), from the details of which it appears that each type of climber we have described reaches its highest perfection, greatest variety in detail, and most exaggerated expression in the tropics. An interesting structural feature with a mechanical bearing may be mentioned. The twiners are, as we have said, twisted on their own axes as well as wound round the support. This, combined with the unilateral pressure against the support, leads to most remarkable anomalies in the secondary thickening of woody forms. The wood is never continuous. The necessary pliability is attained most simply by the wood being split up into wedges which may be again lobed or cleft. In other cases, it occurs in isolated clumps. The cambium is sometimes renewed many times outwards in the cortex so that a series of concentric zones of wood separated by rings of ground tissue is produced. As the stem is often narrowly elliptical in section these wood bands may be only arcs of a circle. Finally, the stem may be deeply cleft into rounded segments, each with its own cylinder of wood. The approach to the structure of cordage composed of separate twisted strands is, in such cases, very remarkable. The great length of the internodes may be noted. The vessels of the tropical lianas are the longest and widest known; this is related to water transport to great heights through a narrow stem (Fig. 47).

Into further details we cannot go; descriptions may be found in Schimper and Warming, and in the works of Schenck and Haberlandt already quoted. We may conclude this account by a quotation from Haberlandt, the description of a rambling palm, which, with some bamboos, may be taken as the highest type of rambler, " . . . the climbing or Rotang palms, which have neither twining stems nor sensitive tendrils, which nevertheless mount to the tops of the highest trees in the thickets of the primeval forest, and, thrusting above the foliage of the wood, let their glittering leaves wave in the wind. . . . If I move from the footpath it may well happen that at the first steps my hat is torn from my head, that the hooks cast out on every side catch my clothes, and that bleeding tears on cheek and hands warn me once and for all to beware. Looking at this gin, near which I have come, I see that the stalks of the graceful feathered leaves of the Rotang palm are provided with extremely elastic and flexible extensions, one or two metres long, on which are numerous very stiff half-whorls of re-

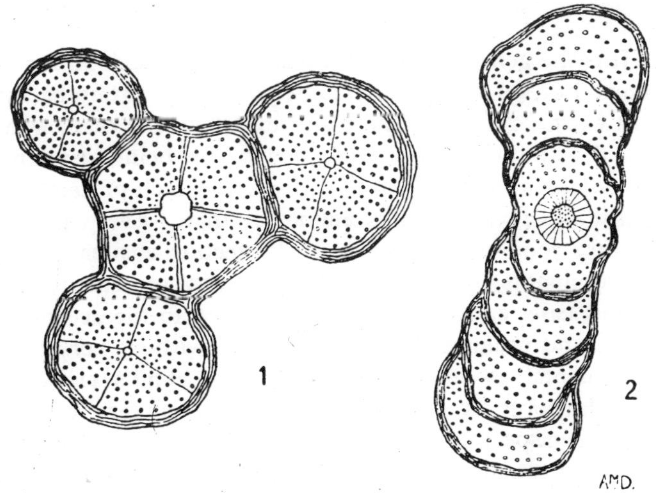

FIG. 47.—Stems of Lianas: 1, cross-section of Paullinia; 2, of Rhynchosia. × 3.

versed hooks. Each leaf runs out into a horrid lash of this nature which lets not lightly go what once it holds. . . . Very flexible, they are waved by the wind to the branches of supporting trees, and there anchor themselves so fast by their numerous hooks that no storm can tear them loose. . . . The smooth snake-like stem reaches in mighty undulations up through the branch work of the tree, creeps over on to neighbouring crowns, and, finally, lifts its youngest leaves over the tips of the supporting tree. It can go no further, for the lashes are whipped round in empty air. But the old leaves die slowly and are cast off. Robbed of its anchors, the smooth stem slips down by its own weight, till the upper lashes are again fast. At the foot of the tree lies, in great coils and loops, the sunken part of the stem as thick as a man's arm. . . . In the forest this mighty cable may reach the length of 200 to 300 metres."

II. Protection against Animals and against Other Plants

It is the general fate of the plant to serve as food for the animal, but, for life on earth to be possible, a certain balance must exist. Nor does the plant, except in the cases where seed distribution is involved, benefit by being devoured. It is common for plants to show features which protect them against the ravage of this animal or that. Every plant is not easy food for any animal. Protection against rodents and ruminants by means of sharp thorns and spines is an obvious case, and its obviousness has led to the acceptance of this and other devices as protective without very much experimental evidence. In a work of Stahl's (1888) we have, however, a monograph which deals very carefully with one aspect of the problem, and also refers to more general questions. A treatment of the subject is also given by Kerner and by Neger.

Stahl points out that the protection of a plant is usually only partial, against some few animals ; but that this may yet be of extreme importance, for one enemy more or less may make all the difference between extinction and survival. He gives as an instance the decimation of the European vine by the Phylloxera introduced from America, a pest unable to do serious damage to the roots of the American species. Plants which are avoided by most animals may be the favourite food of one or a few. Thus *Euphorbia Cyparissias*, the cypress spurge, is never touched by ruminants, rodents, snails, or grasshoppers, but is the only food of the caterpillar of the moth *Sphinx euphorbiæ*. The limitation of many caterpillars to particular plants is a point of great interest.

The fact that plants may be protected in a highly specialised fashion makes it dangerous to draw conclusions from one animal to another. It explains how a particular flora, in a state of balance with the native fauna, may be at the mercy of an introduced species. The ravages of the introduced rabbit and squirrel among our trees are familiar. The introduction of the goat into St. Helena resulted in the practical destruction of the native flora.

Mechanical Protection.—Means of protection may be divided into mechanical and chemical. Of the mechanical the most prominent, though probably not the most important, is the formation of spines, thorns, and prickles, which may perform this function as well as others. Whether protection is the primary function of spines is a matter of doubt. We have already seen that the formation of leaf spines is linked with moisture relations, and that in the development of the individual plant the formation of spines is promoted by good illumination and dry air, and depressed by the opposite conditions. Spinous plants are characteristic of arid regions where thorn scrub and woodland are prominently developed. But the regulation of spine formation by atmospheric moisture, etc., does not necessarily mean that the most important function is not, in fact, protection of the plant against grazing animals, which are numerous in just such country. It is scarcely necessary to give examples, but we may mention the native whin, and the acacias and cactuses of thorn scrub and desert formations. The thistle and holly are examples in which the spines are exaggerated leaf serrations.

Less conspicuous are stiff hairs, such as occur on *Galium aparine* and other bedstraws, and on many Boraginaceæ. As we have seen, these may be of importance to rambling plants in securing their position. Stahl looks on them as an important defence against the attacks of snails, both because they make it difficult for the animal to creep on the plant, and because they wound the delicate mouth parts.

On the other hand, they offer no protection against grasshoppers.

Specialised hairs, only partly mechanical in their action, are the stinging hairs such as those of the nettle. The long tapering point of the unicellular hair is hardened by a deposit of chalk, the rounded tip by silica ; just below the tip is a weak spot. Pressure breaks off the tip and the fine point enters the skin of an animal touching the plant ; at the same time the soft bulb-like base of the hair is compressed and the contents injected into the wound. The well-known inflammation follows quickly. The nature of the toxic substance causing it is not known ; it is probably a protein. The similar stinging hairs of some tropical plants belonging to the Urticaceæ, *e.g.* Urera, Laportea, and to the Loasaceæ, *e.g.* Loasa, Blumenbachia, are much more violent and may have dangerous results.

The silicification of cell walls may make it difficult for an animal to eat a plant. This is well shown by many grasses and sedges, where the epidermal cells along the leaf margin are strongly silicified and also rough. The blade of *Aira cæspitosa* or that of *Phragmites communis* cuts the fingers if drawn through them. The leaves of many bamboos and other grasses of warmer zones cut like razors, inflicting deep wounds.

Perhaps the most important protection against snails and slugs is the presence of *raphides*, the bundles of needle-like crystals of calcium oxalate deposited in the cells of many plants, as, for example, the rhubarb, the cuckoo-pint, and the wood hyacinth. The needles lie parallel in the cell, but are scattered if the tissue is chewed and penetrate the delicate mouth parts of the snail in every direction, choking them effectively. Stahl, who has investigated this point exhaustively, finds that raphide-bearing plants are never touched by snails or slugs, and are avoided by other animals. If swallowed the raphides may have the fatal effect of powdered glass.

Chemical Protection.—Of chemical means of protection the most widespread are the bitter and astringent tannins.

These, and chemical protection in general, may be much more effective than mechanical protection. Stahl states that usually in arid regions with sparse vegetation grazing animals avoid juicy, apparently unprotected, plants, and may be seen wandering round bushes bristling with thorns, nibbling here and there, picking with care and trouble each available leaflet. In Algeria the sharp-needled *Juniperus oxycedrus* is trimmed round by goats and sheep, while the softer leaved *J. phœnicea* is left severely alone; the latter is poisonous. In Britain the sheep-nibbled whin bush is familiar, while the broom, which contains bitter substances and a poisonous alkaloid, is avoided.

Acid sap is a protection against snails, as in *Oxalis Acetosella* and *Rumex Acetosa*. In many plants alkaloid poisons protect against grazing animals—coniin in the hemlock; in the foxglove there is a poisonous glucoside, digitalin. The latex of many plants is poisonous for at least some of their enemies. We have already mentioned the case of the cypress spurge, which is protected by bitter latex against all animals except the caterpillar of one moth. The latex of the para rubber tree does not protect against boring beetles. Ethereal oils are said to be protective, as in the Labiatæ.

Inflammation may be set up, not only by the contents of stinging hairs, but by substances exuded by glandular hairs, as in *Primula sinensis*, or liberated when the leaf is crushed as in *Rhus toxicodendron*, the poison ivy of America. The nature of these very active poisons is obscure. Systematic accounts of poisonous plants are given by Long (1917) and Pammel (1911).

Special Cases.—Three very doubtful cases may be referred to. In tropical South America great damage is done to foliage by leaf-cutting or parasol ants, which use the material thus obtained in the preparation of their fungus culture beds. It has been long supposed that some trees are protected against such enemies by maintaining an army of fighting ants which drive off the invaders. These fighting ants live in hollow stems, *e.g.* in *Cecropia adenopus*, the imbauba, or trumpet tree, or in hollow stipules, *e.g.* in *Acacia sphærocephala*. They gather albuminoid food bodies modified glands, from the leaf tips of the Acacia; on the upper surface of the petiole of Cecropia, hidden in a mat of hairs, they find similar succulent, stalked glands, rich in proteins and fats. It was first suggested by Belt (1874) that the imbauba profited by harbouring the ants, which live in its stem hollows and feed on the glands. He observed that trees, which, for some reason had no inhabitants, were stripped of leaves, while the inhabited trees were immune. He supposed that the food bodies were borne in relation to the needs of the ants; and that these had easy entrance to the stem cavity provided for them by a thin place situated above the insertion of each leaf, which is readily gnawed through. The hypothesis of a close inter-relation between ant and plant was elaborated by many subsequent investigators, of whom the chief was Schimper, in whose book an account of the subject will be found. More recently doubts have been cast on the use of the ant army to the plant and on the idea that any structure in the plant is really related to the attraction of these guests. We may mention the work of Ihering (1907), Ule (1906), and Bailey (1922, 1923), and the critical summing up by Neger. The drift of the criticism is that the clever ant simply makes use of any feature in a plant which may be made to serve its need for shelter or food. As regards the plant it receives little or no advantage from the ant. " *Cecropia adenopus* can get on as well without Azteca as a dog without a flea." The hollow stem or stipule is a common feature of all sorts of plants which have no relations with ants. The thin place through which the ant bites its way is a mechanical result of the pressure of the axillary bud—though this does not explain the openings into the hollow which are formed without the aid of the ant in some plants of the eastern tropics, such as *Humboldtia laurifolia*, and *Ficus inæqualis*. Even food bodies are found in some plants which have no relations with ants, as in the pearl glands of species of Vitis. Leaf-cutter ants are unknown in the eastern tropics where, nevertheless, ant-inhabited trees are common. In America the imbauba tree is regularly inhabited in regions subjected to flooding, an obvious advantage to the warrior ant, though in such regions the leaf-cutters with their subterranean nests are absent. Species of Cecropia are known which do not harbour ants and which are not attacked. The ravages of the leaf-cutters are said to have been exaggerated; in fact, the ants are most destructive to cultivated plants, damage to which is most likely to be noticed—and resented. Ridley (1910) supports the protection theory, and points out that plant enemies other than leaf-cutters may be attacked by the warrior ants. On the other hand, Ihering states that the ant *Azteca Mülleri* is so completely dependent on *Cecropia adenopus* that it cannot survive without its host tree, while the imbauba can grow perfectly well without the ant.

We may sum up by saying that recent opinion tends to emphasise the ability of the ant to make use of any opportunities, and to cast doubt on the reciprocal advantage gained by the plant. There may, however, be a danger in denying all use to the plant in the possible protection afforded against various animal enemies. We may here note that the extra-floral nectaries found on some European plants, as on the backs of the leaves of the cherry laurel, and on the stipules of the bush vetch, have been supposed to attract ants which in their turn protect the flowers from insect damage. There is little evidence that the plant really benefits.

Among the many functions which have been assigned to the violent folding of the leaflets of *Mimosa pudica* when the plant is shaken, is the scaring away of grazing animals. Jost quotes a letter of Heinricher descriptive of the behaviour of the sensitive plant in the botanic garden in Penang: " It was there I first met *Mimosa pudica* as a thick growing, vigorous weed, a third of a metre high and more. If one walked into the low thicket there appeared, as a result of the very prompt reaction, an empty, burnt-out looking space enlarging with every forward step. The same result must follow the entrance of a cow or other similar animal . . .

and I believe that the plant largely escapes grazing through this lack of inviting appearance." Stahl found that goats were in fact prevented from eating the shrub. Goebel points out the danger of drawing conclusions from the behaviour of the plant in the east, where it is not native, or towards an animal like the goat, which does not occur in South America, the plant's original home. Humboldt's observation that horses and cows refused to eat it in South America must also be applied with caution, since these animals are again introduced. Good evidence that the movement is protective in natural conditions is therefore wanting, and, on the other side, we must place Goebel's discovery that plants subjected to frequent shock are much retarded in their development. There is even less evidence in favour of the theory that the movement enables the plant to shake off marauding insects.

Finally, we may refer very shortly to the galls formed by so many plants in relation to the attacks of definite insects. These are teratological structures formed on leaf, stem, root, flower, or fruit after deposition in the tissues, or on the growing point, of an insect egg. They are often complex in structure, and are closely related to the requirements of the larva, which may even have at its disposal a special nutritive tissue, and a preformed opening for its escape at the proper time. How the plant comes to form such a thing, apparently without use to itself, and exactly fitted to the needs of the animal, is a question on which even speculation is silent. It is possible that the origin of the gall and its primary significance lies in a limitation of the activities of the larva, and is thus protective. Every one knows the destruction caused by burrowing larvæ in the leaves of such garden plants as the chrysanthemum, the nasturtium, and the marguerite. If the grub were restricted in its wanderings the leaf would benefit, and in such restriction by means of the formation of a special resistant tissue the gall may have taken its origin.

Protection against Parasitic Plants.—The means of protection against plant parasites have been little investigated. No explanation exists of the limitation of the three races of the mistletoe to their respective groups of host trees. We do not even know whether this is a case of prevention by the inappropriate hosts, or of lack of some intimate factor on the part of the parasite ; these two aspects of the problem are of course inter-related. We may assume that a strong corky covering is unfavourable to the penetration of the primary sucker. This might, for example, partially explain the rarity of mistletoe on the oak. It certainly does not take us far. Gertz (1915) shows that plants are immune from the attacks of the dodder from a variety of causes—the oak because of mechanical resistance, the wood sorrel because of acid sap, the poppy because of latex, the thorn apple because of alkaloids, eschscholtzia because of ethereal oils.

In the case of fungal parasites a strongly cutinised leaf may resist the penetration of those numerous species which pierce the epiderm. Many fungi, however, enter the leaf by the stomata, and through these openings entrance must be free for all. The rusts enter the leaf in this way, and they are remarkable for their high degree of specialisation, and for the numbers of biological races, confined to particular hosts, which exist within the bounds of many species. The case of the rust-resistant strains of wheat may here be cited. The spores of the rust germinate on their leaves and hyphæ penetrate the stomata ; inside the leaf the hyphæ die, or the leaf cells are killed ; in either case the further advance of the parasite is barred. We may compare this with the behaviour of orchid seeds to different strains of the symbiotic fungi ; the proper strain enters into symbiotic growth with the germinating seed ; weak strains are killed ; strong strains kill the orchid. We are evidently here in the presence of obscure protoplasmic reactions, which may have an analogy in the phenomena of immunity which have come into so much prominence in the study of human pathology.

CHAPTER V

REPRODUCTION AND DISPERSAL

§ 1. General Considerations

By reproduction we understand the process of giving rise to a new individual by a parent organism. As we shall see later the meaning of the term *individual* is not at all easy to define, especially in the plant kingdom. Here we may take it that when an organism produces, in a specialised fashion, a cell, which, under suitable conditions, develops into a new organism similar to the parent, a new individual has arisen. To such a process we would confine the term *reproduction.* In the reproductive cell the inheritance of the race is gathered and a fresh start is made. To the increase in the number of independent plants which takes place, both naturally, and especially in gardening practice, by the rooting of parts detached from the parent we would apply some different term such as *vegetative multiplication,* or, conveniently, the gardener's word *propagation.* In such cases the new plant does not start afresh from the beginning.

Reproduction in plants takes place in two ways, *asexually*

by the formation of *spores*, and *sexually* by means of sex cells or *gametes*. The male cell is the *sperm*, the female is the egg or *ovum*. As is well known, in most plants, and always, normally, in the higher land flora, both types of reproduction occur at definite points in a regular life cycle. A sexual stage, or *gametophyte*, reproduces by means of gametes. The single cell formed by the fusion of egg and sperm, the *zygote*, develops into an asexual stage, or *sporophyte*. The spores produced by this give rise to the gametophyte. This cycle is called the *alternation of* (sexual and asexual) *generations*. There may be only one type of gametophyte producing both eggs and sperms, as in many homosporous ferns. In the flowering plants there are male and female gametophytes. In these there are also two types of spore, a *microspore* which produces only male gametophytes, and a *megaspore* which produces female gametophytes. Both types of spore may be produced by the same sporophyte or there may again be two types of sporophyte.

Intimately connected with this alternation of generations in the higher plants is an alternation in the constitution of the nucleus. The nuclei of the sperm and ovum possess a number of chromosomes, definite in any species; their fusion results in the formation of a zygote nucleus with double that number, and this double set is maintained in all the cells of the sporophyte. In the formation of the spores the penultimate nuclear division is unique in nature, and results in one-half of the chromosomes passing to each of the daughter cells, so that in these, and in each of the two spores to which each may give rise, only the single set of chromosomes is found. This number is retained in all the cells of the gametophyte. The single set is spoken of as the *haploid* number, the double as the *diploid*.

The angiosperm plant is a sporophyte. It produces two types of spore. Numerous microspores, the pollen grains, are formed in each of the four sporangia normally borne by the stamen. The nucellus, the megasporangium, normally brings to maturity only a single megaspore, the embryo sac. The gametophyte generation is reduced to three cells in the pollen grain, and to eight in the embryo sac, including the gametes in each case. Of the two male gametes one fuses with the egg cell, and forms the zygote, which immediately develops into an embryo, the early stage of a new sporophyte generation. The gametophyte generation in the angiosperms is thus reduced to a very few cells and has only a transient existence. In the gymnosperms it is rather better developed, but in these, too, it passes its existence entirely within the parent spore. Only by comparative studies with lower types has it become perfectly clear that we have, in the flowering plants, a definite alternation of generations of the same type as that found so much more obviously in the mosses and ferns.

In the moss the gametophyte is the moss plant, the sporophyte is the capsule with its stalk. The sexual generation is the dominant one. The fern plant is a sporophyte, but the gametophyte leads an independent though modest existence, as the prothallus. As we follow the changes upwards through the evolutionary scale we find the gametophyte becoming more and more dependent. In Selaginella, a heterosporous lycopod, the female gametophyte just protrudes from the bursting spore. In the gymnosperms it is entirely contained within the spore, and, further, the megaspore remains embedded in the tissue of the sporangium, the nucellus, nourished and protected throughout by the parent plant. In the Cycads and in Gingko we still find free-swimming ciliate sperms, but already in the higher gymnosperms the sperm has lost its power of independent locomotion and is carried by the pollen tube to the embryo sac. At the most it may have some power of amœboid movement there. The cases mentioned are not stages in one phylogenetic line, but they may be taken as illustrating steps in the reduction of the gametophyte generation.

This process of reduction has taken place during the evolution of a flora more and more suited to life on a land surface, and the advantage of a reduced and dependent gametophyte to a land plant is easily understood. The prothallus of a fern is extremely delicate, only one cell thick over most of its area, and quite incapable of withstanding desiccation. It can live only in moist stations. In the Lycopodiaceæ and Ophioglossaceæ the gametophyte is tuberous, and is in many cases saprophytic and subterranean with a mycorhizal fungus. For good development it requires moist soils rich in humus. An independent gametophyte therefore at once limits the plant to stations where the moisture relations are very favourable. The evolution of a type of gametophyte with drought-resisting capacity could be imagined; but it has not occurred. Moreover, water is necessary for another reason. The sperm can swim to the ovum only in free water. Fertilisation cannot take place unless surface water is available. Only by living in a constantly moist situation can the plant be assured of a supply of water at the critical point in its life cycle. Now, as the sporophyte generation is a sedentary organism growing where the zygote, embedded in the tissues of the gametophyte, has started to develop, it follows that the sporophyte, too, however marked drought resisting capacity it might possess, would, from its individual point of origin, be limited to moist stations. In fact the abundant remains of the once dominant flora of ferns and their allies which persist to the present day are characteristic of damp situations, the shade of woods, marshes, and slow streams. A few exceptions like the bracken are known to multiply for the most part vegetatively by branching rhizomes.

The alternative to the evolution of a drought-resisting gametophyte is the reduction of the sexual stage. In the plants where the process is complete we find the utmost possible emancipation from the necessity of a steady or abundant water supply. Reduction is the primary condition, and it has been accompanied by the evolution of the specialised absorbing, conducting, and economising systems of the sporophyte which we have already studied.

§ 2. Sexual and Asexual Reproduction

We may now put a more fundamental question. The mosses, the ferns and their allies, the gymnosperms, and the angiosperms all show normally an alternation of generations; they have two distinct types of reproduction. What different functions do these serve? Why should not sexual reproduction alone suffice, or asexual reproduction alone, for by either a multitude of new individuals may arise? To answer this completely is not possible, for the essential nature of sexual reproduction is not fully understood.

If we try to trace the stages in the evolution of sexual reproduction in the plant kingdom we find that we must go back to the algæ, for in the land flora the gametes are throughout highly specialised. In the genus Chlamydomonas, unicellular, free-swimming, green algæ, reproduction occurs by internal division into a number (2 to 8) of zoospores, each of which is just like the mother cell. When these spores escape by the rupture of the mother cell wall they increase in size, secrete a cell wall, and so reach the adult state in a very simple fashion. The mature cell may also give rise to gametes.

These may be indistinguishable morphologically from the zoospores, or they may be smaller, with slight structural differences, as many as sixty-four being formed in a single cell. On escaping, the gametes swim about freely but do not grow into mature individuals. They fuse in pairs, form a resting zygote, with a thick wall and reserve food. The zygote germinates by dividing internally into four zoospores. Here we have typical sexual reproduction in which the reproductive cell proceeds to develop only after union with, or fertilisation by, another reproductive cell. It is on a low level because there is no difference between the two gametes, and little difference between the gamete, the zoospore, and the vegetative cell. Yet this simple condition does not obtain in all the species of the genus.

In *Chlamydomonas grandis* the two gametes are equal and naked; they come together point to point and fuse either thus or side to side. In *Chlamydomonas media* the equal gametes have walls, and before union the plasma contracts and escapes from the wall. In *Chlamydomonas Braunii* the two gametes come together while still enclosed in walls ; the contents of the one slip inside the wall of the other, where fusion takes place. Here we have a difference in the *behaviour* of the two gametes. In *Chlamydomonas coccifera* a definite ovum is formed by a vegetative cell casting its cilia and increasing in size. Sperms are formed by the internal division of another cell in the more usual way. Compared with the egg they are quite small and they are motile. The egg is fertilised by one of the sperms. Here we have a difference in *structure.*

Within the bounds of a single primitive genus we have, then, an advance from the pairing of equal gametes, *isogamy*, to the fertilisation of a large egg by a small motile sperm, *oogamy*. Similar series may be traced in other families of the green algæ, *e.g.* in the Volvocaceæ and the Ulothricaceæ. What we wish here to emphasise is that while the fundamental characteristic of sexual reproduction is the fusion of two gametes, the trend of evolution is towards the differentiation of a large motionless egg and a small motile sperm, though the motility of the sperm has been subsequently lost, for secondary reasons, in the higher members of the land flora.

In many of the higher algæ a new feature is found in the retention of the egg in organic connection with the parent plant, and this condition is constant in the land flora, where the egg is produced in a special organ, the archegonium, which from its universal occurrence in all forms in which the reduction of the gametophyte has not resulted in its obliteration, must be taken to be a structure very favourable to the nourishment and protection of the delicate egg cell. The enlargement of the egg may be regarded as advantageous in making possible an accumulation of reserve food substance with which the zygote may start its career. The retention of the egg in the parent plant makes for its better protection and nourishment and also benefits the zygote.

Now it requires but little consideration to see that this type of reproduction is ill suited to secure the dispersal of the offspring. For its success it depends not on dispersal of the gametes, but on their concentration in a limited neighbourhood. The chances of meeting of these minute cells, scattered by water currents, must in any case be small. In Chlamydomonas and other free-swimming organisms the necessity of linking dispersal with reproduction does not exist; and this is true of most animals which are characteristically motile. The plant at an early stage of its evolution, however, became sedentary; its lack of mobility is one of its most striking attributes. In all sedentary plants dispersal can take place only when reproductive bodies are cut loose from the parent. Even when the egg cell is set free it is unsuited to dispersal; when it remains attached, dispersal by it is impossible. This is the condition in the land flora. Some other means of dispersal is therefore imperative, and there is no doubt that this, along with the possibility of more rapid multiplication given by cells which do not require to pair, is the fundamental function of *asexual* reproduction. In the algæ it is carried out by zoospores, or by non-motile spores which drift in water currents. In the mosses and ferns the spores are never motile. Power of motion, however, ceases to be important when the varying, and often violent, air currents of the land surface are available. Spores are minute bodies and are carried, literally like dust, over great distances. With loss of motility the spore has evolved a positive character of the utmost importance, the double wall which gives the possibility of resistance to desiccation, and of lying dormant for long periods. The spore is thus rendered independent of an immediate supply of water or of a supply at any particular time. The reproductive body of the land plant must be capable of resisting drought and of air dispersal, and the spore of the moss or fern fulfils these conditions.

Significance of Sex.—We have thus a plausible explanation of the importance of asexual reproduction, which, however, does not assist us to understand why sexual reproduction, the maintenance of which throughout the vegetable kingdom forces us to believe in its usefulness, should be necessary also. We here enter one of the most difficult regions of biology, for, as we have said, we do not really know what is implied in the sexual process.

(*a*) The ovum is a reproductive body which normally can develop only after a suitable stimulus has been applied to it. This stimulus is normally the entrance of the sperm. If the essence of sexual reproduction lies merely in a stimulation to development, it is not clear what advantage the organism, or the race, gains in clinging to a practice as precarious as it is expensive. Further, the stimulus is not of so highly specialised a character as to make essential the co-operation of the second gamete. Parthenogenesis—the development of an unfertilised egg—is a widespread phenomenon in the animal kingdom, and also occurs among plants. Here the normal development of the egg cell is determined by some stimulus or condition which, whatever its nature, is not that of fertilisation. Further, artificial parthenogenesis has been brought about both in plants and animals by subjecting the gamete to a variety of abnormal external influences, such as wounding, the action of salt solutions and of organic acids. The stimulation of fertilisation is not therefore unique—it may be due to some simple effect, such as the admission of oxygen—nor does it in any case explain the biological significance of sexual reproduction.

(*b*) It has been suggested that after a certain span of vegetative existence the organism becomes weakened or worn out; the span of life is, in fact, definitely limited in some plants as in many animals. It is possible that the process of sexual fusion is in some way the means of rejuvenation, that it starts the new generation of individuals with an equipment of revived protoplasm. How this should come about by the union of two exhausted cells is not easy to see. The theory is strengthened if it is made broader, and the increased vitality is made dependent on the union of gametes derived from different individuals or stocks; though this introduces new factors. Undoubted weakening does occur in some species if close inbreeding is persisted in through many generations. This is the case in human beings and, among plants, in the maize. In this plant the crossing of two races leads to the production of vigorous offspring; it is a case of "hybrid vigour." The factors involved have been investigated recently by East and Jones (1919). They have shown that vigour is produced by a number of independent dominant hereditary factors: the greater the number of these present in an individual the better its growth. When two individuals of different races are crossed an addition of such factors takes place. Theoretically it is possible to breed an individual containing all these dominant factors in homozygous condition, and such an individual if inbred would show no weakening. Practically this is extremely difficult or even impossible because of linkages which are only rarely broken (cp. Morgan, 1919). Other cases of hybrid vigour in plants are described by Darwin (1876), and these are probably to be explained in the same way. There are many plants and animals in which inbreeding is the rule, and in these no degeneracy results. This more advanced type of sexuality which is represented by outbreeding is therefore without a primary effect on stamina, and this makes it all the more difficult to believe in a rejuvenation resulting from the more primitive fusion of two sister cells, or of cells derived from the same individual. We shall return to the question of the possibility of prolonged vegetative existence in another connection, and may here note simply that the evidence in favour of the rejuvenation theory of sex is not convincing.

(*c*) The case of the maize, however, brings us to a third possible explanation of the importance of sexuality. The zygote is remarkable in that it contains a contribution from each of two cells, and very often these two cells are derived from different individuals or even races. In the zygote the characters of two parents are brought together, and the individual which arises from it has also a dual set of characters. Now if we consider the case of a species which occasionally produces, by mutation, an individual with a new character, and if we imagine that this species has lost the power of sexual reproduction, it will be clear that each new character will be confined to the individual in which it originated, and to the descendants of that individual; and that a new character appearing in one individual will never have a chance of combining with another new character in a second individual. Let us suppose that in the course of time ten new characters have appeared in ten different individuals; we have then ten new types with no power of intercombination. If, however, the species possesses the power of sexual reproduction and the new types are free to cross, then combination becomes possible and with it the possibility of the origin of a vast number of new types from a small number of individual changes. With 10 distinct characters there is a possibility of 1023 distinct combinations. The chance of new and successful races being produced is enormously increased and the rate of evolution must be accelerated. The importance of hybridisation in the production of garden varieties, and improved races of cultivated plants, needs no emphasis, and may indicate a similar importance in natural evolution. If it does nothing else, sexual reproduction would therefore seem to have an evolutionary function; indeed, hybridisation has been made by Lotsy (1916) the basis of a complete theory of evolution.

This is the best explanation we have of the significance of sexual reproduction, but we must keep in mind a difficulty which stands in the way of its acceptance. Every character of an organism must stand the test of natural selection through competition with its fellows and the action of adverse conditions, and, if it be definitely disadvantageous to the individual, the race which shows it may succumb. We must believe that sexual reproduction, especially in its lower grades, is a precarious process; we do not know that it in any way benefits the individual possessing it or arising through its action; the *rôle* that we have assigned to it lies rather in the possibility it gives of producing fresh types. It is not easy to see why such a property should survive. We should look in some other direction for an explanation of its origin and power of persistence in primitive forms.

(*d*) An interesting suggestion has recently been made by Jones (1918). He points to the resemblance between the sexual fusion of egg and sperm, and a parasitic attack by the latter on the former. It is quite possible that sexual fusion originated in a mutual parasitic attack of feebly assimilating, semi-starved, primitive organisms. If the habit of such parasitic fusions were once deeply stamped on the organism it might have given rise to the whole structure of sexual reproduction, which would in the course of evolution have thus changed entirely its significance for the organism.

§ 3. Seed Formation

Taking once more the fern as a type of a land plant, we may resume the foregoing discussion by saying that, in its spores, it has a means of reproduction capable of giving rise to great numbers of new individuals, of scattering the offspring widely, and of resting through periods of adverse conditions ; through its gametes it secures the advantages of crossing. The definite alternation of the two in one life cycle ensures the regular recurrence of the two kinds of benefit.

The simple condition of the fern is lost in higher plants. First the gametophyte was reduced until it became confined to the spore, and a marked differentiation of the spores took place. Then the megaspore remained attached to the parent plant, embedded in the sporogenous tissue. This was an important change, for at this point the megaspore lost the two chief characters and functions of the spore—the possibility of dispersal and the power of drought resistance. This is the condition in the flowering plants, gymnosperms as well as angiosperms. The details are different in these two great groups but they resemble each other in this, that after the fertilisation of the egg cell, the development of the new sporophyte starts at once and proceeds until an embryo is formed; development then ceases temporarily and the embryo loses water, while changes in the integuments of the ovule result in its enclosure in a more or less resistant envelope. Thus is formed the seed. The seed typically contains a store of food, fat or carbohydrates, and proteins. These may be stored in the cotyledons, in the hypocotyl, or in a special storage tissue. The storage tissue may be derived from the nucellus of the ovule, the megasporangium, when it is called the *perisperm*, a rather uncommon case of which the pepper is an example. Most frequently it develops as the result of a second fusion occurring in the embryo sac, in which three nuclei are concerned; one of these is a sister of the egg cell, the second is one of the group of four antipodal cells, which may be regarded as the last vestiges of the vegetative part of the female gametophyte; the third is the second sperm cell which enters from the pollen tube. The triple fusion nucleus divides many times, and after cell formation has taken place the nutritive tissue known as the *endosperm* is formed. It is well shown in the cereals. The food store of the gymnosperm seed is an endosperm derived from gametophytic tissue which is well developed within the embryo sac.

The seed is thus a body of complex structure and of multiple origin. The seed coat belongs to the parent sporophyte, as does the perisperm if present. The endosperm is gametophytic in origin in the gymnosperms, while in the angiosperms it is a unique body in a sense homologous with the new sporophyte. The embryo is the new sporophyte generation. The whole may thus include three different generations in its constitution. It is not a reproductive body in any strict sense of the word. The reproductive bodies, in the seed plants as in others, are the spores and the gametes. The seed is the consequence of two separate reproductive events, and is, in fact, simply a young plant in which development has been arrested, and which has at this stage been cast loose from the parent, enclosed in a peculiar covering and well stocked with food. It is capable of dispersal and also of lying dormant and resisting desiccation.

We may look on the seed habit as the logical conclusion (for the time being) of an evolutionary trend shown by plant life on the land surface, a trend which was initiated by the reduction of the gametophytic generation. Perfect protection of the female gametophyte and independence from the necessity of external water are possible only when the gametophyte is enclosed in the tissues of the parent. In early stages the megaspore was not enclosed, but its retention in the sporangium, and the prolonged attachment of the sporangium itself gave further advantages in the way of nourishing the ovum. The typical properties of the spore were lost and the functions of dispersal and dormancy were transferred to the new organ, the seed. To begin with, the seed was cast loose after fertilisation, but before the development of the embryo. A longer attachment to the parent organism, however, brought with it the possibility of better protection and nourishment of the embryo, and of its provision with a food store, and so the modern seed has been evolved.

§ 4. Parthenogenesis and Apogamy

The process of reduction of the gametophyte has not stopped at the stage we have indicated as being characteristic of the angiosperms. Many cases are now known in which an embryo is developed without fertilisation having taken place. This happens regularly in many plants, but whether it is merely an abnormality, or whether it is a comparatively new evolutionary departure in the angiosperms, we do not precisely know.

Three different types exist.

1. **Parthenogenesis** is the development of an unfertilised ovum. It is known in the Compositæ, *e.g.* in Antennaria and Hieracium, in the Rosaceæ, *e.g.* in Alchemilla, in the Ranunculaceæ, *e.g.* in Thalictrum, and in some other

families. In some species it is habitual and in others it exists alongside typical sexuality. In all cases the genera which have parthenogenetic species have also normal species. In Alchemilla (Murbeck, 1901) the spore mother cell does not undergo reduction division and becomes itself the megaspore; from it there develops an embryo sac, perfectly normal except that the nuclei have the diploid number of chromosomes. The ovum develops into an embryo without fertilisation. In *Hieracium flagellare* (Rosenberg, 1906) some flowers of a head show normal reproduction, while in others parthenogenesis takes place. Here, however, reduction division occurs and an embryo sac is formed in the ordinary way. This disintegrates and a cell of the integument enlarges and gives rise to a second embryo sac with diploid nuclei, the egg cell of which develops parthenogenetically. This is a case of parthenogenesis combined with a second abnormality—*apospory*, that is asexual reproduction without true spore formation, a cell of the sporangial tissue taking the place of the megaspore.

2. **Apogamy**: the egg cell is dispensed with and the embryo is formed (parthenogenetically) from another, vegetative, cell of the gametophyte. Among the flowering plants this is an uncommon occurrence. In *Alchemilla sericata* the embryo sac is derived from a megaspore with diploid nuclei, and sometimes an embryo is developed in it from one of the synergids (the two cells which lie beside the ovum) as well as from the egg cell.

3. **Adventitious Embryos.**—A third case is that in which embryos are formed directly from the tissues of the nucellus without the intervention of the gametophyte generation at all. The development of several such embryos leads to the occurrence of polyembryony in the seed. The best known example is the orange; another common plant with polyembryony is the garden *Funkia ovata*. The embryos in these cases do not develop unless fertilisation of the normal egg cell has taken place. In Cœlebogyne (Euphorbiaceæ) adventitious embryos are formed and seeds ripen without fertilisation of the egg cell. The production of

these adventitious embryos from the tissues of the sporophyte must be regarded rather as a case of vegetative multiplication than as a modification of the ordinary reproductive process, for neither spore nor egg cell is concerned. For details the monographs of Winkler (1920) and Ernst (1918) should be consulted.

Origin of Parthenogenesis.—In parthenogenesis, apogamy and nucellar budding, we might be tempted to see a new evolutionary departure. That these methods have arisen recently in the angiosperms is indicated by the fact that neighbouring species of the same genus, or even other individuals of the same species, show the normal mode, but as they are found in algæ and ferns, it is clear that the possibility of such modes of reproduction has existed throughout the history of the plant kingdom. There is, however, good reason to believe that these processes are abnormal.

It is well known that in crosses between species which can produce healthy hybrid offspring, these are often more or less completely sterile. This sterility is due, in the higher plants, to failure to form normal spores; in many cases irregularities occur in the reduction division. The amount of sterility varies; it may be complete, or some good seed may be produced. So pronounced is the tendency to produce imperfect spores that the presence of a proportion of bad pollen, shrivelled grains, has been used as a means of identifying specific hybrids in nature. Ernst (1918) claims that it is just in such natural specific hybrids that apogamy and parthenogenesis are found. The abnormal relations which induce parthenogenesis are due to the irregularities in cell and nuclear division which occur in hybrid races. Ernst supports his theory with a formidable mass of evidence and it seems to fit the case in the plant kingdom remarkably well. Winkler (1920) has criticised it chiefly on the ground that it does not explain similar phenomena among animals. Ernst's theory receives support, and a physical explanation, through recent work of Haberlandt (1922). He refers the stimulus which causes renewed embryonic growth near wounds—callus formation, cork formation, and the production of roots and other new organs—to the action of special substances produced in the wounded cells, which he calls "wound hormones." He supposes that the stimulus to parthenogenetic development lies in wound hormones produced as the result of the abnormal divisions and disintegrations to which we have referred. He has in fact been able to initiate the development of the egg cell, and to induce the formation of nucellar embryos by mechanical injury, squeezing and pricking, of the ovules of Œnothera. He extends his theory to include the stimulus to development by normal fertilisation, the wounding being here the result of the entrance of the sperm. Ernst's theory, then, which has evidently much to recommend it, supposes that those abnormal modes of reproduction in which the sexual fusion is omitted occur not in normal species, but in interspecific hybrids. There are three possibilities in regard to the fate of these hybrids. The abnormality may become permanent, a new mode of reproduction. It may result in the ultimate disappearance of the hybrid. It may disappear, being replaced by normal reproduction, and a new and normal species may thus arise.

§ 5. Sex Distribution in Flowering Plants

The "essential" organs of the flower are the stamens (collectively the andrœcium) and the carpels (collectively the pistil or gynœcium); both are, of course, sporophytic. Each stamen has four embedded sporangia which give rise to numerous microspores, the pollen grains. The carpel contains one or more stalked and integumented sporangia, the ovules, in each of which a single functional megaspore, the embryo sac, is produced. It is customary to refer to the stamens as male, and to the carpels as female organs. This is done as a matter of convenience and tradition. It is not really legitimate, for these are not sexual reproductive organs. They are organs of the sporophyte, and produce spores within which the sexual gametophytes are formed.

But as the male gametophyte is always derived from the spore produced by the stamen it is not unnatural to extend the term male to the stamen ; and so for the carpel. Strictly speaking the sexes in the flowering plants are always separated and occur on two different individuals. When we talk, however, of sex distribution in the flowering plants we mean the distribution of the mega- and micro-sporophylls; we use the word sex in the looser sense which carries over the sexual character to the spores, to the organs which produce them, and finally to the sporophytes themselves.

The distribution of sex in this sense in the flowering plants is very complex and varied, and it is not possible here to do more than outline the various conditions which occur. We must in the first place distinguish between flowers which are *hermaphrodite* with both stamens and carpels, and those which are *unisexual* and have only the one or the other. In the angiosperms the hermaphrodite flower is the commonest type, and it probably represents the primary condition in this group. Unisexual flowers have arisen by the suppression of one or the other essential organ, as is indicated by the numerous cases in which the suppression is normally or occasionally incomplete. Suppression may go so far that both stamens and carpels disappear and a flower results which has attractive functions only, as in the marginal flowers of the inflorescences of many Umbelliferæ and Compositæ. In the gymnosperms unisexual flowers (cones) are the rule, and this condition is primary. Only in Welwitschia is a rudimentary ovule present in the male flower, though hermaphrodite cones are known as abnormalities in some conifers.

A species or individual with hermaphrodite flowers is termed *monoclinous*, in contrast with a species having male and female flowers, which is *diclinous*. In the simplest cases all the flowers are either hermaphrodite or unisexual. Diclinous species with male and female flowers on the same individual are *monœcious*, *e.g.* the hazel ; if the male and female flowers occur on different individuals the species is *diœcious*, *e.g.* the willows. In diœcious species the separation of the sexes is frequently incomplete, and a few male flowers may be found on the female plant, and *vice versâ*, *e.g.* in *Mercurialis annua.*

More complex are the relations in those species with hermaphrodite as well as unisexual flowers. We must distinguish :—A. Species all individuals of which have the same kinds of flower : (*a*) all individuals have both hermaphrodite and female flowers, are *gynomonœcious*, *e.g.* Parietaria; (*b*) all individuals have hermaphrodite and male flowers, are *andromonœcious*, *e.g.* Æsculus ; (*c*) all individuals have hermaphrodite, male, and female flowers, are *trimonœcious*, *e.g. Poterium Sanguisorba.* B. Species in which different individuals have different types of flowers : (*a*) *gynodiœcious* species, of which some individuals have hermaphrodite and others female flowers, *e.g.* many Labiatæ ; (*b*) *androdiœcious* species, of which some individuals have hermaphrodite and others male flowers, *e.g. Caltha palustris ;* (*c*) *triœcious* species, of which some individuals have hermaphrodite, others male, and yet others female flowers, *e.g. Empetrum nigrum. Polygamous* is a general term applied to species with three types of flowers.

This does not exhaust the possible or actual combinations. We may give as an example of further types, *Silene inflata* with male, andromonœcious, female, gynomonœcious, and individuals with hermaphrodite flowers. Moreover these types must be regarded to a certain extent as ideal, for distribution is rarely quite clean cut. Thus irregularities occur even in monoclinous and simply monœcious and diœcious species, where the constancy is greatest. Thus Stout (1919) has shown that in *Plantago lanceolata*, the flowers of which are normally hermaphrodite, individuals may be found showing every gradation between this condition and complete suppression of pollen production. Davey and Gibson (1917) have studied *Myrica Gale*, a species usually described as diœcious, and they find individuals, (*a*) with male and female catkins, (*b*) with catkins bearing male and female flowers, (*c*) with hermaphrodite flowers. The willows are very regularly diœcious, but Schaffner

(1919) in a study of *Salix amygdaloides* found 9 per cent. of the individuals to bear catkins which are male below and female above, with a zone of hermaphrodite flowers in between; in *Morus albus*, he found 20 per cent. of intermediate individuals. We have already mentioned the case of *Mercurialis annua* in which male plants occasionally bear female flowers and *vice versâ*. It is probable that all diclinous species may at one time or another exhibit such irregularities. In species with more than one type of flower on the individual the intergrades are most frequent. A general review of this subject will be found in Yampolsky (1920).

§ 6. Determination of Sex

The ordinary type of angiosperm with hermaphrodite flowers must possess in the inheritance of each individual the potentialities of both sexes, since it does in fact bear both male and female organs. Here the separation of the sexes, to whatever it may be due, is somatic, and occurs when the stamens and carpels come to be laid down at different points of the meristematic zone of the same flower rudiment. Even there derangement may set in; teratological flowers are known in which stamens bear ovules, *e.g.* in Sempervivum, or carpels produce pollen, *e.g.* in Begonia. In simply monœcious plants the same thing is true. Each individual can and does produce both types of sexual organ; the separation may take place when the flowers or inflorescences are developing, or it may occur earlier in the individual development in plants where one type of flower constantly occupies a definite place on the axis. In Sagittaria the male flowers are always borne higher than the female; in the maize the male inflorescence is terminal, the female is axillary.

In diœcious species we might be tempted to believe that some individuals inherit only maleness and others only femaleness; or, in the more complex cases, that some inherit one sex and others both. A little consideration shows that so simple an explanation is not sufficient. It does not explain the occurrence of the intergrading individuals.

An illuminating example is that of *Lychnis dioica*, the red campion. It is strictly diœcious and intermediates are rare, having been observed by one investigator only (G. H. Shull, 1910). It is subject, like many other Caryophyllaceous plants, to the attack of a smut fungus, *Ustilago violacea*, which invades the anthers and fills them with violet spores. If the smut infects a female plant, which bears no stamens, its presence acts as a stimulus to the production of these organs and the spores are formed in the anthers as usual. Here is an individual, normally pure female, which can yet produce male organs under suitable conditions. It must in reality be potentially bi-sexual. It is difficult to avoid the conclusion that if this is so in an extreme case of unisexuality like Lychnis, then it must be true of all individuals of all flowering angiosperms. Maleness and femaleness *as such* cannot be inherited alternatively; they must exist as potentialities side by side in the same individual and flower, their expression determined by other factors.

These determining factors may in some cases be external to the plant. In monœcious species it is frequently possible, by change in the external conditions, to modify profoundly the types of flower borne by the individual. Thus Riede (1922) found that conditions favouring assimilation promote the development of female flowers in the monœcious maize, while increased absorption of mineral salts favours the male flowers. In *Arisæma triphyllum*, a diœcious aroid with many intermediate individuals, Schaffner (1922) found that plants cut back and kept dry tend to produce only male flowers in the following year, while plants well watered and dunged produce only female. Stout (1923) has shown that in *Cleome spinosa*, which goes on flowering for two to three months, production of male flowers alternates regularly with production of female and hermaphrodite flowers. This looks like a nutritive effect comparable to the two other cases cited.

It seems to be quite easy to modify the sex distribution in species the individuals of which produce two or more kinds of flowers. It is scarcely possible to change the sex of an individual with only one type of flower. A striking example of the difficulty of this is *Satureia hortensis*, which has gynomonœcious and female individuals. The individuals with hermaphrodite and female flowers are easily influenced. Under bad conditions they produce only female flowers; with specially favourable nutrition the proportion of hermaphrodite flowers is much increased. But the individuals with only female flowers can by no means be made to form hermaphrodites. Indeed Schaffner's experiments with Arisæma are so far the only indication we have that it may be possible to influence such individuals, and this plant normally has many intermediates. In strictly diœcious plants no means are as yet known by which the sex distribution may be controlled. It may be altered by *Ustilago violacea* in the case of Lychnis, and a slight degree of intergrading may exist. The potentialities of both sexes must be present in each individual, but the natural determination is precise and firmly fixed. It requires a profound and intimate influence, which we cannot yet imitate, to change it.

The question then arises whether the determination of the potentiality which is to appear is caused by some inherited factor. There is good evidence that, in some cases at least, this is so. Correns (1907) first obtained evidence from the behaviour of the diœcious *Bryonia dioica* when crossed with *B. alba*, which has hermaphrodite flowers. Whichever plant is used as pollen parent, the offspring are all unisexual, which shows that, in the first place, *Bryonia dioica* has some factor which prevents the formation of a hermaphrodite flower. If pollen of *B. alba* is used on *B. dioica* the offspring are all females. If *B. dioica* is used as the pollen parent, half the offspring are males and half females. This behaviour is explained on the hypothesis that the male *B. dioica* has an inherited factor which suppresses femaleness; but, as it is effective in half the offspring only, it is necessary to assume that the male plant is heterozygous for the factor in question. A diœcious plant of this nature really consists of two races living side by side, differing in the possession by one of them of a factor responsible for the suppression of femaleness and therefore for the appearance of maleness. When the heterozygote male breeds with the homozygote female (which is, of course, the only possible mode of reproduction within the species) equal numbers of the same two classes are always produced. The case may be illustrated in the usual Mendelian notation :—

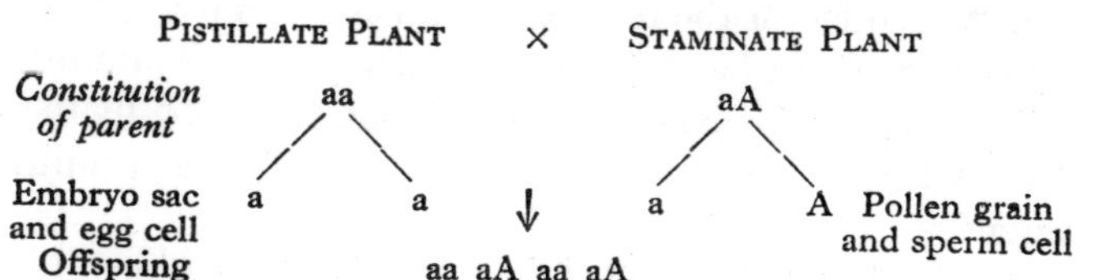

If Z be used to indicate the factor repressing hermaphrodite flowers, then the two crosses with *B. alba* may be represented :—

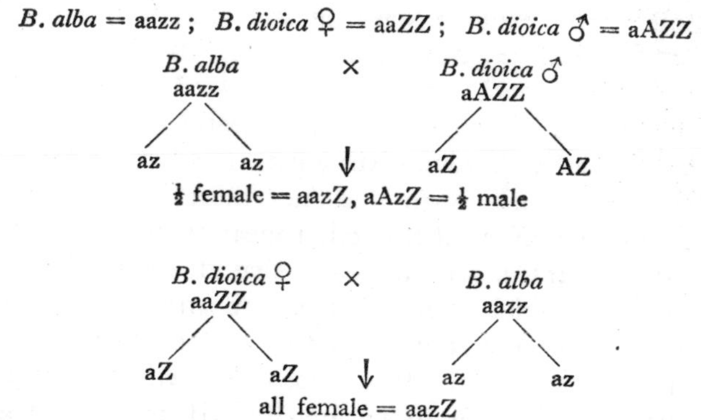

The hybrids are unfortunately completely sterile and further proof of the correctness of the theory by breeding the second generation is not possible. Correns (1907)

obtained the same result for the cross between the diœcious *Lychnis alba* and the hermaphrodite *Silene viscosa*. Shull (1910, 1911), using hermaphrodite individuals of *Lychnis dioica*, of which he found 6 among 8000 individuals, obtained similar but rather more complex results.

In these three cases, and perhaps in the majority of diœcious plants, the appearance of the male sex is determined by the presence of a single dominant Mendelian factor, which suppresses the female potentialities. This gives a reasonable explanation of how the potentialities of both sexes may be present though only one appears: in the absence of the special factor the female potentiality is most powerful. As the determining factor is, of course, subject to the influence of external conditions, its action may be modified by these, as when the pistillate flower of the red campion produces stamens.

Diœcious plants do not, however, all behave alike in this respect. Strasburger (1909, 1910) obtained results with *Mercurialis annua*, confirmed by Bitter (1909) and Yampolsky (1919), which do not admit of this explanation. The female plants of this species occasionally bear male flowers, and the male plants female. If the female flower is fertilised by pollen from male flowers on the same plant, the offspring are all, or nearly all, female. If the occasional females on a male plant are fertilised by pollen of that plant, the offspring are all, or nearly all, male. The normal cross fertilisation gives equal numbers of male and female. The male plant does not therefore behave as a heterozygote; all its germ cells appear to bear male determinants. Strasburger supposed that these were of two "strengths." Half the male plants bear a "weak" male determinant = M I, the other half a "strong" = M III. All the female plants have a female determinant expressed as F II. When an M I sperm cell meets an F II ovule the female tendency predominates and a female results; when a strong male sperm, M III, meets an F II egg cell the male tendency predominates and a male results. Something of this sort may occur in cases of complex distribution.

Correns (1904, 1905) found that the female *Satureia hortensis* always gives rise to females, and the plant with hermaphrodite and female flowers produces almost exclusively this type, and this is also the case in *Silene inflata*. The theory of varying potency of the male determinant carried by the pollen grains makes it easier to understand the possibility of influencing the sex distribution by external conditions, and also the occurrence of intergrades. The recent work of Goldschmidt (1923) on sex determination in the gipsy moth makes use of similar ideas of varying potencies.

On either theory the separation of the factors which control maleness takes place in the reduction division preceding pollen formation, so that, of each four pollen grains produced from a mother cell, two should be of one sort and two of the other. Endeavours have been made to show that this is the case. The most interesting results so far obtained are those of Correns (1917, 1922) on the modification of the sex ratios in *Lychnis dioica* and in *Rumex Acetosa*, obtained by varying the amount of pollen used. He found that in the campion the proportion of female plants produced could be considerably raised if a large amount of pollen were applied to the stigmas. With a very scanty supply of pollen he obtained 737 females to 555 males; with abundant pollen the figures were 895 to 381. The same result was obtained with the sorrel. In one series of experiments he obtained, with abundant pollen only 12·6 per cent. males, and with little pollen 29 per cent. The explanation is that the pollen tubes of the grains carrying the male determinant grow more slowly than the others; thus with abundant pollen the chances are that a disproportionate number of ovules will be fertilised by "female" pollen grains. With a small amount of pollen a greater number of the available grains will be required to fertilise all the ovules, and a larger proportion of "male" grains will become effective.

We have not as yet any general scheme which will cover the mode of sex determination in all plants. Indeed, it seems from the available evidence that, although much is yet obscure, more than one method of determination exists.

The mode of determination of the different types of individual in the more complex cases of distribution have not been elucidated.

§ 7. Changes in Sex Distribution in the Course of Evolution

The changes in the mode of sex distribution during the evolution of the land flora are of great interest. In the ferns proper the gametophyte is usually hermaphrodite, though this condition is easily modified. The horsetail gametophyte is diœcious. With the advent of heterospory diœcism in the gametophyte generation became fixed, and this has remained unaltered through all subsequent changes. Heterospory was probably at first united with bisexuality of the individual, as in the heterosporous pteridophytes which still exist. These are not primitive forms, and it is possible that in other lines heterospory may have been early associated with a separation of the sporophyte individuals into two classes, male and female. In the earliest seed plants, the pteridosperms, there may have been only one type of sporophyte. From these may have arisen, on the one hand the diœcious cycads, and on the the other the mesozoic Bennettites with its hermaphrodite flower. At the end of some such evolutionary line come the angiosperms. The gymnosperms are all diclinous, but about 75 per cent. of existing angiosperms have hermaphrodite flowers. The hermaphrodite flower is characteristic of the primitive angiosperms and may be taken as primitive for the group. In the course of further evolution, however, it has undergone, more than once, modification towards unisexuality, which is pronounced, if not rigidly fixed, in a great many flowering plants. It is very likely that the great range of intermediate conditions represent stages in the trend towards complete and simple diœcism.

Beginning, therefore, with a union of the two sexes in both gametophyte and sporophyte, the initial evolutionary change was the separation of male and female gametophytes.

Then separation of sexual characters in the sporophyte took place, at first limited to the spore, spreading to the sporangia, sporophylls, and finally, on the lines which lead to the various gymnosperm families, to the cones or even to the whole plant, as in the cycad type of pronounced diœcism. A critical change was the evolution of the hermaphrodite flower, from which modern diclinous species of angiosperms have been derived by suppression. We must, however, mention the theory of von Wettstein, that the primitive angiosperms are to be looked for in the Cupuliferæ, a group with marked dicliny, a theory which has not found very general acceptance.

Such a history must leave us in doubt with regard to the fundamental importance of cross fertilisation or out-breeding, which is most easily and decisively secured by diœcism. When sex distribution has undergone so many changes during the course of evolution it appears unlikely that any particular type has great superiority. Either cross fertilisation is not of great importance, or it is secured with sufficient frequency, no matter how the sexes are combined or separated. The importance of sex is, however, emphasised by its maintenance through this long and varied history. Parthenogenesis and apogamy crop up here and there, but have never been permanently established.

§ 8. Secondary Sex Characters

In connection with this discussion of sexual reproduction reference may be made to a phenomenon of great interest in the animal kingdom, though relatively quite unimportant among plants, the occurrence of secondary sexual characters. Every one knows the secondary differences between a cock and a hen, but it is not easy to get striking differences between male and female plants, apart from the difference between stamens and ovaries. It is easy to pick out a male or a female willow at a distance, but what catches the eye is the brilliant yellow of the stamens due to the presence of pollen.

A true secondary distinction has already been touched

on, in the difference in growth rate between the " male " and " female " pollen tubes. In one species which shows this, *Rumex Acetosa*, the male plants are markedly smaller and less hardy than the females. The same is true of the hemp. There is sometimes a marked difference between male and female inflorescences, but differences between pistillate and staminate flowers are scarce. The facts have been collected by Goebel (1910 and *Organographie*), and one or two examples may be given. In the maize the male inflorescence is terminal and freely branched, the female is lateral and simple. In *Mercurialis perennis* the female inflorescences are short and have a few short-stalked flowers, the male are longer with many sessile flowers. The orchid Catasetum is a unique case ; the form of the perianth in male, female, and hermaphrodite flowers is so different that they were formerly referred to distinct genera ; the occasional occurrence of two forms on one plant enabled them to be identified. In our native *Sagittaria sagittifolia* the female flowers are borne low down on the inflorescence, their peduncles are twice as thick as those of the staminate flowers, while the perianth of the latter is much larger. In most cases it is impossible to assign a functional significance to such differences. Goebel gives the case of *Eriocaulon nautiliforme*, in which the pistillate flower has the posterior perianth segment inflated to a bladder, which later acts as a float for the fruit.

§ 9. Pollination—The Stamens and the Pollen

Pollination, the transference of pollen from stamen to stigma, is the necessary preliminary to fertilisation and seed production. The structure of the flower, especially in the angiosperms, is intimately related to the accomplishment of pollen transference. Except where reduction has taken place the flower of the angiosperm possesses, in addition to the " essential " organs, an envelope, the perianth, or two envelopes, an outer calyx, typically protecting the flower in the bud stage, and an inner corolla, typically bright and showy, a mark for the visiting insect. In the great majority of cases pollination is carried out by insect agency or by the wind. In wind-pollinated flowers calyx and corolla are usually reduced or absent. We may first consider some properties of the stamens and pollen.

The Stamens.—The structure of the stamen or microsporophyll is in ground plan very constant, though profound modifications sometimes occur. Typically it consists of a stalk, the filament, bearing a head, the anther. This has four internal sporangia, elongated sacs which when ripe are full of microspores, the pollen grains. At maturity the anther opens, or dehisces, by two longitudinal slits, due to the tearing of the external wall tissue between each two sporangia away from the dividing wall; the sporangia thus open in pairs. The number of sporangia may be increased by partitioning or decreased by abortion. The form of the anther is subject to modification. The stamens may be united more or less completely, or individual stamens may be divided into numerous parts. Union of the stamens with other floral organs, *e.g.* the style, may take place. Staminodes are sterile stamens which may be rudimentary, or may perform other functions.

In the gymnosperms the structure of the microsporophyll is much less uniform. In the Cycads very numerous microsporangia are borne on the lower side of large microsporophylls, which are, in Zamia, peltate. In Taxus about half a dozen sporangia are borne on the lower surface of a peltate scale. In Pinus two sporangia are borne on the lower surface of a scale sporophyll. In Ginkgo the stalk-like sporophyll bears two sporangia at its tip. In Welwitschia three, in Gnetum and Ephedra several sporangia are borne on a stalk. In Ginkgo and the Gnetales the microsporophyll is stamen-like, but in the other gymnosperms, especially in the Cycads, the departure from the angiosperm type is very wide. The stamen, though strictly homologous with the gymnospermous microsporophyll, is so uniform in its general plan that it may be regarded as a new type of organ, the stabilised end product of a long series of evolutionary changes. It has one other important and constant difference from the microsporophyll of the gymnosperm in the mechanism of its dehiscence.

At maturity the pollen grains lie, usually completely separated, packed together in the sporangial cavities. As we have said, the stamen usually opens by longitudinal slits. In the gymnosperms the opening is also by slits though their direction is more varied. The actual opening is partly a passive process, due to the weakening and dissolution of the cell walls along the line of dehiscence; partly it is due to the mode of drying out of a special layer of cells of the anther wall, the walls of which are peculiarly thickened. This opening mechanism is very ancient; it is already highly specialised in the fern sporangium, with its annulus. In the gymnosperms, the specialised cell layer belongs to the epiderm of the sporangium; it is an *exothecium*. The sub-epidermal layer in the angiosperms is effective, and is termed an *endothecium*.

In a good many angiosperms the active layer appears to be an exothecium, but in such cases we are dealing with a secondary reduction phenomenon and not with a primary condition homologous with the gymnospermous exothecium. In the Ericaceæ a series of forms may be traced with typical endothecium at the one extreme, *e.g.* in Clethra, and an apparent exothecium at the other, *e.g.* in Erica. The details of the reduction are treated of by Goebel and by Staedtler (1923). In some submerged aquatics the endothecium fails to develop and special modes of opening are found. In Zannichellia a swelling of the inner cells of the wall bursts the epiderm open.

The endothecium consists of a layer of cells with their long axes at right angles to the surface of the anther wall. The inner walls of these cells are strongly thickened, and thickened bands run up the side walls tapering off and ceasing at the outer wall (Fig. 48). Steinbrinck (1906) has elucidated the way in which this peculiar type of thickening aids dehiscence. As the cells dry out the cohesion of the diminishing water tends to produce collapse. At the inner side this is restricted by the thickened wall, but outwards the radial walls are drawn together in folds between the thickened strips. This process occasions a relatively much greater shrinkage of the outer surface of the anther wall than of the inner surface; the wall contracts and also, after rupture has taken place, bends outwards and the opening gapes; the wall may ultimately be rolled quite back. This explanation has been combated, *e.g.* by Schipps (1914), but the evidence in its favour seems to be very strong. It will be noted that the mechanism is in principle the same as that which causes the opening of the fern sporangium by the annulus. The well-known springing back of the fern sporangium wall, which is responsible for jerking out the spores, does not occur in the anthers, because, when air ultimately enters the drying cells and the cohesion tension is thereby released, the cell does not resume its original shape and size. This is due to the pleats of the delicate wall sticking together.

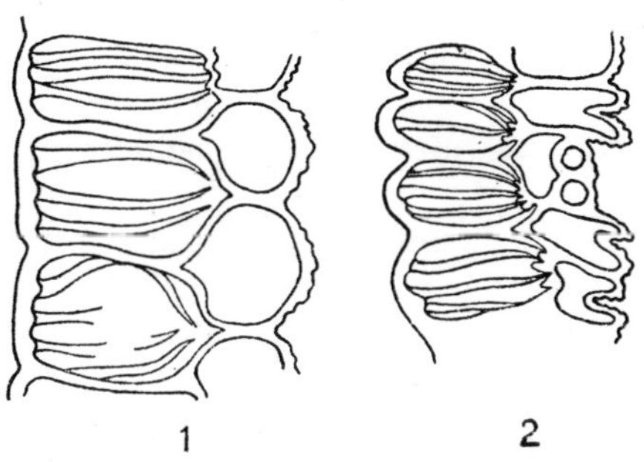

FIG. 48.—Endothecium of *Lilium candidum*: 1, from unopened anther; 2, from opened anther. (After Steinbrinck.)

In the Cycads the mode of thickening of the exothecial cells is strongly reminiscent of the fern annulus. In the other gymnosperms the thickening is less marked and regular. The mechanism of opening is the same throughout. It has been retained, an extremely conservative feature, through the whole terrestrial flora, though in the angiosperms it is carried out by a tissue which is not homologous with that in the lower forms.

One or two departures from the typical opening by slits may be mentioned. We have already alluded to the absence of an endothecium in submerged plants, where indeed it could not function. In the Ericaceæ, where the endothecium is much reduced, dehiscence occurs by pores instead of by slits. These pores, two in number, open at the apex of the anther by the dissolution of a special tissue. In some cases, *e.g.* Rhododendron, the pollen is squeezed out by the contraction of the anther ; in others, *e.g.* Calluna, it lies in the anther and is shaken out by visiting insects. In the Berberidaceæ, Lauraceæ, and some other families, dehiscence is again by pores which open by the shrinkage of valve-like lids, to which, in these cases, the endothecium is confined. The sticky pollen is carried out attached to the lid.

Pollen is sometimes shaken out of the anthers, especially in wind-pollinated plants. Goebel states that only in Ricinus is this due to a mechanism of the anther. In Ricinus the thickening of the walls of the exothecial cells is different near the future opening and on the opposed side of the anther. In the former the walls are spirally and annularly thickened, and tend to resist the opening movement of the latter ; when dehiscence actually occurs the resistance is suddenly released and the wall flies violently back, throwing out the pollen in a cloud.

In all other cases the sudden movement which ejects the pollen is not connected with the opening of the anther, but is due to a tension in the filament. The most famous case is that of *Parietaria officinalis*, the wall pellitory, in which the elastic filaments are held arched in by the anthers sticking in the boat-shaped perianth segments. At a certain stage of the opening of the male flower, often as the result of a slight shock to the plant, the anthers come free with a jerk ; they fly back and the pollen explodes into a little puff.

The Pollen.—The pollen grains vary enormously in shape, size, and other characters. In size they range from 0·0025 mm. in diameter in Myosotis to 0·25 mm. in Cucurbita. They are usually round or oval, but may have sharp angles. Like fern spores, they possess a delicate inner wall, the intine, and a stout outer wall, the extine, in which thin spots are often found, through which the pollen tube, an extension of the intine, makes its way. Of peculiar interest are the grains of Zostera and of some related marine monocotyledons. They are long and thread-like and coil round the stigmas of the female flowers. The elongated form is probably favourable to flotation, but is not found in all genera. In plants with water-borne pollen the extine is usually absent.

The extine may be quite smooth or may be sculptured in various ways, bearing warts, ridges, spines, or delicate bristles. Such roughened grains are found only in insect-pollinated plants. The rough surface makes the grains stick together or to other bodies, such as the insect's legs. In some few cases, notably in certain Onagraceæ, the grains have several long delicate threads of viscid substance which bind them together. In most of the Orchidaceæ, and in many Asclepiadaceæ and Mimoseæ, pollinia are formed. In the first two families these may consist of the whole mass of pollen of two neighbouring sporangia glued together; in the Mimoseæ and in many orchids the number of pollinia is greater. The pollinia are transported by insects, to which they become attached by a variety of means. The advantage of the pollinium is specially obvious in the orchids where each ovary contains innumerable ovules. The deposition of a single pair of pollinia on the stigma ensures the fertilisation of large numbers of ovules.

Germination of the pollen grain takes place by the protrusion of the intine through the extine and its growth into a long slender pollen tube. In some cases, as in Malva, several pollen tubes are formed, though only one is functional. The conditions under which germination occurs are very varied. In the angiosperms the pollen may be capable of resting in the dry state for considerable periods without losing its power of germination. Kerner relates that the Arabs, who artificially pollinate the date palm, "put aside some of the pollen from year to year, so that in the possible event of the male flowers not developing, they may ensure a crop of dates." He gives the extent of viability for a number of other plants, ranging from 3 days in *Hibiscus*

Trionum to a month in *Ajuga reptans* and 2 months *Pæonia tenuifolia*. Molisch (1893) tested 26 different species and found the life of the pollen grain to range from 12 days in *Trifolium hybridum* to 72 days in *Narcissus poeticus*. Hayes and Garber (1921) state that in the maize the pollen does not survive two days after leaving the anther. Anthony and Harlan (1920) find that in the barley fertilisation can be secured with certainty only if the pollen is transferred straight from the stamen to the stigma. There are evidently very great differences in this respect, and more exact and extended information would be of interest, especially in regard to the relative conditions in closely related wild and cultivated species.

The pollen of many plants will germinate with ease in sugar solutions, on gelatine, or even in water. Rayner (1916) finds that the pollen of *Echeveria retusa* germinates in an hour in 15 per cent. sugar solution. Adams (1916), in a study of the germination conditions of a number of fruits, found that the pollen of the apple would germinate in sugar solutions from 2·5 to 10 per cent., even after the pollen had been stored for three months. That of the strawberry germinated in 8 per cent. sugar, and that of the olive in 16 per cent. Tischler (1917) found pollen of Cassia to germinate in a 70 per cent. sugar solution. Anthony and Harlan (1920) found that, while on the one hand the pollen of the barley is extremely sensitive to desiccation, on the other it bursts at once if placed in pure water, and this seems to be the case, too, in the wheat and rye. Jost (1905) found that the regulation of the water supply was most important for the germination of grass pollen. Martin (1913) considers that in *Trifolium pratense* the regulation of the water supply is the most important factor in securing conditions favourable to germination, and that it is this relation which determines the suitable concentration of sugar. Molisch (1893) obtained germination, in sugar solution with the pollen of 12 species. Many tolerated a very wide range of concentration ; *Robinia pseudacacia* germinated in solutions from 10 to 40 per cent., *Galanthus nivalis* from 1 to 7 per cent., *Allium ursinum* from 3 to 5 per cent. Only the pollen of the primrose germinated in water as well as in sugar solution. Lidfors (1896), however, obtained germination in water with plants belonging to 61 families, for example, with Lysimachia, Lobelia, Urtica, Parietaria. Germination occurred only in distilled water, not in water containing salts.

There are, however, many cases known in which germination does not occur under such simple conditions, and where the presence of substances excreted by the stigma seems to be essential. Burck (1900) found that the pollen of Mussaenda germinated only when a piece of the stigma was placed in the water; the influence of the stigma could be replaced only by lævulose which was effective in traces. Here the regulation of water supply could not explain the effect, nor could the necessity of food substances. The case is as yet unanalysed. The chief interest lies in the possibility of some regulation of the germination of the pollen grain. Molisch was unable to obtain germination under any artificial conditions with the pollen of certain Compositæ, *e.g. Taraxacum officinale*, *Chrysanthemum Leucanthemum*, Urticaceæ, *e.g. Urtica dioica*, *Cannabis sativa*, Malvaceæ, *e.g. Althæa officinalis*, *Malva sylvestris*. Pollen of the Ericaceæ, *e.g.* Rhododendron, Azalea, germinated only in acid solutions, especially in malic acid. The pollen of *Pavetta javanica* can germinate only on the stigmas of that species or of the closely related *P. fulgens*, but not on those of other species of the genus.

It is evident that in many species pollen can germinate under a great variety of conditions, and it is probable that such pollen frequently germinates on strange stigmas. Whether fertilisation follows depends on other factors, in the first place of course on the closeness of the species to each other. In other cases pollen can germinate only in special conditions, which, in nature, are provided by the stigmas of the particular species in question. Even where only a definite range of concentration of sugar is required the chance that germination may occur before the pollen reaches the stigma is avoided. It may be doubted whether the more extreme cases of specialisation bring any further advantage, for it is unlikely that the pollen grain often gets a second chance if it has once been deposited on the wrong stigma.

We may mention here a subject to which we shall return in another connection—those cases of self-sterility where failure is due to the fact that pollen will not germinate on the stigma of the flower which produced it.

Protection of Pollen.—We have seen that barley pollen is readily damaged by water, and the same is true for many plants. Lidfors (1896, 1899) found that in 38 families out of 80 which he investigated, species were found with pollen damaged by water. All degrees were found, from the extreme sensitiveness of the barley, where instant bursting occurs, to pollen which germinates in pure water.

Such sensitive pollen must be affected by rain, and it is natural that certain features in the floral structure and behaviour should have been interpreted as offering protection to the pollen against this danger and against the grosser one of being washed away.

Many pollen grains are covered by a thin film of oil, which makes wetting difficult. Especially in zygomorphic flowers the structure of corolla and calyx is frequently such as to protect the stamens. It is only necessary to think of the arched hood of the sage, or monkshood, of the pursed-up corolla of the toad-flax or of the snapdragon, of the enclosed stamens of the broom or whin, of the sheltering scales which close the corolla tube of the forget-me-not, of the narrow tube entrance of the centaury, to see how frequently and well pollen may thus be protected from wet. In hanging flowers the pollen is again shielded, as in the wood hyacinth and harebell. It is not necessary to regard these structures as primarily related to pollen protection, probably they are not; yet their protective action may be none the less effective.

Floral Movements.—It is rather more difficult to assess the usefulness in this respect of flower movements. In *Anemone sylvatica*, *A. japonica*, and other species, the peduncles droop in the evening so that the flower, erect by the day, is pendent at night when the danger of wetting by dew is greatest. More general and striking are the movements carried out in the flowers of many plants (by the whole flower head of the Compositæ) which result in the corolla closing at night and opening, for a shorter or longer period, through the day. The tulip, the crocus, the daisy, the dandelion, the goat's-beard, the flax, the ice plant, are common and familiar examples. These movements are classed along with the "sleep" or nyctinastic movements of leaves. Unlike the majority of leaf movements they are due to differential growth rates on the two sides of the petals. Depending on the life of the flower, they occur only once or several times. The factor which usually causes opening is rise of temperature. This is certainly effective in the tulip, crocus, flax, scarlet pimpernel, and many others. In the daisy and marigold (*Calendula officinalis*), change from dark to light is the causal factor. Closure may take place as the result of a fall in temperature, as in the crocus and tulip; it is probable that it is more generally due to an automatic reverse at a definite interval after opening. Such a reverse is seen even in the tulip and crocus, though in the former it does not lead to complete closure. This automatic reverse leads to the early closure of such flowers as the goat's-beard, which shuts when the sun is at its zenith.

In the majority of cases, just as with leaf movements, a rhythm underlies the movement, which is induced, accentuated, or modified by the external factor. This rhythm may be due to a summing up of the effects of successive changes in external conditions, or it may be inherent; in different flowers its nature may differ. At present it is not possible to decide definitely which explanation is correct. The scarlet pimpernel opens for the second time, only if the temperature is raised about 24 hours after the first opening; a suitable rise of temperature before this is quite without effect. Normally the opening occurs about 9 to 10 a.m. By suitable treatment the first opening may, however, be made to occur in the evening ; in such a flower the second opening can take place only in the following evening. Here we have a rhythm, not indeed causing opening (or only to a very slight extent), but regulating the time at which opening may take place under the influence of the suitable change of external conditions. In the marigold, fully investigated by Stoppel (1910) and by Stoppel and Kniep (1911), opening is caused by illumination. The flowers also open in constant darkness, and in this condition carry out a periodic opening and closing every 24 hours. Illuminated by night and darkened by day the opening is shifted 12 hours from the normal time. Illuminated and darkened in 8-hour periods, the rhythm is speeded up to suit the new period of external change. Illuminated and darkened in 4-hour periods, the flowers open and close fully only *once in the day*, but slighter movements show the influence of the external change. Illuminated and darkened in 2-hour periods, only the daily movement takes place, the external change has no visible effect. Here we have an internal rhythm which may be influenced, but only to a limited extent, by external change. In those flowers which open at an approximately definite hour of the day we may suppose that the regularity is partly the effect of internal periodicity and partly due to the regular onset of the external conditions, favourable to opening, at a definite time. A factor which may have an important effect in regulating such movements is the diurnal alteration in the conducting power of the atmosphere for electricity (see p. 204).

The flowers mentioned and many others remain closed through the night, and are thus protected from the deposition of dew on the essential organs. In cool, dull weather opening does not take place through the day, or is only partial, so that protection is again secured. But passing showers on a sunny day, accompanied as they are by lowered temperature and illumination, do not, as a rule, cause closure, and against these the flower is unprotected. The tulip, the crocus, and the pheasant's eye are exceptions which close rapidly on a fall of temperature. It has been suggested, too, that the nocturnal closure protects against over-cooling, but of such an effect we have no good evidence. Some flowers—the tobacco, the evening primrose, the night-scented stock—are closed by day. They are pollinated by night-flying moths and are open, and most strongly fragrant, when these are abroad. The significance of the day closure we do not know, nor has its causation been cleared up.

It is clear that some types of floral architecture, and those movements which result in night closure, must protect the pollen, but the relation of structure and movement to the necessities of a particular plant does not seem to be a very close one. It is significant that no case is known in which the movement is controlled by change in atmospheric humidity. Hansgirg (1904), indeed, refers a large number of floral movements to moisture changes, but in no case with sufficient proof. As a number of the cases given by him, *e.g.* the daisy, the scarlet pimpernel, the anemone, are certainly related to temperature or light his entire work must be taken as requiring revision.

Protection of Pollen and Sensitiveness to Damage.—Lidfors (1896) has correlated the sensitiveness of the pollen to damage by rain with the degree of protection existing in the flower. He examined representatives of 80 families; 55 of these included species where unprotected pollen was resistant, while 23 had species with unprotected sensitive pollen; 23 families had species with protected pollen which was damaged by water, while 6 had species with protected and resistant pollen. He notes other features which are seen where unprotected pollen is sensitive; thus in the grasses the pollen germinates very quickly and is produced in great preponderance over the ovules, though this is, of course, related to the mode of pollination. He also gives some interesting contrasts between related species. In Rumex the pollen is resistant and quite unprotected; in the closely related Polygonum the pollen is protected and is sensitive. His conclusion is that on the whole unprotected pollen is resistant, sensitive pollen is protected.

Hansgirg (1904), on the other hand, cites a large number

of cases to prove that there is no general correspondence. His criterion of damage is, however, the power of germinating in water and not the power of germinating in a suitable solution after exposure to water, and this vitiates his results. In the same work he finds no relation between floral movement and sensitiveness.

There is evidently a wide field for research here, and it is likely that results of more value will be obtained by intensive study of a few carefully chosen types than by cursory examination of such large numbers as are dealt with by these two investigators.

§ 10. Pollination and Fertilisation

The pollen grain of the angiosperms is received on the receptive surface of the stigma which is often roughened by papillæ and so tends to hold the grains. It is often viscid, too, a condition which both holds the grains and promotes germination. This occurs without delay, and the pollen tube grows into the tissue of the stigma, down through the style, and so into the ovarial cavity. The tube may actually pass through the cells, dissolving the cell walls by enzymatic secretions, or it may grow through the spaces of a special loose tissue, or through a well-defined canal. The direction of its growth seems to be determined partly by negative aerotropism, and partly by positive chemotropism. Pollen grains germinating on gelatine grow into the medium away from the oxygen of the air. It has been shown by Molisch (1893), Miyoshi (1894), and Lidfors (1899, 1909), that the tubes grow towards pieces of stigma or ovules. The active substances are usually sugars. In *Narcissus Tazetta* and many other plants, Lidfors showed that a protein was the directing chemical. In the ovarial cavity the tube grows along the placenta and enters the ovule by the micropyle; it pierces the nucellar tissue and the wall of the embryo sac; its own wall is broken down and the conditions for fertilisation are realised. Incidentally we may note that all this cell destruction gives ample occasion for the formation of the wound hormones postulated by Haberlandt and regarded by him as responsible for the stimulus leading to development of the ovum which accompanies fertilisation.

In a number of plants, notably among the Cupuliferæ, the pollen tube, instead of entering the micropyle, passes up the funicle and enters the ovule by the chalaza—*chalazogamic* as opposed to *porogamic* fertilisation. There are great variations in the exact route followed by the pollen tube in these cases. The cause and significance of this procedure are obscure. One other abnormal case may be noted. In cleistogamic flowers, *e.g.* in the sweet violet, where self-pollination occurs without the flower opening, the pollen grains may germinate in the pollen sac and the tubes pierce the walls of anther and ovary.

In the gymnosperms the conditions are quite different. The ovule is naked and the pollen is received directly by the micropyle. The pine may be taken as an example. At the time when the pollen is shed in spring, a year after the microspores are fully formed, a drop of mucilaginous fluid is excreted by the nucellus through the micropyle, and in this the pollen is caught. The fluid dries up, retracts, and the pollen is pulled through the micropyle, where it germinates and begins to penetrate the nucellus; slow growth continues till checked by winter. Only in the following spring do the tubes reach the egg cells, several of which may be fertilised. In gymnosperms, other than the conifers, while the growth of the pollen tube is always slow, fertilisation takes place within a few months of pollination. In the Cycads, in Ginkgo, and in the Gnetales a pollen chamber is formed at the tip of the nucellus, and in Ginkgo it may even project beyond the micropyle. In the Gnetales the pollen is received by a long protruding micropylar tube. In the Cycads and in Ginkgo the sperms are ciliate, the last case of active sperms in the plant kingdom. In the Cycads a branched pollen tube penetrates the nucellus, but functions only as a haustorium. This suggests that the primary function of the pollen tube was nutritive, it was a haustorium, and that the transport of passive male cells to the embryo sac is secondary ; in the course of evolution the organ has taken on a new function. Branched pollen tubes are also found in some angiosperms.

We have already noted one ancient character in the microsporangium in the mechanism of dehiscence. We may here refer to two ancient spore characters retained by the pollen grain in many cases, though lost in the embryo sac. The spore is a reproductive body capable of rest and of dispersal. Both features are shown by the pollen grain. The period of rest is, however, generally short, and has significance only in that it permits pollination to take place for a period after the anthers have dehisced. Dispersal no longer means a scattering of the new generation ; it is specialised and provides for the transference of the pollen to the stigma and the bringing together of male and female cells.

§ 11. Pollination—Agencies

The study of the ways in which pollination takes place may be said to date from the publication in 1793 of Christian Konrad Sprengel's book *Das entdeckte Geheimniss der Natur im Bau und in der Befruchtung der Blumen*, a pioneer work of the first importance, and a model of exact investigation. Since then a colossal literature has arisen round the subject, in which the names of Darwin and Hermann Müller are conspicuous. In Knuth's *Handbook of Floral Biology* we have a standard compendium of knowledge. Pollination is dealt with in all text-books of botany, and many of the more striking cases are familiar, so that we may here confine ourselves to a statement of principles with a few illustrative examples.

Pollination may be carried out by a number of agencies. The most important are insects (*entomophily*), and the wind (*anemophily*). Much less common is pollination by birds (*ornithophily*), or by snails (*malacophily*), or by water (*hydrophily*).

From a different point of view, that of the origin of the pollen, we distinguish between *autogamy*, in which the flower is pollinated with its own pollen, and *allogamy*, where the pollen comes from a different flower. Allogamous plants may be subdivided; by *geitonogamy* is understood pollination from another flower on the same plant; by *xenogamy* from a flower of a separate plant; by *hybridisation* from a flower of a different race or species.

Finally, we may distinguish those flowers, the great majority, where opening precedes pollination—*chasmogamous*—from those where autogamy takes place without opening—*cleistogamous*.

I. *Entomophily*

Origin of Insect Pollination.—This is perhaps the most important type of pollination, and not alone because the majority of species are entomophilous, but because it is in relation to insect visits that the angiosperm flower, the gaily-coloured, scented, nectar-bearing " flower " of the popular sense, has evolved. These three properties, often found together in a single flower, have no conceivable use apart from insects.

The early seed plants were wind-pollinated, a condition that persists in the modern gymnosperms with the exception of Welwitschia, and some Cycads; but it is likely that insect visits, perhaps very irregular, may have occurred even in the pteridosperm stage. The great insects of the carboniferous strata had biting jaws, and were not nectar suckers, but they may have found food by gnawing the fleshy parts of the sporophylls, or by eating the microspores. It is possible that the origin of attraction by brilliant colours was the production of red anthocyans as by-products. The female cones of fir and larch, the larch roses, are brilliant red, although insects do not pollinate them. It is also possible that the yellow colour of the pollen grains was the beginning of this feature. In the secondary rocks, where in Bennettites we have an approach to the angiosperm flower, insects of modern orders were well represented, nearly all, however, with biting jaws. In the tertiary epoch, where the angiosperms predominate, and representatives of many modern families occur, almost all the modern types of insects existed. The Lepidoptera and Hymenoptera arose with the angiosperm flora, and the one is unthinkable without the other.

Floral Classes.—Many details of floral structure can be brought into relation with the structure and habits of particular insects, and with the way in which they visit the flower. Partly according to structure, and partly according to their insect visitors, Müller divided the entomophilous flowers into nine classes :—

Po, Pollen flowers, without nectar, but with a superabundant production of pollen which is gathered for food, especially by bees. Examples are the poppy, the roses, and the rock-roses. In some flowers special "food pollen" which has lost, partially or completely, the power of germination, is formed in special stamens, as in Cassia.

A, Flowers with exposed nectar, such as the Umbelliferæ, many Saxifragaceæ, the maples, the elder, the lime. The flowers are often small, wide open, and with abundant nectar. They are chiefly visited by short-tongued flies, ichneumons, and beetles, though the maples and lime are visited by bees.

AB, Flowers with partly concealed nectar, such as the buttercups, many Cruciferæ, the willows. The nectar is partly concealed in short corolla tubes, by hairs, or by scales, but can be easily reached by dipterous flies and short-tongued bees.

B, Flowers with fully concealed nectar, such as the eyebright, thyme, mint, whortleberries, heaths, forget-me-nots. In these the nectar lies at the bottom of a fairly long tube, and is accessible only to insects with fairly long tongues —bees, butterflies, moths, hover-flies. It will be noted that in this class are included for the first time typical zygomorphic flowers, a form which allows of regulation of the mode of visit.

B′, Inflorescences with concealed nectar, especially characteristic of the Compositæ. The only reason for separating this from the previous class is the flower-like appearance of the infloresence, which acts as a single flower so far as attraction is concerned.

H, Hymenoptera flowers, visited almost exclusively by bees. Here we have a class in which specialised structure permits the visits of only one particular insect or of a limited group. In some the nectar is so deeply placed that only an insect with a very long tongue can reach it, as in the red clover, where the nectar, 9 mm. from the mouth of the flower, is available only to the humble bee. In others only a heavy insect can open the flower, as in the snapdragon or broom. In markedly zygomorphic flowers like the sage, monkshood, violets, snapdragon, broom, and orchids, a convenient landing stage for the bee is combined with deeply concealed nectar at the base of a long corolla tube, or in special spurs. Where such flowers have wide tubes the access of small flies may be prevented by hairs, scales, etc.

F, Lepidoptera flowers have the nectar deeply concealed in narrow tubes or spurs. They are less massive than bee flowers, and often do not have landing stages, for the butterfly and moth continue fluttering while sucking. They are frequently marked by a peculiar aromatic scent, the quality of which may best be indicated by reference to the scent of the clove pink. To the same class belong the flowers pollinated by night moths, such as the night-scented stock, the honeysuckle, and the Nottingham catchfly. They are characterised by their stronger fragrance in the evening, and by pale tints easily visible in the twilight. The famous example of the Madagascar orchid, *Angræcum sesquipedale*, may be mentioned (Fig. 49). It has a spur nearly a foot long in which the nectar is produced. On the strength of the existence of this flower, Wallace, in his "Essays on Natural Selection," predicted that a sphingid moth with a tongue of the same length would be found; such moths have since been described from Brazil (Müller, 1873) and East Africa (Wallace, 1907).

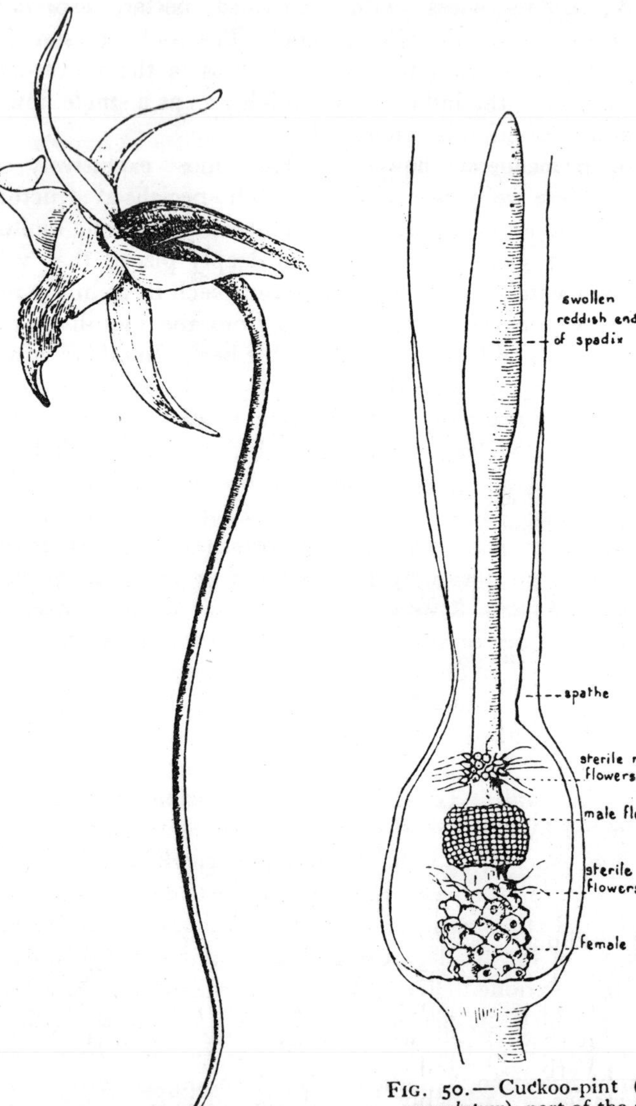

FIG. 49.—*Angræcum sesquipedale*, single flower showing spur. ½ nat. size.

FIG. 50.—Cuckoo-pint (*Arum maculatum*), part of the spathe cut away to show the structure of the inflorescence. (From Kirkwood, "Plant and Flower Forms.")

D, Diptera flowers, visited chiefly by flies, and usually less highly specialised than the last two classes. The small, slightly zygomorphic flowers of the speedwells are visited by syrphid flies which can reach the nectar in the very short tubes. There are also cases of very peculiar specialisation. In the cuckoo-pint, *Arum maculatum*, flies are attracted to the inflorescence, enclosed in its spathe, by the lurid red colour and rather putrid odour of the terminal portion of the spadix axis. When they have crawled in they are trapped among the female flowers at the base of the inflorescence by a ring of hair-like abortive stamens (Fig. 50). They escape only when these have withered, and must pass up through the male flowers, the anthers of which are now dehiscing. The Aristolochias, which again have a reddish colouring, also trap flies in a curiously shaped perianth.

K, Small insect flowers, pollinated by small bees, flies, and beetles often not more than 2 mm. long; *Herminium monorchis* is an example.

Food of Visiting Insects.—The substantial advantage which the insect derives from the visit to the flower is, in almost all cases, a supply of food, typically nectar, sometimes pollen. In some flowers neither is available; in such highly specialised flowers as some of our native orchids, *e.g. Orchis latifolia*, and *Orchis morio*, which are visited by bees, there is no nectar, and the pollen, cemented in pollinia, cannot be collected for food. The spurs are lined with juicy cells the delicate walls of which are pierced by the bee, which sucks the contents. Bees learn to pierce the wall from the outside, thus obtaining the sap without carrying off the pollinia, a habit that is also acquired in connection with some nectar flowers like *Erica Tetralix*. In *Pinguicula alpina* the nectarless spur is lined with special "fodder" hairs which are filled with a sugary sap. Similar hairs occur in Verbascum and in some orchids.

Nectaries are probably derived from glands of the active hydathode type, which, on leaves, excrete water and mineral salts; the presence of extra-floral nectaries on leaves has already been mentioned. In the flower the nectary, associated with a variety of organs, may occupy a number of positions. In the orchids the nectar is secreted by the walls of the perianth spur. In the violets it is contained in a corollar spur but is secreted by two staminal spurs. In the Umbelliferæ and many others, the nectaries form discs or protrusions on the upper surface of the ovary, while in Allium they lie on the outer ovarial wall. In many Rosaceæ the nectaries are on the inner surface of the cup-shaped receptacle; in the Cruciferæ they form discs at the apex of the receptacle. In Ranunculus they lie at the base of the petals, sometimes covered by a scale; in Helleborus the cup-like nectaries are modified stamens. These few examples show that the secretion of nectar may be performed by almost any organ of the flower.

The nectar is a watery fluid containing about 25 per cent. of glucose. It should not be confused with honey, the product manufactured by the bee, of which it is only the raw material. Many of the features which may protect pollen from rain also serve to prevent the nectar from being washed out. Whether the danger of this is great may be doubted, for in a multitude of flowers the nectar is freely exposed. It is of interest to know that Sprengel's first observation on floral biology was the presence of hairs in the throat of the corolla of *Geranium sylvaticum*, which he decided had the function of preventing the nectar from being washed out. From this beginning sprang his great work on pollination.

Colour.—Colour and scent are the two chief means by which the insect is guided to the flower, and a good deal of discussion has taken place as regards their relative efficiency. For us the flower's most striking character is, in most cases, its colour; it does not follow that the same is true for the bee, butterfly, and fly. The question turns on the colour sense of insects, and can only be answered experimentally, though there are suggestive facts open to observation.

It is significant that in the different classes of flowers detailed above different colours prevail. There are many exceptions, but, speaking generally, in class A the colour is white or yellow, often of a dirty tinge, or dull red or green ; in class AB yellows and whites predominate ; in class B blues and purples are also found ; in class H, the bee flowers, blue and purple are predominant, though yellows are common. In the butterfly and moth flowers of class F, the colours are varied, though pale tints of pink or purple predominate. The fly flowers of D are frequently lurid red or dirty green. There is a very distinct tendency for the less highly specialised flowers to be white or yellow, for bee flowers to be blue or purple, for pronounced fly flowers to be reddish-green, for moth flowers to be pale. Pure reds are extremely rare. It is generally assumed that the blue tints have appeared at a later stage of evolution than the yellow and red. The very fact that flowers related to different classes of insects have different types of colouring should be a warning not to generalise from the behaviour of one insect to that of others, and still less from our own colour-sense to that of the insect.

The importance of colour in the entomophilous flower is also emphasised by its absence from anemophilous flowers ; in these pigment, if it is present, is a red anthocyan, as in the larch cones or stigmas of the hazel. It has nothing to do with pollination, and it is doubtful if it has any significance at all. It is the colour conspicuous by its absence from the insect-visited flower. In the floras of Oceanic islands bright flowers are scarce, and this is related to the absence of insect life. Wallace, in his *Tropical Nature*, writes, "... the Galapagos Islands, which ... with a tolerably luxuriant vegetation in the damp mountain zone yet produce hardly a conspicuously coloured flower ; and this is correlated with, and no doubt dependent on, an extreme poverty of insect life, not one bee and only a single butterfly having been found there." Finally, we may repeat the statement that the great advances in the evolution of insects and of brilliant flowers were contemporaneous.

Colour Sense of Insect.—Such facts make it certain that the bright colouring of the flower is related to the insect visit, but they do not decide whether colour is more important than scent, and they leave it a possibility that the pigment of the flower is effective simply by reason of its brightness, and not because of its colour as such. A great deal of inconclusive work has been done on these points, *e.g.* Plateau (1895, 6), Giltay (1904, 1906), Andreæ (1903), Lovell (1920). The questions, so far as the honey-bee is concerned, have been settled recently by von Frisch (1914). V. Hess (1913) had made an interesting comparison between the colour sense of insects and other invertebrates and that of colour-blind human beings. That very rare individual, a totally colour-blind man, can distinguish no colours, though he can distinguish between different light intensities. When he views the solar spectrum he sees it shorter than we do—the red end is merged in darkness ; he sees the brightest light in the yellow-green region, while normal vision sees it in the yellow. The relative intensities of different spectral regions as seen by colour-blind and normal vision may be measured and expressed as graphs, which differ in characteristic ways. Von Hess found that this graph for bees and other invertebrates, measured by making use of phototropic response, corresponds with that of totally colour-blind men. He drew the conclusion that these animals, too, were totally colour-blind and could perceive only differences in light intensity. If such were the case, the flower would attract because it is bright and not because it is, for example, blue.

Von Frisch started out from the idea that this was not a legitimate deduction from the experimental results, and carried out experiments which seem to give conclusive evidence in favour of a limited colour sense. If the bee has no colour sense but distinguishes only intensities, then it should be unable to distinguish between a colour and that shade of grey which, to its senses, is of the same degree of brightness. What that particular shade of grey in any given case will be we cannot tell, but we can offer the insect a series of greys, very finely graded, from which to choose. Von Frisch used a series of 30 greys, in all shades from white to black ; this grading was, as a matter of fact, finer than necessary, and in some experiments a smaller number was used. His honey bees were first drilled to visit a piece of blue or of yellow paper placed among the grey papers, but with a watch-glass of sugar-water on it to act as an attraction. The drilling lasted two days. After that the bees were offered the series of papers arranged at random, and without any sugar on the coloured slip. To avoid any influence that a difference in smell between the coloured and grey papers might occasion, the whole was covered with a glass sheet. The bees settled only over the coloured paper, and never on any of the greys. This held for blue and for yellow. For red it was found that the bees settled on black or on dark grey as readily as on the colour ; for blue-green it was found that the bees settled with equal readiness on a medium dark grey. It was further found that the bee could not distinguish between yellow, orange-reds, and yellow-greens, or between blues and purples. If bees were drilled to a green or orange-red, they subsequently settled rather on yellow. Drilled to a purple-red paper, they would settle on a blue, but not on a red.

Von Frisch concludes that the bee can distinguish the colours yellow and blue, but not red or green. When it is presented with yellow-greens or orange-reds it sees in these only the yellow constituent ; when it is presented with purples it sees only the blue constituent. Its colour sense therefore approaches that of a red-green colour-blind person (the commonest kind of colour-blindness), though it is not exactly the same.

The bee can therefore see yellows, blues, and purples, and the *colour* of the flower affects its senses. We have here an experimental explanation of the facts that bee flowers are characteristically blue or purple in colour, and that yellow is also frequent. When we say that these colours *attract* the bee, we do not mean that the bee *likes* these colours ; we mean that it can distinguish them, and that it comes to associate them with nectar, so that it uses them in the search in nature exactly as it does in the experiment.

Honey Guides.—This gives a sound basis for work on other insects, and for the investigation of the usefulness of particular colours and colour arrangements. For the possibilities of guidance by colour do not stop with the production of a single colour. We have already mentioned the first of the train of ideas which led Sprengel to his discoveries ; the second step was the posing of the problem as to what was the use of the yellow eye of *Myosotis palustris*. Sprengel concluded that it served to guide insects to the entrance of the nectar-containing corolla tube, after they had settled on the flower. Only then did he see that the blue colour of the corolla might guide the insect, from a distance, to the flower. His first idea concerned the guiding of the already alighted insect ; he termed the yellow eye a *honey guide* (saftmal). By Sprengel, and by many biologists since his day, various markings on the general ground colour of the corolla have been described as such honey guides. We need only mention as familiar the dark lines running towards the base of the petals in the violets (Fig. 51) and eyebrights. The usefulness of such guides has been the subject of a good deal of scepticism. A general review of the types of guide has recently been made by Kraepelin (1920) ; he concludes that they may assist the insect to make its landing, but that their significance in helping to locate the nectar is small. Von Frisch has given some attention to this subject. He classified the colours of 94 European flowers with

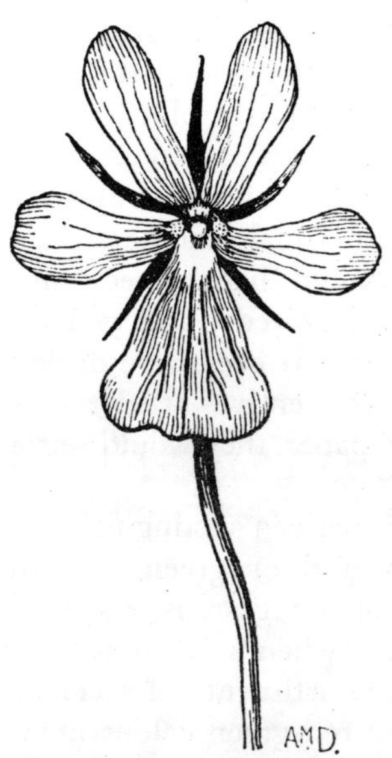

FIG. 51.—Honey guides on the petals of *Viola cornuta*. The flower is pale blue-purple with dark purple lines and a cream-coloured eye. × 1½.

distinct honey guides and obtained the results given in Table XXXIV.

TABLE XXXIV

COLOUR CONTRASTS OF HONEY GUIDES

Number of flowers.	Colour contrast for human eye.	Colour contrast for eye of bee
16	Yellow—blue	Yellow—blue
6	Yellow—violet	,, ,,
10	Yellow—purple-red	,, ,,
1	Orange-red—blue	,, ,,
33		
14	White—yellow	White—yellow
1	White—orange	,, ,,
15		
11	White—blue	White—blue
7	White—violet	,, ,,
11	White—purple-red	,, ,,
29		
3	White—red	White—black
6	Purple-red or yellow with black or brown	Blue or yellow with black
8	—	Different tones of one colour

From this it appears that, in fact, the honey guides are always such that the eye of the bee can distinguish them from the ground colour of the corolla. The frequency of the yellow-blue contrast is very striking, and, along with white-blue and white-yellow, makes up the great majority of combinations as seen by the bee's eye.

Further interesting experimental results were obtained. It was found that bees could distinguish between, for example, a yellow circle surrounded by a blue ring, and a blue circle surrounded by a yellow ring. The former is the design of a forget-me-not flower. It can distinguish between different designs in a single colour. Thus a pattern of blue or yellow rays is distinguished from a pattern of four blue or yellow arms; the former gives the appearance of a composite flower, the latter of a gentian flower. There is thus, in addition to a sense of colour, a very distinct sense of form; it is limited, however, to the recognition of forms which have some affinity to those of flowers; purely geometrical and artificial patterns are not distinguished. Von Frisch holds that the form sense, including the power of distinguishing between colour patterns, is chiefly important in enabling the bee to pick up the particular flower it is visiting from others of similar colour. The habit of visiting only one species is very strongly marked. The honey guide may help the insect to probe the flower more easily, but it is more probable that it forms a colour pattern which makes identification of the flower to be visited easy, and that this is its chief function. It is clear that there is plenty of room for experimental work here.

Using the results and methods of von Frisch, Knoll (1921, 1922 *a* and *b*) has investigated the colour sense and relation to flowers of two other insects. In the flower of *Muscari racemosum* the tips of the perianth segments form a white circle surrounding the dark opening, and standing out from the violet-blue ground colour of the perianth. The plant is visited by various insects, of which Knoll chose the humming-bird hawk-moth (*Macroglossum stellatarum*) to work with. He imitated the colour, but not the form, of the flower with suitably coloured paper; through a hole in a white circle on this sugar-water was available. The moths were drilled with this artificial flower, and were then presented with the same appliance without the food and covered by a glass plate. They flew to the white ring, and the traces of their tongues on the glass sheet were all grouped round the ring (Fig. 52). They did not visit the other parts of the violet paper, nor did they visit white rings on yellow or grey paper. Violet paper without the white ring had relatively little effect. This is a good case of a honey guide. The experiment does not show that the white ring helps the insect which has reached the flower to find the nectar; it does show, however, that the impression which guides the insect to the flower is that of a combination of white circle and violet ground. It also demonstrates the colour sense of the moth for blue. This moth also visits the flowers of *Linaria vulgaris*, and individuals which frequent this flower are found to react to yellow instead of blue. Like the bee, the moth distinguishes a yellow group of colours and a blue group; but an individual which frequents a particular flower becomes "drilled" to the colour of that flower (exactly like the bees in the experiment) and always reacts to that colour. The flower-fly (*Bombylius fuliginosus*) has the same colour sense, distinguishing yellow and blue. In both these insects the sense of smell is of very subordinate importance.

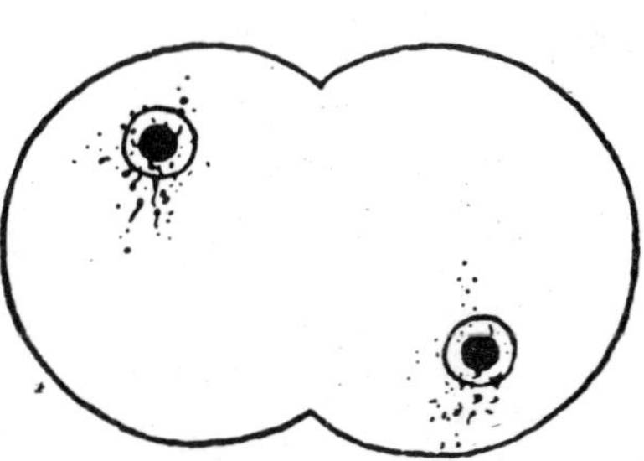

FIG. 52.—Knoll's artificial flower; a black dot and white circle on a purple ground; the small spots are the traces of the tongue of a visiting moth. (After Knoll.)

Special Cases. — A puzzling case may be mentioned here. There are three orchids, *Ophrys apifera*, *O. muscifera*, *O. aranifera*, the flowers of which, to our eyes, bear a close resemblance to a bee settled on a rose flower, to a fly, and to a spider respectively. It has been supposed that these resemblances frighten away visiting bees, which perceive that the flower is already occupied, and that the bee is an unwelcome visitor—an explanation rejected by Darwin (1862). Detto (1905) has shown that when one of these flowers is pinned into the flower or inflorescence of another plant habitually visited by bees—*e.g.* the peony—bees approach, and then, apparently perceiving that the flower is already occupied, swerve off, just as they do if a real insect is already in possession. Detto inclines to believe in the unwelcome guest theory. But there is no evidence that the Ophrys flowers are avoided in this way on their own spikes; they are simply not visited. The result for *O. muscifera* and *O. aranifera* is that very little seed is set. Many counts have been made by Darwin, Detto, and Eckhardt, and the number of capsules with seed ranges from less than 1 to about 17 per cent., and usually lies between 5 and 10 per cent.; *O. apifera* is completely fertile, but this is due to the fact that it is self-pollinated. In view of our new knowledge of the colour sense of the bee, it seems very likely that the reason for the lack of visits is that the dark red blossom is invisible or indistinguishable from green. It seems that we are here dealing with three flowers which have definitely disadvantageous characters; the resemblance to insects is quite accidental and visible only to us. It is dangerous to use such a resemblance as the basis of hypotheses as to supposed advantages to the plant. This danger is illustrated even better by another example recently described. Möbius (1912) states that the flowers of certain species of Delphinium, *e.g. D. elatum*, have a very strong resemblance to a flower with a bee entering; it is even stronger than in *Ophrys apifera*. But bees visit the Delphinium industriously, and the bee-like appearance is explained as being useful. "Perhaps the bees are tricked to believe that others of their species have visited the flowers, and that they are therefore worth visiting; so that they then search for unvisited flowers on the same stalk, and on coming closer are convinced that the flowers are still free and realise their error." The author's use of the word "perhaps" is well justified.

Scent.—The importance of scent is peculiarly difficult to estimate. Experimental work is difficult, because there is no objective method of classifying scents or of measuring their strengths. We can distinguish between "aromatic," "sweet," and "nauseous" odours in a rough sort of way (Linnæus set up seven classes on such a basis, and Henning (1916, see also Parker, 1922) has proposed a similar classification), and we can distinguish between strong and weak by direct comparison, but that is all. It is certain that our sense of smell is very coarse compared with that of many animals ; the example of the fox-hound will at once occur. Even among normal human beings the sense of smell is much more variable than are the senses of hearing and sight. Fabre has shown that the olfactory sense of moths is infinitely more delicate than ours. Moths liberated at a distance of several hundred yards from, and out of sight of, a honeysuckle bush flit straight to it.

Such exact knowledge as we possess is again due to the work of von Frisch (1919), and is again confined to the case of the honey bee. The rather unexpected result of a long series of experiments was that the bee's sense of smell is essentially the same as our own, both in its acuteness and in its power of distinguishing between different scents. Bees which had been drilled to oil of orange could pick out this scent with complete accuracy from forty-three other ethereal oils. In addition to the oil of orange, they were attracted only by oil of citron and oil of bergamot, two essences of similar derivation, which, to our sense also, have scents similar to that of the orange. There are certain pairs of substances which, with very different chemical constitution, have for us similar scents, *e.g.* isobutyl benzoate and amyl salicylate ; such pairs tend to be confused by the bee too. In one particular the bee's sense seems to be sharper than ours ; the insect can pick out a particular scent from a mixture better than we can. Von Frisch was unable to show that certain inconspicuous flowers, which are visited by bees, and are to us scentless, such as those of the wild vine or the red currant, are scented for the bee. Visits to such flowers seem to be made easy by the fact that they always occur in masses. When bees were drilled to colour and scent simultaneously, and then offered the two attractions separately, colour only was perceived from a distance, even when the scent was very strong and a breeze carried it to the bee. The general conclusion is that colour is the guide to the flower, and that scent is useful in enabling the bee, flying among many flowers of similar colours, to pick out the species it has formed the temporary habit of visiting ; in this it is supplemented by the sense of form.

These results cannot be taken to apply to insects in general. It is very probable that flies, beetles, and possibly butterflies and moths are more strongly affected by scents than the bee. The prevalence of rather unpleasant odours in fly flowers, the strong aromatic scent of butterfly flowers, the stronger evening scent of those flowers pollinated by night moths, tell in favour of a greater importance of the sense of smell in such cases. But the hawk-moths investigated by Knoll resembled the bee in their relation to colour and scent. The proper method of attacking the problem has now been worked out, and we may hope that von Frisch's work will be extended to other insects, and that this very interesting aspect of the relation of the flower to the insect will soon be entirely cleared up.

Special Relations.—In a very small number of cases the insect visits the flower to deposit eggs in the ovary. Thus Kerner states that certain Caryophyllaceæ, *e.g. Silene nutans*, *S. inflata*, *Lychnis Flos-cuculi*, are pollinated by owlet moths which lay eggs in the ovary. The caterpillars devour many, but not all, of the ovules and developing seeds before they gnaw their way out. Small blue butterflies have a similar relation to the flowers of *Anthyllis Vulneraria*, *Colutea arborescens*, and *Sanguisorba officinalis*. A much closer relation is that between some American Yuccas, *e.g. Y. filamentosa*, and a pollinating moth, *Pronuba yuccasella*. The female moth first collects a little ball of pollen from the anthers with special maxillary appendages ; it then lays its eggs in the ovary between the ovules with the help of a long ovipositor ; and finally, clambering down the style of the pendent flower, it packs the pollen ball into the stigmatic grooves. The plant is said to be completely sterile in the absence of the moth, as when it is cultivated in Europe, and the caterpillars can live only on the developing seeds, many of which may be destroyed.

Perhaps the most remarkable of all inter-relations between an animal and a plant is afforded by the mode of pollination of the genus Ficus. The unisexual flowers are borne on the inside of a hollow inflorescence axis, a *synconium*, which opens to the outside by a constricted apical pore. The swollen and fleshy infructescence is the " fruit " or edible fig ; each " seed " is in reality a fruit, the product of a separate flower. Pollination is carried out by various small wasps. Of the 600 species of the genus the cultivated fig of Mediterranean countries is best known. It has been a subject of investigation from the time of Aristotle and Theophrastus. Much of our exact knowledge is due to Solms-Laubach (1882) ; more recently Tschirch and Ravasini (1911) have given an account, new in many details, based on the examination of very extensive Italian material. We follow the account of Tschirch (1911). The pollination of several other species has been described by Solms-Laubach (1885) and Cunningham (1889).

The wild fig, *Ficus carica*, the " fico sylvatico " of the Italian peasant, is still found in Italy, sometimes in communities, as on the walls of Monteriggioni, sometimes as isolated individuals. It bears three generations of flowers and fruits in the year. The first, the " profichi," are formed in February ; the inflorescences contain numerous male flowers, just inside the mouth, and, lower down, numerous " gall flowers." The gall flower has a short style, with an open canal, and a single rudimentary ovule incapable of forming a seed. Female wasps (*Blastophaga grossorum*) enter the synconium and deposit eggs in the ovules of the gall flowers, one in each. In the ovule the larva is hatched out, feeds, and undergoes metamorphosis. The male wasps gnaw their way out ; approach gall flowers containing female wasps, pierce the ovarial wall, and fertilise the female within ; they then die without leaving the synconium. By this time the fig is ripe, though still tough and bitter, and the male flowers are shedding their pollen. The female wasps leave their abodes and crawl out of the synconium, becoming liberally dusted with pollen on the way. They are lazy and fly but little, crawling about the tree in search of young inflorescences. These they find in the second generation, the "fichi," now developing about the end of May. These contain only normal female flowers with long styles. The wasp tries in vain to lay its eggs in the ovaries, at the same time pollinating the stigmas. The fichi ripen about the end of September, and are fleshy and edible. Meantime the third generation, the "mamme," is developing, and the gravid wasps ultimately find their way into the synconia and lay their eggs in the gall flowers which alone are present. In these the larvæ pass the winter, escaping in spring to repeat the cycle. Fig and wasp are entirely dependent the one on the other.

The cultivated fig, in all its numerous varieties, is derived from this wild species. It exists in two races, the fig (*Ficus carica, domestica*) and the caprifig, or goat fig (*Ficus carica, caprificus*), which never produces edible fruit. Each bears three generations of flowers. The wasps pass the winter in the mamme of the caprifig, and, escaping about March, enter the profichi of the caprifig and also the "fiori di fico" of the fig. In the former they lay their eggs in the gall flowers; the latter contain sterile female flowers only, in which eggs cannot be laid. The fiori di fico ripen in some varieties, and are edible, but usually they fall off. From the profichi the gravid females escape in June, becoming dusted with pollen as they make their way out. They then enter the "mammoni" of the caprifig and the "pedagnuoli" of the fig. In the former they find gall flowers in which they lay eggs, in the latter they pollinate the female flowers which alone are present. The pedagnuoli ripen into edible figs from August to December, and form the main crop of all varieties. From the mammoni a new generation of gravid female wasps escapes in September, and these, sparingly dusted with pollen from a few male flowers, pass to the mamme of the caprifig, in the gall flowers of which the larvæ pass the winter. They also enter the "cimaruoli" of the fig, in which only female flowers are present, and pollinate these. The cimaruoli of some varieties produce a crop of edible figs in winter. The cultivated fig may thus bear two crops of figs, or very rarely three, in the course of a year. The wasp seems to go through three generations, though this is not quite certain.

From the earliest times it has been known that the presence of the caprifig is necessary for the production of a crop on the fig. " Caprification " is an established feature of fig cultivation in many places. The peasant grows caprifigs among his fig trees, or grafts shoots on the figs, or even hangs branches in the bearing trees. When the Smyrna fig was planted in California it was found necessary to introduce the caprifig, with the Blastophaga, before a crop could be obtained. In the north of Italy caprification is not practised ; the varieties of fig grown there are parthenocarpic, and produce swollen fruit without fertilisation. These do not keep well and cannot be dried ; the dried figs of commerce always contain seed.

The caprifig bears gall flowers and male flowers ; the fig bears only female flowers, which in one generation are sterile. We have here a unique case in which a monœcious wild plant has been changed by selection (unconscious doubtless, since it occurred in very early times) into a diœcious cultivated form. The caprifig is essentially male, the fig female. Only in cultivation can this condition be maintained, for the seeds of the fig revert to the wild species ; artificial propagation by cuttings and grafting is necessary to carry on the cultivated fig.

II. *Ornithophilous Flowers*

This mode of pollination is of considerable importance in some countries, *e.g.* in Patagonia, West Australia, South Africa, and Brazil. The birds which are active are small honeysuckers, sun-birds, and humming-birds, often not bigger than moths. A recent paper by Werth (1915) gives an interesting account. In general, there is no such marked difference between a bird flower and an insect flower as between either and a wind flower. Many bird flowers are also visited by insects. In both the colours are brilliant, but in bird flowers reds and especially scarlets are frequent.

This is interesting in view of von Hess's (1917) investigations on the colour sense of the bird. The retina of the eye contains yellowish oil drops, and the effect of these is that of viewing an object through a pale orange glass; the bird is blue colour-blind, but can distinguish red and green. Of 159 ornithophilous flowers listed by Werth, 84 per cent. are red, 8 per cent. white, 5 per cent. yellow, and 2·5 per cent. are blue. Brilliant colour contrasts are frequent; in *Strelitzia regina*, for instance, orange and blue. The flowers are scentless and produce nectar in great quantities, so much, in some Australian Proteas, that the natives find it worth while to collect it. The nectar is thin and watery. A striking feature is the rigidity of the styles, stigmas, and filaments which are frequently lignified, evidently in relation to the vigour of their visitors. In some cases a stout platform is provided, as in Strelitzia, where a bract serves. The birds usually sip, however, while hovering, and the platform is usually absent. This is well seen in comparing related ornithophilous and entomophilous species. Thus *Salvia pratensis* has a large lower corolla lip on which the bee alights, while the ornithophilous *S. aurea* has no lower lip at all. In the leguminous *Erythrina indica* the keel and wings, so characteristic of the family, are much reduced, the stamens and style being fully exposed and rigid (Fig. 53). Exposure of these organs is a common feature.

Sargent (1918) has made observations on the ornithophilous flowers of West Australia which support the conclusions of Werth. Most West Australian plants are visited by birds, which may do very serious damage to the blossom, *e.g.* of species of Erica and Arbutus. The woody stamens and style prevent this damage to some extent, so that the feature may be regarded as protective, rather than as directly related to pollination. In other cases there is a direct relation between structure and the visit of a bird. The flowers of *Loranthus aphyllus* are too large to be pollinated by an insect. The small delicate flower clusters of *Acacia celastrifolia* are pollinated by birds brushing against them, while taking honey from an extra-floral gland on the phyllode, in the axil of which the inflorescence is borne. The nectar-sipping habit may have arisen, Sargent thinks, from the birds sipping dew from the flowers and other parts of the plant in arid regions. It is also possible that the association of bird and flower began with insectivorous species taking insects from the flowers.

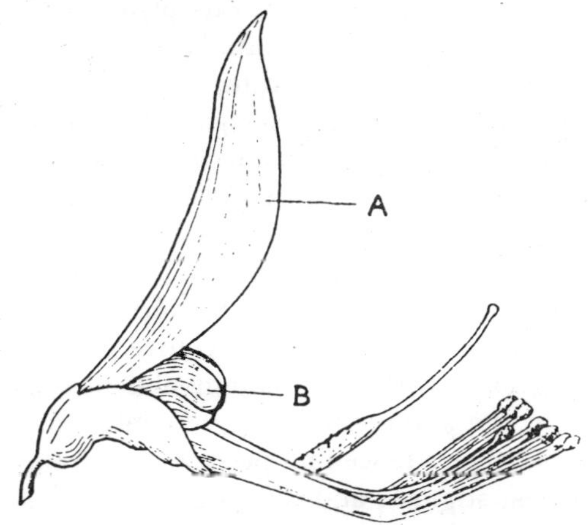

FIG. 53.—*Erythrina indica*, a bird-pollinated flower: A, standard; B, the much reduced wings and keel. Nat. size. (After Werth.)

III. *Malacophilous and Chiropterophilous Flowers*

Pollination by snails has been described for *Chrysosplenium alternifolium*, the golden saxifrage, on the flowers of which Müller saw snails crawling about and leaving pollen in their tracks. As the said snails were engaged in eating the stamens, their beneficent influence is open to doubt; the flowers are also visited by many beetles and other insects. Pollination of a few tropical trees by bats has been described, *e.g. Frycinetia sp.* in Java, and *Bauhinia megalandra* in Trinidad. The bats seem to visit the flowers to catch insects, and do much damage by tearing the corollas.

IV. *Anemophilous Flowers*

These are most strongly marked by negative characters. They lack scent, they have no nectar, they have lost the brilliantly coloured floral envelopes. This may be related to wind pollination in two ways. It may be regarded as a loss of characters which have become unnecessary and therefore wasteful. It also favours wind pollination positively by getting rid of organs which impede the free scattering of the pollen from the stamens, and its easy access to the stigmas. Many anemophilous plants have certainly been derived recently, in the evolutionary sense, from insect-pollinated ancestors. Thalictrum, the wind-pollinated meadow rue, is an example. Though belonging to a family remarkable for brilliant flowers and advanced specialisation for insect visits, it possesses no corolla. We cannot be sure that in all cases wind-pollinated plants have had entomophilous ancestors. The angiosperms may be polyphyletic in origin, and it may well be that such families as the Gramineæ and Cupuliferæ have never, in the course of their descent, passed through a gaily flowered entomophilous stage. Wind pollination in the gymnosperms is certainly primitive.

On the positive side we find peculiarities in the structure of the stamens, stigmas, and pollen grains. These last are generally small in size, quite smooth in surface, and produced in enormous quantities. Near pine woods sheets of water may be covered with a yellow sulphur-like dust when the microsporangia dehisce in spring. Any one who has brought a spray of hazel catkins or of elm or ash blossom into the house knows how thick the dust of pollen settles on the table where it stands. Compare this with the behaviour of another plant in which the pollen is also abundant and conspicuous. The scarlet *Lilium bulbiferum* has brilliant orange-scarlet pollen, and the large anthers brim over when they dehisce, but the grains stick about the anthers or fall in lumps on to the petals. They are not emptied in the air even when the flower is shaken. This is characteristic of the insect-visited plant. In the wind-pollinated plant the separation of the grains, the absence of any tendency to stick together, due to the absence of sculpturing, is most important, for, along with small size, it increases the time during which the grains will float, like

PLATE VI

WIND- AND INSECT-POLLINATED FLOWERS.

Above, *Garrya elliptica*, wind-pollinated; below, *Buddleia yunnanensis*, pollinated by bees and butterflies.

finest dust, in the air, and consequently the extent and thoroughness of their dispersal. The sufferer from hay fever knows too well how ubiquitous is grass pollen in the air in early summer. The pollen grains of the pine have two little flotation bladders formed by the inflation of the extine at the opposite ends of the grain.

The way in which the pollen flies into a cloud, especially when the plant is shaken, draws attention to the free exposure of the stamens. The floral envelopes being absent, the stamens stand or hang freely in the air. This may be emphasised by unusual length in the filaments. The attachment of the anthers to the filament is often very slight, so that the faintest breath of wind sets them shaking, an arrangement admirably seen in the grasses, where the attachment to the middle of the anther makes this still less stable. The same thing is achieved in a different way in the birch and the hazel, where the male inflorescence, the catkin, is a pendent and easily swung tassel. In the hazel the pollen falls from the stamens into the slightly hollowed back of the bract next below; out of this it is shaken only when the wind sets the catkin swaying; that is, it passes into the air only when conditions are such as to secure its dispersal. A more advanced type of mechanism is shown by those flowers already mentioned in which the stamens, at some point of the flower's opening, are flung violently back and the pollen explodes, as it were, into the air, *e.g.* in the wall pellitory.

The stigmas, like the stamens, are freely exposed. The conspicuous and beautiful crimson stigmas of the hazel project in groups of three from the bud-like female catkins. Frequently the stigmas are long and feathery, as in the grasses.

In trees and shrubs the access of pollen is frequently facilitated by early flowering, so that pollination takes place before the leaves can act as a screen to catch the drifting grains. The hazel, the ash, the elm, are examples. In the grasses the need for this does not arise, as the foliage is less obstructive and the flower spikes rise above its general level.

In windy situations the percentage of anemophilous plants tends to increase. We have already quoted Wallace's remarks about the Galapagos Islands. Knuth reckoned that of the flora of Germany as a whole the anemophilous plants make up 21·5 per cent.; in the wind-swept plains of Schleswig-Holstein the figure is 27 per cent.; in the North Friesian Islands it is 36·25 per cent.; and in the very low Halligen Island it is 47 per cent. There is not here a positive correlation between anemophily and wind, for strong wind is not favourable to pollination, but a negative correlation, due to the difficulty of life for flying insects in such conditions.

Some estimates have been made of the distance to which pollen may be carried by the wind. It is said that pollen of the pine may travel for over 500 miles, and "pollen rains" have been observed from 30 to 40 miles out to sea. It does not seem that transport to a distance is of much importance, since wind-pollinated plants are habitually gregarious—forest trees and meadow grasses.

V. *Hydrophilous Flowers*

The number of plants with water-borne pollen is not very large. In a great many aquatics the inflorescences stand above the water, and the flowers are pollinated by insects, as in the water lilies, water plantains, arrow-head; or by the wind, as in the bur-reeds, bulrushes, and pondweeds. This happens even with plants which are otherwise completely submerged, as with the water milfoil and some water buttercups.

A famous case which stands between air-borne and water-borne pollen is the much investigated *Vallisneria spiralis*. A recent paper by Wyllie (1917) clears up some obscurities and corrects some errors of earlier descriptions. The solitary female flowers are carried to the surface of the water by an elongation of the peduncle, which may reach a yard in length. They open, and lie, with their stigmas recurved, in a little depression of the water. The male flowers, which are only 1 mm. in diameter, are formed in hundreds in a single inflorescence. Each consists of two stamens enclosed in a perianth of two large segments and a small one. They are detached under water and rise slowly to the surface, where they open, the perianth segments curving back, and supporting the flower on a little tripod which is very stable. They are drifted about on the water. In the immediate neighbourhood of a female flower they are drawn to it by surface tension, and cluster round it, tipping over so that the stamens touch the stigma. If the female flower is momentarily submerged, as often happens with small waves, the attendant males are inverted over it in an air-bubble, and the chances of pollination are improved. After fertilisation the peduncle coils up and the female flower is drawn once more under water, where the fruit ripens.

In *Elodea canadensis*, also investigated by Wyllie (1904), the solitary male flowers are detached under water and shoot to the surface, where they burst open, scattering pollen on the water, where it drifts to the stigmas.

Finally, we have the case of completely submerged flowers like those of Zostera and Zannichellia. The pollen is shed under water and drifts, submerged, to the filamentous stigmas. We have already noted the tendency to elongation in such pollen grains; this culminates in the thread-like pollen of Zostera, and is correlated with the absence of the extine. The secondary nature of the aquatic habit in flowering plants is nowhere more evident than in the very general retention of flowers with sub-aerial pollination, and the small number of cases of completely submerged flowers.

§ 12. Pollination—Floral Mechanisms

We have been concerned so far with the agencies effecting pollination, without entering, in most cases dealt with, into the exact method of pollen transfer. Of wind- and water-pollinated flowers not much more need be said—pollination is random. In entomophilous plants, on the other hand, the pollen is often removed and deposited in a very definite way. In flowers of class A, which are open, and visited by a variety of small and large insects, the transference is carried out simply by the " guests " scrambling at random over stigmas and stamens. It is in the higher, and particularly in the zygomorphic flowers, that more exact mechanisms are found. We may take as an example the meadow sage, *Salvia pratensis*, as described by Müller (Fig. 54). The tube of the corolla is horizontal and contains at its base a drop of nectar secreted by axial glands. The corolla runs out in front into the broad

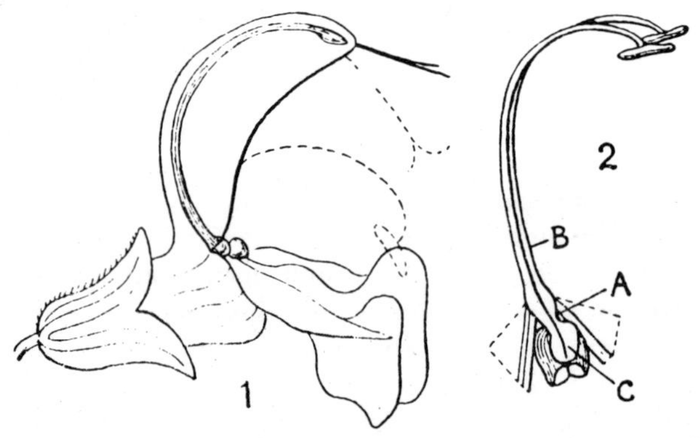

FIG. 54.—Pollination of sage (*Salvia pratensis*); 1, diagram of flower with the corolla hood supposed transparent, to show position of stamens: dotted lines show position of stamens when depressed and of style in older flower; 2, the stamens: A, lower, and B, upper half of connective, C, lower halves of anthers forming the plate. Magnified. (After Müller.)

lower lip, which forms the landing-stage for visiting bees, and the arched and hooded upper lip, under which lie the stamens and young style. There are only two functional stamens, the anthers of which are borne on short, stout filaments. The lower half of each anther is sterile, and, with its neighbour, forms a broad plate which blocks the entrance to the tube; from it the connective, enormously elongated, arches up under the upper lip of the corolla, and carries at its tip the fertile half of the anther. The connective is not rigidly fixed to the filament, but is pivoted, so that, if anything strong enough pushes against the plate formed of the two sterile anther lobes, this moves inwards and the fertile anther halves swing down. This is what happens when a bee pushes into the flower seeking nectar, with the result that the insect's back is dusted with pollen at a definite spot. As the bee leaves, the stamens swing back into their former position. At this stage of the flower's development the stigmas, just projecting from the upper lip, are not receptive, and are not touched by the bee. Later, when the stamens have shed their pollen, the style grows out and down so that the stigmas occupy the position reached by the anthers when they touch the back of a visiting bee. If a bee visits the flower at this stage the stigmas will touch its back where it is dusted with pollen from other flowers. Pollination can thus take place only in one particular way, and it is carried out only by heavy bees sufficiently powerful to work the mechanism.

It will be further noted that only cross-pollination can take place; this is ensured both by the floral mechanism and by the fact that the stamens and stigmas are mature at different times. The number of methods by which cross-pollination is more or less certainly effected is very large. This, taken along with the breeder's experience of hybrid vigour, is the chief support of the opinion cautiously expressed by Darwin in the *Origin of Species*, that " it is a general law of nature (utterly ignorant though we be of the meaning of the law) that no organic being self-fertilises itself for an eternity of generations; but that a cross with another individual is occasionally—perhaps at very long intervals—indispensable."

The investigation of floral mechanisms in relation to pollination, particularly by insect agency, has been carried out largely under the influence of this belief, which has perhaps tended to influence conclusions unduly. Along with some further examples of floral mechanism we may consider this problem of cross- and self-pollination.

§ 13. Cross-Pollination

Dicliny and Dichogamy.—In strictly diœcious plants only xenogamy is possible. This is seen, for example, in the hydrophilous Vallisneria and Elodea, in the anemophilous poplars and junipers, in the entomophilous willows. In monœcious plants, on the other hand, autogamy is impossible, but geitonogamy may take place. This state is seen in the majority of the conifers, in the beech and the hazel, most sedges and many palms among anemophilous plants, in the arrow-head, and the arums among entomophilous, and in the grass-wracks among hydrophilous plants. In monœcious plants, however, a separation of the sexes in another way almost invariably occurs; the flowers of the one sex mature before those of the other, and, in fact, the pistillate flowers ripen several days before the stamens shed their pollen. This separation of the two sexes in time, which is also frequent in the individual hermaphrodite flower, is termed *dichogamy*. We distinguish between *proterogyny*, the condition in all monœcious plants where the stigmas are ripe first, and *proterandry* where the stamens are ripe first.

The interval between the receptiveness of the stigmas and the dehiscence of the stamens is, in general, two or three days, but may be longer, *e.g.* nine days in *Alnus viridis*, and perhaps in the hazel. Proterogyny would seem to be an absolute safeguard against self-pollination. We have not exact information, however, as to the length of time during which the stigmas remain receptive in absence of cross-pollination. Further, different shoots of the same tree may be ripe at different times. So that there is a possibility of occasional geitonogamy, though it must be unusual.

The various more complicated schemes of sex distribution are, on the whole, less suited to secure cross-pollination than is diœcism, or even monœcism. They need not be discussed in detail; for their effectiveness can be readily estimated by considering the different types already described. In such cases, however, dichogamy is usually present. Thus, for example, the ash, which has andromonœcious, gynomonœcious, female, and monoclinous individuals, is also proterogynous.

Dichogamy is very frequent in hermaphrodite flowers, and in these we find both proterandry, *e.g.* in *Epilobium angustifolium*, *Ruta graveolens*, the Salvias, and almost all the Compositæ and Umbelliferæ, and proterogyny, *e.g.* in Plantago, Helleborus, Thalictrum, and others. A great many anemophilous species, perhaps the majority, are diclinous and markedly dichogamous. Hermaphrodite wind-pollinated flowers are found in the grasses and in some smaller groups, *e.g.* the genera Thalictrum, Potamogeton, Plantago, and Rumex (some species of which are diclinous). It is of interest that Plantago, Thalictrum, and Rumex are closely related to entomophilous genera or families, and the inference may be drawn that the anemophilous habit is a recent acquisition (Fig. 55).

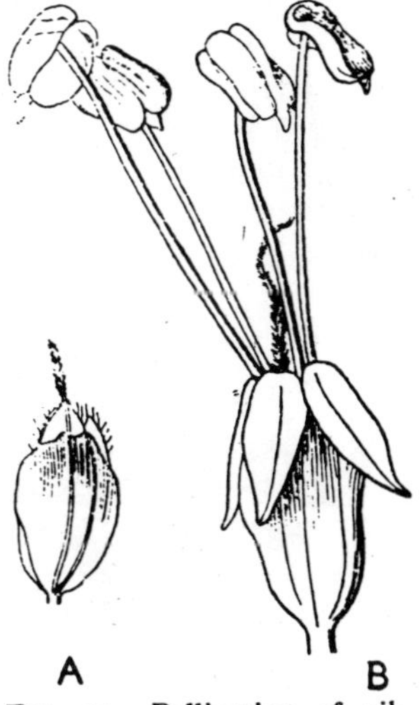

FIG. 55.—Pollination of ribwort plantain (*Plantago lanceolata*): A, flower in female stage with petals and stamens still enclosed in the calyx; B, older flower in male stage. Magnified. (After Müller.)

The grasses require special consideration. The predominant condition throughout this great family is hermaphroditism, though some, *e.g.* the maize, are monœcious, and some show other types of sex distribution. The flowers may be classed as fugacious, as they open, and the stigmas are functional, for a few hours only, often early in the day. The opening—that is, the separation of the pales—takes place by the rapid swelling of the lodicules; at the same time the filaments undergo very rapid growth, so that the anthers may almost be said to tumble out, emptying their pollen in the air, either at once or very soon after. The feathery stigmas likewise protrude, standing out above the pendent stamens. Most grasses follow this scheme of opening; in some genera, *e.g.* Phleum, Anthoxanthum, Nardus, the pales do not open, and the stamens and stigmas push forcibly between their edges. Some grasses are described as proterandrous, some as proterogynous, and many as homogamous. It is obvious that, even where dichogamy obtains, it can be little marked when the open stage of the flower lasts only a few minutes or hours. The stigmas may be exposed alone for a few minutes, or the stamens may have shed their pollen shortly before the protrusion of the stigmas, but the separation can never be very great. Many

FIG. 56.—Flowers of oat; left with stigmas exposed, right with stamens dehiscing. × 2. (After Kerner.)

grasses are habitually self-pollinated before the flower opens. This is notably the case in the oat, wheat, barley, and rice; to what extent it is a property of cultivated races does not seem to be known (Fig. 56).

In wind-pollinated plants, then, dicliny and dichogamy predominate. With pollen scattered broadcast sex separation is the only available method of securing cross-pollination. In the grasses, however, we have a great and very successful family in which the flowers are hermaphrodite; where dichogamy does occur it is so little accentuated that its effectiveness may be doubted in absence of experimental proof to the contrary.

Amongst entomophilous plants, too, we have seen that diclinous and dichogamous species are found; but if we look over the common flowers in any garden or meadow we see at once that the hermaphrodite condition predominates. In a great many hermaphrodite flowers pollination is nearly as indiscriminate, though carried out by insects, as it would be with wind carriage. This is the case in the floral classes A and PO, and to a less extent B and AB. In such cases, where small insects wander at will over the flower, indirect autogamy must be as frequent as cross-pollination. The situation is altered if the flower is dichogamous. Thus proterandry is common in the saxifrages, *Chrysosplenium oppositifolium* is proterogynous, and the Umbelliferæ are usually proterandrous. In the more highly specialised flowers dichogamy is frequently combined with a relation of the essential organs to each other and to the form of the corolla, which determines how the flower is visited, and is such that cross-pollination is favoured.

Further Examples of Floral Mechanism.—In the case of *Salvia pratensis*, already described, we have a good example of a fine floral mechanism combined with proterandry. In the monkshood, *Aconitum Napellus*, the posterior sepal forms an arched hood over the stamens and carpels, and, with its brilliant blue colour, is the most showy part of the flower. The other four sepals, like it, resemble petals and are also brightly coloured; the two lower form a landing-stage for the humble bees which alone pollinate the flower. The two narrow petals curve up under the hood and function as nectaries. The bee, entering the newly opened flower, brushes against the upturned stamens and the lower side of its body is dusted with pollen. Later on the stamens wither and curl down, their place being taken by the stigmas, which are now receptive. A bee visiting the flower at this stage covers the stigmas with pollen from another flower. The genus Aconitum is of special interest, because its distribution lies wholly within that of the humble bees, on which it is completely dependent for pollination. The northern limits of the monkshoods and the humble bees coincide throughout Europe, Siberia, and America (Fig. 57).

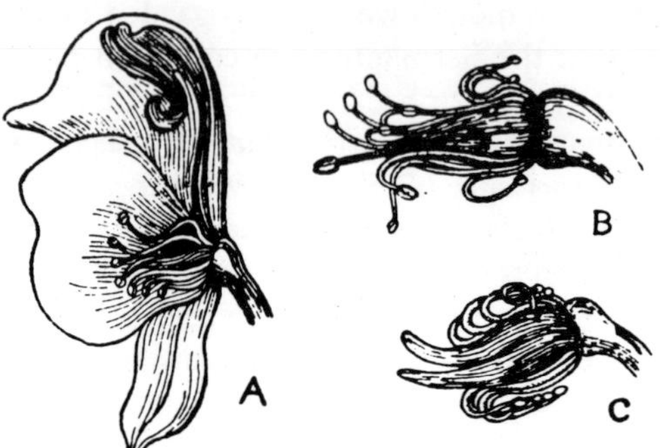

FIG. 57.—Pollination of monkshood (*Aconitum Napellus*): A, section through flower; B, stamens and carpels, young flower; C, stamens and carpels, older flower. Nat. size. (After Müller.)

The small bells of *Erica Tetralix*, the cross-leaved heath, are visited vigorously by bees and many other insects. The club-like stigma lies in the mouth of the flower, the stamens hang a little further in. They dehisce by apical pores. From the basal end of each anther two spurs stick out, reaching the sides of the bell. A bee, clinging to the inflorescence, first touches the stigma; as it pushes its proboscis past the anthers these, or their appendages, are jarred, and pollen is shaken on the insect's head. Self-pollination may follow, the pollen falling on the stigma. The bell is often just too long for the proboscis of the honey-bee, which then pierces the base of the flower from the outside and so obtains the nectar (Fig. 58).

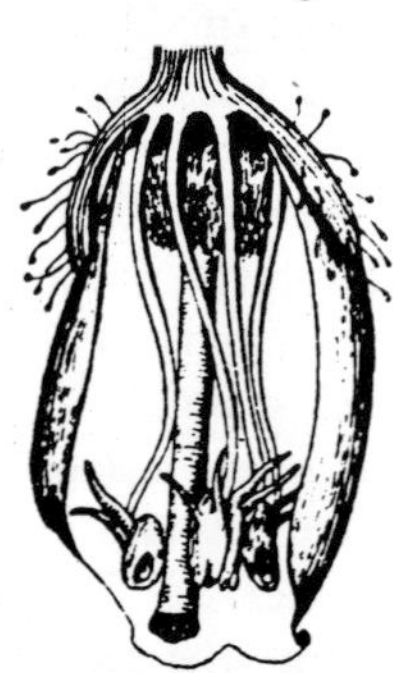

FIG. 58.—*Erica Tetralix*; section through flower to show relative position of stigma and stamens, the latter with anther spurs. × 5. (After Müller.)

The honeysuckle, *Lonicera Periclymenum*, is a moth flower with a long corollar tube and a strong evening scent. On the night of its first opening the five stamens project horizontally, well beyond the mouth of the corolla, and the moth, touching them as it sips nectar in its flight, receives pollen ; the style, with the rounded stigma, is depressed out of the insect's way. On the following night the stamens have withered and sunk down, while the style has curved up so that the stigma is now touched by the moths and is pollinated from other flowers. On the second night the corolla is yellower in colour, and its segments are rolled back. Knuth suggests that moths may visit the newly opened flowers first because of their brighter colour, but he was unable to confirm this because of the rapidity of the insect's flight.

In the yellow flag, *Iris Pseudacorus*, the nectar lies in the perianth tube to which access may be obtained by bees under the broad stigma, one lobe of which lies over each perianth segment. As the bee pushes in it rubs the receptive flap of the stigma, and later brushes against the stamens which are epipetalous within the tube. As it retreats the bee pushes back the receptive flap so that this does not receive pollen from the same flower. A form of this plant has the stigmatal lobes so closely pressed to the perianth that bees cannot push their way in ; it is pollinated by syrphids. The bee flower is also visited by small insects which drink nectar, but do not transfer pollen.

In *Linnæa borealis*, a moth flower, the bracts, inferior ovary, and sepals are thickly clad with viscid, glandular hairs which are said to prevent ants and other small crawling insects from stealing the nectar. The flower has a peculiar aromatic fragrance. The tube of the flower slants downwards so that protection of pollen and nectar from rain is ensured. The stigma stands out from the mouth of the corolla so that it is touched first ; the anthers lie further back, and pollen can scarcely fall on the stigma.

The floral mechanism of the orchids in all its variety has been extensively investigated, particularly since the publication of Darwin's monograph (1862). Except in Cypripedium and its relatives the visiting insect brings off, stuck to its head, the pollen of the single fertile stamen in the form of the pollinia. As it flies the pollinia sink on their stalks by their own weight, so that when the insect visits another flower they are pressed against the stigma; this lies below the stamen, separated from it by the beak or rostellum; the rostellum also usually prevents the pollinia from reaching the stigma of its own flower.

Heterostyly.—The primroses are the classical examples of the arrangement termed heterostyly. In *Primula acaulis*, the primrose, and *Primula veris*, the cowslip, for example, some individuals have flowers in which the five stamens stand halfway down the throat of the corolla and a long style carries the rounded stigma to the mouth. Other individuals, about one-half of the whole population, have the stamens at the mouth and a short style bringing the stigma about halfway up the tube (Fig. 59). As a result, the visiting insects, probably moths, transfer pollen from the long-styled, or "pin-eyed," flowers to the stigmas of the short-styled, or "thrum-eyed," flowers, and *vice versâ*. The larger pollen grains of the thrum-eyed flowers also fit better between the coarser stigmatal papillæ of the pin-eyed. There is little chance of pollen of one type of flower reaching a stigma of the same type, and still less of self-pollination. In *Lythrum Salicaria* there are three types of flowers, long-, short-, and intermediate-styled; the first has short and intermediate stamens, the second long and intermediate, the third short and long, so that pollen is transferred from any one type to the other two.

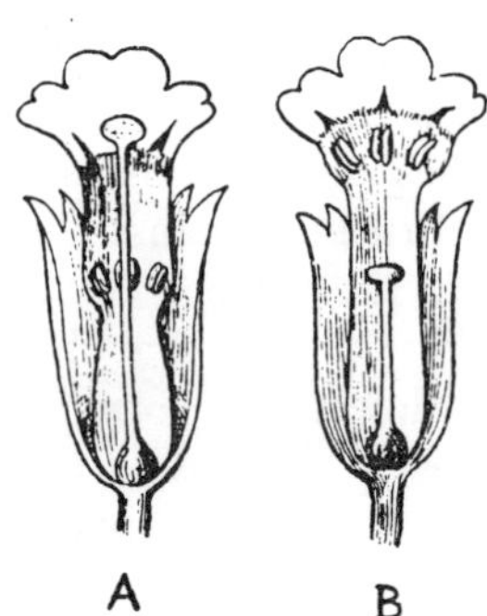

FIG. 59.—*Primula veris*; sections through, A, long-styled, and B, short-styled flowers. × 2. (After Hildebrandt.)

Darwin (1876, 1877) showed that the "legitimate" pollination of thrum-eyed stigma by pin-eyed pollen habitually produces more seed and more vigorous offspring than the "illegitimate" pollination of thrum-eyed stigma by thrum-eyed pollen, or of pin-eyed stigma by pin-eyed pollen, a result confirmed by other workers. Bateson and Gregory (1905) were able to prove that the two types of flower are determined by a single Mendelian factor. In *Primula sinensis*, if the long-styled flowers are pollinated by the same type only long-styled flowers are found in the progeny. Short-styled flowers selfed give either short-styled offspring only, or short- and long-styled in the proportion 3 : 1. The " legitimate " pollination gives either half and half short- and long-styled or only short-styled. This proves that the short style is due to a single dominant factor. Long-styled plants are homozygous recessives. Short-styled plants are either homozygous or heterozygous dominants. In the former case they give only short-styled progeny, either when crossed or when selfed ; in the latter case they give a mixture. In nature, where only legitimate pollination occurs, the short-styled plants are all heterozygotes, and the primroses thus consist of two distinct races in each species, which can exist only by continual crossing. It will be seen that this is closely analogous to the relation between the male and female plants of such diœcious species as *Bryonia dioica*.

Sensitive Stigmas and Stamens.—A type of floral mechanism, the meaning of which is sometimes very obscure, is that exhibited by flowers in which the stigmas or stamens respond by rapid movements to the stimulus of mechanical shock (*seismonastic* movement, like that of *Mimosa pudica*). The two lobes of the stigma of a Mimulus or an Incarvillea close together in a second or two if one is lightly touched. This has been said to be of use in protecting the stigma from deposition of " own " pollen when a visiting insect is leaving the flower. As, however, the insect in leaving brushes the *back* of the stigma, such deposition is in any case unlikely and the movement has probably not much practical significance. Newcombe (1922) has shown that in general sensitive stigmas which reopen quickly remain closed for longer if pollination has taken place, and show a second and permanent closure a few hours later.

Only on the closed stigma does the pollen germinate unless the air is very moist.

Sensitive stamens are more common. A familiar example is afforded by the barberries. A touch on the inside of the filament causes it to move towards the ovary. In many Compositæ (Small, 1917), and particularly in the Cynareæ, a rapid contraction of the filaments, by as much as 20 per cent., draws the staminal tube down over the style, and brushes out the pollen. In the Cistaceæ the bunches of stamens move outwards when the filaments are bent. In all cases the movements are reversed after a short rest period. The use of the contraction of the filaments in the Compositæ is evident. As the same result is achieved by the majority of species by the growth of the style, and without sensitive filaments, it cannot be said that the contractile filament adds much to the efficiency of the floral mechanism.

The movement of the barberry stamens is described by Knuth as ensuring that the visiting insect, which stimulates the filament, shall receive pollen. As the larger insects, such as bees, to which alone the explanation could apply, would certainly receive pollen without any movement, the advantage is again questionable. Still more doubtful is the case of the Cistaceæ. We may take as an example *Cistus salvifolius*, recently described in detail by Knoll (1914*b*). The numerous stamens surround the stigma and pollen falls from them on it. The stamens, if vigorously bent inwards, move, after the lapse of about a second, rapidly outwards till they lie against the petals; after a few minutes they resume their former position. Knoll maintains that by this means the flower presents alternately " male " and " female " conditions to the visiting insect; in the former the visitor is more likely to receive, in the latter to deposit, pollen. It is not proved that the chances of cross-pollination by a large insect rambling about the flower are materially improved (Fig. 60).

§ 14. Self-Sterility

The most perfect means of preventing autogamy in a hermaphrodite flower is self-sterility. This seems to be a

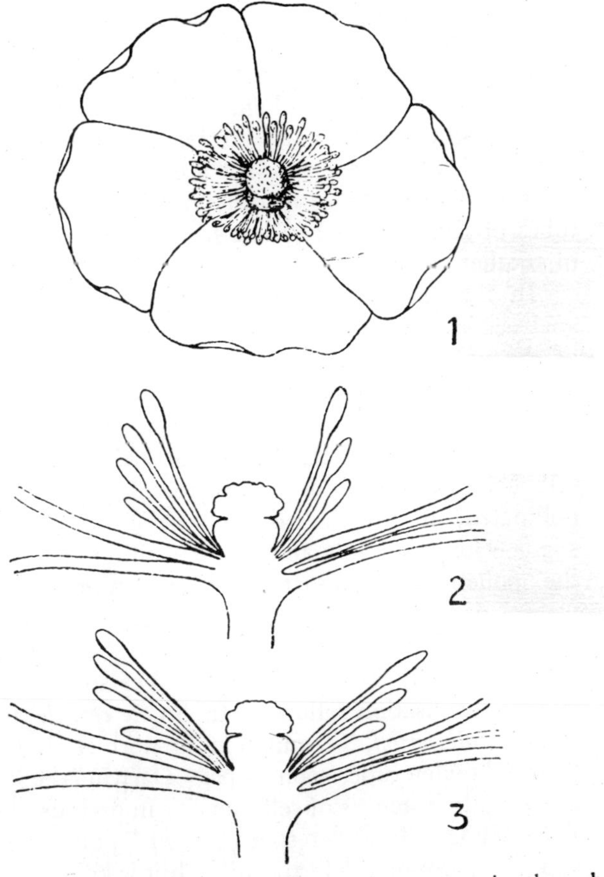

Fig. 60.—*Cistus salvifolius*; 1, flower from above; 2, section through flower, stamens in normal position; 3, stamens after stimulation. 1 nat. size, 2 and 3 × 2·5. (After Knoll.)

not uncommon phenomenon, though much carefully controlled work must be done before we can have a true idea of its frequency. Knuth states that in most cases there is no difference as regards fertility between the results of autogamy and allogamy. He gives a list of about 150 plants which are said to be self-sterile, but notes that this is not exhaustive. Sterility of apples and pears within a race is well known, and mixed plantings are resorted to in practice to secure the setting of fruit.

Even when a plant is not strictly self-sterile, the pollen from another individual may be more active than, or prepotent over, " own " pollen. The tubes may grow more quickly, so that the ovules are reached first by the foreign, if " own " and foreign pollen are placed on the stigma at the same time. Cases of differential growth rate in pollen tubes of Rumex and Lychnis have been already mentioned in another connection.

In the cases of self-sterility which have so far been analysed the cause lies in some failure in the germination of the pollen grain, or in the growth of the pollen tube. Jost (1907) has shown that in the laburnum, *Cytisus Laburnum*, pollen does not germinate unless the stigma has been wounded. If the flower is artificially or naturally selfed the grains do not germinate. If the flower is insect-pollinated, in which case cross-pollination takes place, slight wounds are inflicted by the visiting bee, which enable the pollen to germinate. If the stigma is artificially wounded, then " own " pollen will lead to fertilisation. In *Corydalis cava* germination of " own " pollen takes place if the stigma is crushed, but the tubes soon cease to grow, and no fertilisation follows. In *Secale cereale*, the rye, and *Lilium bulbiferum*, germination is normal, but growth of " own " pollen tubes soon stops. Darwin (1868) gives some remarkable instances of self-sterility in orchids, described by Fritz Müller. In eleven species " own " pollen is not only incapable of producing fertilisation, but is killed by the stigma. If pollen of *Oncidium flexuosum* is placed on the stigma of the same flower, or on that of another flower of the same plant, it becomes brown and dies in five days, while pollen from a distinct plant on the same stigma is perfectly fresh. In Notylia, " own " pollen is killed in two days and the flower may fall. We have thus a graded series showing different degrees of interference with " own " pollen, while, in every case, pollination from another individual of the same species is followed by fertilisation.

It has frequently been assumed that pollen from *any* other individual of the same species is effective, but, as Bateson (1913) writes, this " has always seemed to me a self-evident absurdity, for it would imply that there can be as many categories as individuals." In fact, we now know for two cases that sterility exists not for each individual, but within definite races of the species. Correns (1913) investigated *Cardamine pratensis*, the cuckoo flower. " Own" pollen just germinates on the stigma, but gets no further, while pollen from another plant is effective. Correns crossed two parents, B and G, and from the progeny selected sixty plants. Each of these he crossed back with each parent. Half were sterile with B, half sterile with G, and half were fertile with G, half fertile with B. The relation to one parent was completely independent of that to the other, so that the daughter generation could be divided into four equal classes, W fertile with both parents, X fertile with B only, Y fertile with G only, and Z fertile with neither. The species therefore consists of races, probably many in number, inside each of which complete sterility exists, while seed is set when pollination takes place from another race. The attempt to explain the results by the inheritance of two different inhibitors is not completely satisfactory. The second case, that of *Veronica syriaca*, described by Lehmann (1918, 1922), is similar, though apparently more complex.

The actual cause of the failure of the pollen grains to germinate, or of the tubes to grow, is not known. It may be due to the presence of an inhibiting agent as Correns believes, and in support of this the behaviour of the orchids is strong evidence. Jost, on the other hand, thinks that the failure is due to the lack of some essential growth factor. He points out that in no case does a pollen tube attain its normal length in a culture solution. Compton (1913), in a general review of the subject, makes a comparison between the inhibition of the pollen tube growth and the phenomena of immunity to disease in the animal kingdom. There is the possibility that in the cells of the stigma an anti-body inhibiting further growth is formed as a reaction to the entrance of the pollen tube. The subject awaits exact investigation.

§ 15. Self-Pollination

Cross-pollination is thus widespread amongst flowering plants. Sometimes it is, from one cause or another, obligatory; sometimes it is more or less favoured. Sometimes, however, the chances as between cross- and self-pollination are about equal or inclined to the latter. Self-pollination may be favoured by the structure of the flower, and it, too, may be obligatory.

Over open flowers like those of the buttercups, poppies, and brambles many insects wander at will, the flowers are homogamous, and, unless self-sterile, as in *Ranunculus acris*, *Papaver Rhœas*, *Rubus odoratus*, there is nothing to prevent autogamy taking place, and, indeed, it is likely to be the rule. No special mechanism exists, but the chances of indirect autogamy must be strong. Even when special mechanisms which assist cross-pollination exist, there is often a very strong chance of geitonogamy. Bees in particular tend to restrict their visits to a particular flower during considerable periods. Any one who has watched a bee busied about a sage bush knows how other plants are neglected and flower after flower of the sage is tried. Even though the flowers are proterandrous, it is clear that the chances of a stigma being pollinated from another flower on the same plant are great. The same must be true of monoclinous and monœcious wind-pollinated plants, where dichogamy is not complete. In a spike of the ribwort plantain, *Plantago lanceolata*, the stigmas are produced from the upper flowers while the stamens hang from the lower. In the monœcious proterogynous sedges the stigmas may still be receptive when the anthers are dehiscing. Even the presence of a complex floral mechanism may not therefore be much safeguard against geitonogamy, and it is not clear that this differs essentially from autogamy. Rare cases have been described in which the inheritance of the reproductive cells at different levels on the axis is different, as in the rogue peas investigated by Bateson and Pellew (1920). We do not know that such cases have any general application. Without definite proof to the contrary we must regard those plants in which geitonogamy may occur as belonging to the same category as those which are often autogamous.

More than a chance of autogamy exists in many crucifers, such as the wallflower. The anthers of the long stamens surround the stigma and dehisce directly on to it. Insect visits may result in the deposition of foreign pollen, but, unless this is prepotent, autogamy is certainly favoured. According to Knuth, this condition is particularly common in small annual plants.

In some flowers, where the anthers originally lie below the stigma, they grow up later so as to come in contact with it, as in *Adoxa Moschatellina* and many saxifrages. In *Hypericum perforatum*, *Lysimachia nemorum*, and *Azalea procumbens*, the anthers at first stand away from the stigmas and later bend or are bent towards it. Here autogamy is postponed till allogamy has failed. These examples may suffice, but we may note that Knuth distinguishes twenty different ways in which autogamy is secured.

Cleistogamy.—The most extreme case of autogamy is offered by the cleistogamous flowers, in which the perianth never opens and pollination takes place, as it were, in the bud. About 150 species are known with cleistogamous flowers. The most familiar example is the sweet violet, *Viola odorata*. The sweet-scented spring flowers are not very conspicuous, but are visited by a variety of insects. Presumably the visits are not frequent, for seed is not often set. Later in the summer the cleistogamous flowers are formed hidden deep among the leaves. They are bud-like, never open, but set abundant seed (Fig. 61).

Cleistogamous flowers have been regarded as specialised structures which insure the plant against the chance failure of the open flowers to set seed. Goebel (1904 and Org.) has shown, however, that essentially they are flowers in which development is inhibited at an early stage. In the

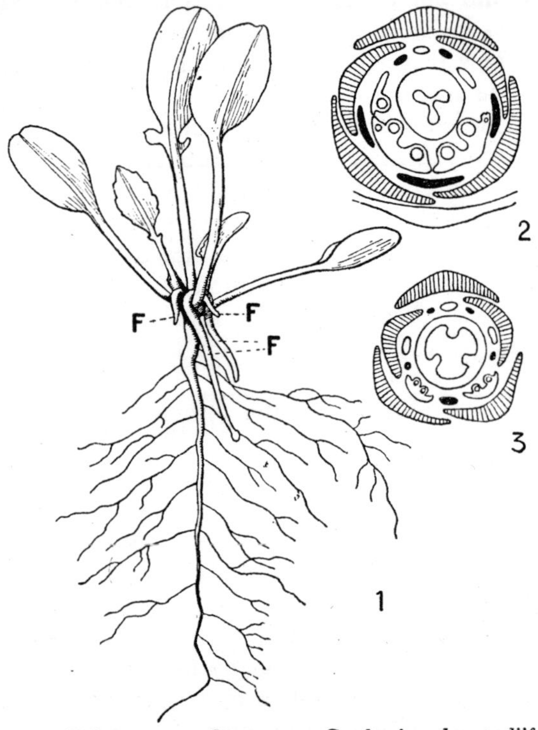

Fig. 61.—Cleistogamous flowers: 1, *Cardamine chenopodiifolia*, at F the cleistogamous flowers penetrating the soil; 2 and 3, *Viola sylvatica*, cross-sections through flowers, sepals shaded, petals black; 2 is an intermediate between the open and the cleistogamous flower, 3. (After Goebel.)

violet the sepals are normal, though small, the petals are represented by five whitish scales, and the spur is not developed. Five stamens are present, but only two sporangia are developed in each; the staminal spurs are wanting. The stigma remains hidden among the stamens. The endothecium is reduced and does not function, the pollen germinating in the anthers and the pollen tubes piercing their walls. The cleistogamous flower, in this case, and in other cases where reduction is more marked, corresponds in the main to an early stage in the development of the open flower ; at this stage, however, functional maturity of the germ cells is reached. The most general feature is the reduction of the corolla, which is completely absent, for example, in *Cardamine chenopodiifolia*. In normal development the corolla generally appears after the other floral organs.

The conditions in which cleistogamous flowers are produced favour this interpretation. They tend to be formed either when growth of the plant is predominantly vegetative, or under conditions of malnutrition. Thus, frequently they appear early, preceding the open flowers, as in *Impatiens noli-me-tangere*. A striking example of this is the Brazilian *Cardamine chenopodiifolia*, which bears cleistogamous, subterranean flowers on plants which have only developed a few pairs of leaves (Fig. 61). The touch-me-not may be forced to produce cleistogamous flowers by growing it in starvation conditions on dry soil. Many plants, such as the pea or the shepherd's purse, form flowers which never open at the end of their flowering period when the supply of food substances is falling low.

There are all gradations between plants bearing only normal flowers, some of which may not open in unfavourable conditions, and those which produce only cleistogamous flowers, *e.g. Salvia cleistogama*. Our native typically cleistogamous plants are all of the intermediate types—*Viola odorata*, *Juncus buffonius*, *Lamium amplexicaule*, *Stellaria media*, *Oxalis acetosella*. Not very far removed from cleistogamy, so far as effect is concerned, are those plants in which autogamy takes place as the flower opens, as in the oat, the barley, and the wheat. Although the formation of cleistogamous flowers is evidently controlled to an important extent by external conditions, we must nevertheless assume the presence of an inherited tendency to their production.

§ 16. Pollination—General Considerations

Looking back over this account of pollination, we see that, in the vast majority of cases, the passive microspore is carried to the stigma (or the micropyle in the case of the gymnosperms) either by insects or by the wind. We cannot say that one of these methods is better than the other. The number of species which are entomophilous is the greater, but probably not the number of individuals. The most successful dicotyledonous family, the Compositæ, is entomophilous, the most successful monocotyledonous, the Gramineæ, is anemophilous. Yet the influence which has directed the course of evolution towards the production of the gay, fragrant blossom, which we naturally associate with the phrase "flowering plant," has been the relation of the insect to pollination.

Not only is pollen transferred, it tends to be transferred in a way which, at the least, makes cross-pollination possible. In anemophilous plants the means are chiefly dicliny and dichogamy; these are employed, too, in entomophilous plants, but much more striking in these are the floral mechanisms. We must admire the variety and apparent ingenuity of these, but we are not yet in a position to evaluate their net importance, efficiency, and necessity. They are very widespread. In some cases they restrict pollination to one particular method, as in the red clover, which sets seed only when visited by the humble bee; when this crop was introduced into New Zealand no seed was set till the humble bee, too, was introduced. On the other hand, Kirchner (1922) has shown that of the hundred odd European orchids, fifteen are habitually self-pollinated. The same is known to be true of about 150 exotic species. These are flowers of the type most highly specialised in relation to insect visits, and the autogamous species appear, in most cases, to be as much suited structurally as the others to insect pollination, and have indeed been frequently described as insect-pollinated. If this is so in a highly specialised flower, it is clear that only careful experimental work can tell us, in most cases, how much the flower really depends on the insect.

Cross-pollination is, almost certainly, the rule. This supports the view that out-breeding is advantageous, and emphasises the importance of this aspect of sexual reproduction. But we do not know much about the relative frequency of geitonogamy, nor of its significance. We know that autogamy occurs occasionally in many flowers, normally in many others, and exclusively in at least a few. This last fact seems to indicate that there is no essential *rejuvenating* action in fertilisation, and makes its *evolutionary* aspect more prominent. We do not know how general conditions, such as are exhibited in the maize, may be in wild species. The caution of Darwin's statement on out-breeding must be maintained, and its confession of ignorance is still applicable.

§ 17. The Seed and the Fruit

The essential result of fertilisation is the initiation in the ovum of the process of development leading to the formation of the new sporophyte, which shortly enters on a state of rest as the embryo of the seed. The critical stage in the transition from rapidly growing embryo to resting embryo is accompanied by a marked loss of water, which falls from round about 70 per cent. to round about 10 per cent. Of the causes governing the change we know little. Kidd (1914, Part II) has recently suggested that the inhibition of growth is connected with the narcotic action of carbon dioxide accumulated by the vigorously respiring embryo. The embryo may be accompanied by an endospermic food store of independent origin.

Subsidiary Effects of Fertilisation.—The effects of fertilisation are not confined to the initiation of growth and division of the ovum; they are felt, too, in various parts of the flower—that is, in the parent sporophyte. Most intimately connected with the embryo are the changes which result in the integuments of the ovule being converted into the seed coats. These changes consist in growth, keeping pace with the growth of the embryo, and, towards the close of development, in the drying out of the seed coats and in the formation of special, mechanically resistant and impervious layers of cells. A hard or leathery seed coat is protective in various ways ; it may protect the seed from mechanical injury, from desiccation, from digestion by animals using the fruit as food. It may also, as we shall see, prevent immediate germination. New structures may appear in connection with the seed coats in the form of an outgrowth from the funicle—*aril*, or from the micropyle—*caruncle*. The aril may be fleshy and brightly coloured, as in the outgrowth which has earned for the seed of the yew

FIG. 62.—Seeds with arils : 1, nutmeg (Myristica) ; 2, African lucky bean (Afzelia). Nat. size.

the courtesy title of " berry " ; or it may be dry and wrinkled as in the mace of the nutmeg (Fig. 62). The tufts of hair of such seeds as the cotton or willow are arillar in origin.

The aril has usually a function in connection with seed distribution, though in some cases it may assist in the opening of the fruit and liberation of the seed.

The Fruit.—After fertilisation has been effected, other parts of the flower, and particularly of the ovary, enter on a new phase of development, which results in the formation of that structure peculiar to the angiosperms—the fruit. In the simplest cases the fruit is derived from the ovary alone ; but many fruits, in the common acceptance of the term, include parts derived from the style, the floral axis, and even the calyx and the peduncle. A strict and uniform application of the term is not easy. Where the pistil is apocarpous the product of each carpel is entitled to be called a fruit, and the collection must also be called a fruit, especially in such cases as the strawberry, where it is set on a fleshy axis; such fruits may be distinguished as polycarpic. Again, in some cases, what is commonly regarded as a fruit is the product of a number of separate flowers, as in the mulberry or fig. These may be called collective fruits. Neither the statement that the fruit is the result of the fertilisation of a carpel or syncarpous ovary, nor that it is the result of fertilisation of a single flower, covers all the common fruits; and we must use the term rather broadly; this does not matter as long as we are aware of what exactly is involved.

The ovary enclosing the ovule, which develops into the fruit enclosing the seed, is the distinctive feature of the angiosperm. Biologically it may be regarded as affording greater protection and superior nutrition to the ovule, as well as greater protection, and the production of more varied means of dispersal, for the seed. The commonest change in the ovary, which usually denotes the setting of seed, is swelling. The ovarial wall becomes the *pericarp* or wall of the fruit. This may be leathery in texture, as in the broom, or hard and stony, as in the hazel nut. Its middle tissue or *mesocarp* may become fleshy, as in the tomato; and, in addition, its inner layer, the *endocarp*, may become hard and stone-like, as in the plum.

In epigynous flowers part of the floral axis forms the outside of the ovarial wall, and is concerned in the wall of the fruit. It is possible that this co-operation of the axis in the formation of ovary and fruit wall is a more perfect arrangement for nutrition and protection of the developing seeds. In berries, such as the gooseberry, and drupes, such as the walnut, the changes following fertilisation spread to the floral axis. More striking are those cases, like the strawberry, where the independent axis of a hypogynous flower swells up and becomes fleshy. In the rose hip the cup-shaped axis of the perigynous flower gives rise to the succulent and showy part of the fruit.

The peduncle rarely takes part in fruit formation, though it is frequently much strengthened in relation to, and perhaps as the result of, the increased weight it must bear. In *Anacardium occidentale*, the cashew nut, the one-seeded kidney-shaped nut is borne on a heart-shaped, fleshy swelling of the peduncle. In the fig the flowers are borne on the inner surface of an urn-shaped axis, from which the fleshy part of the fruit develops.

Usually the style withers away after fertilisation, but occasionally it undergoes development on totally new lines, and takes on fresh functions in connection with seed distribution. The styles of *Anemone Pulsatilla* and of *Dryas octopetalla* grow into long, feathered organs ; the style of *Geum urbanum* becomes mechanically strengthened and provided with a hook ; the styles of Geranium and Erodium also become strengthened with mechanical tissues of peculiar hygroscopic qualities.

The calyx frequently withers ; frequently it persists for longer or shorter periods, sheltering the young fruit. Occasionally it undergoes fresh development, taking on new functions. The pappus of the fruits of the valerians is an outgrowth of the rim which represents the calyx in the flower ; the pappus of the Compositæ may possibly be homologous with a calyx. In the tropical Anisoptera and Dipterocarpus sepals enlarge enormously and form great wings. In a few cases the sepals swell and form a fleshy envelope, *e.g.* in *Dillenia retusa* and *Stictocardia tiliæfolia.*

In many Compositæ the involucre closes in and protects the developing fruits, *e.g.* in the dandelion. In a few, as in Xanthium and Arctium, it persists as a resistant and hooked envelope serving in distribution.

Fall of Corolla.—The corolla and stamens take no part in these changes ; they wither and fall off. Often the individual petal, or the corolla as a whole, falls before withering sets in. The separate petals fall from the poppy, the rose, the rock-rose, the crane's-bill, and the flax; in the pimpernel, the comfrey, and the sage the gamopetalous corolla falls as a whole; in the daffodil, the dandelion, and the honeysuckle withering precedes the fall. It is well known that some flowers last only a single day, or even a few hours, and that withering sets in regularly whether pollination has occurred or not. Fitting (1909, *a* and *b*) states that in the orchid *Phalænopsis violacea* the corolla, which would otherwise remain fresh for a month, withers a day or two after pollination has taken place, while in *Geranium pyrenaicum* and *Erodium Manescavi* the petals fall about an hour after pollination. In these cases the stimulus is connected with pollination, for fertilisation cannot have taken place. The fall may be brought about in the case of the orchid by the application of dead pollen or of pollen of other plants. When the petals fall fresh there is a preformed abscission layer, as in leaves. The male flowers of many diclinous species fall as a whole, *e.g.* Begonia, Mercurialis, Populus (inflorescence). Abscission of female flowers may take place if pollination fails, as in the scarlet runner or the potato. Abscission of flower buds in unfavourable conditions, as when poisoned by coal-gas or dried up, is not infrequent. Abscission, whether of flowers or of inflorescences, may be regarded as the unloading of an organ which, while present, draws on the supply of water and food substances, and which has become useless, either as having performed its functions, or as having failed in the normal span to do so. The rapid withering of petals which do not fall suggests that changes in the water relations may be an important cause.

Post-floral Movements.—Another class of post-floral changes are those which result in specific movements, particularly of the peduncle. The most remarkable case is that of *Arachis hypogæa*, the earth-nut. The floral axis, between the calyx and the ovary, elongates very greatly, reaching a length of as much as five inches, and growing downwards so that the young fruit, which meanwhile remains small, is pushed into the soil. Only there does it begin to grow and mature. How so remarkable a habit can have arisen is difficult to understand. The fruit seems to draw both water and salts from the soil, and can ripen only when buried. This can hardly have been the primary meaning of the process. It is possible that, to begin with, a certain protection from preying animals was achieved, and that subsequently the fruit became more dependent on the soil covering.

In *Trifolium subterraneum* a post-floral movement of the peduncle leads to the burying of the fruiting head, and this is the case with *Voandzeia subterranea* and *Cyclamen europæum*. In the last the burying of the fruit takes place by a spiral inrolling of the peduncle, in the others by a positive geotropic curvature. After pollination the peduncle of *Vallisneria spiralis* coils up and draws the ovary to the bottom of the water, where the fruit ripens.

In *Linaria Cymbalaria*, the ivy-leaved toad-flax, the flowering peduncle is positively phototropic; after pollination the peduncle becomes negatively phototropic. On old walls we may frequently see the flowering peduncle carrying the flowers outwards, clear of the leaves, while the negative reaction after pollination brings the fruit into cracks of the wall where the seeds are shed. Here no actual burying takes place, and we have a case where the advantage secured by liberating the seeds in a place suitable for the growth of the plant is clear. The use of many other similar movements cannot be said to be understood. In *Tropæolum majus*, the garden nasturtium, the flowering peduncle is nearly erect and points forward from the foliage. A few hours after pollination a downward curvature takes place in the peduncle just below the ovary, and within 24 hours a second, very sharp, downward curvature takes place about 2 in. lower. The result is that the fruit is buried in the foliage. This movement has been analysed by Oehlkers (1921), who finds that it is chiefly due to a reversal in the geotropic reaction, taking place about the time of fertilisation, but independent of that process. In unpollinated flowers a slight movement also takes place. Fertilisation acts by inducing renewed growth in the peduncle, and through this the movement is accentuated and accelerated. One might be tempted to read into the burying in the foliage some protection from rain or from birds, but other species of Tropæolum which do not show the reaction do not seem to suffer. In Euphorbia a similar movement of the pedicle of the female flower brings the fruit hanging over the edge of the cyathial cup into a completely exposed position. According to Schmitt (1922) fertilisation is a necessary prelude to the post-floral movements of Althæa and Digitalis.

FIG. 63.—Scarlet pimpernel, showing pre- and post-floral movements of the peduncle. Nat. size.

In the scarlet pimpernel the peduncle before flowering is sharply bent just below the bud, so that the bud nods; it straightens before the flower opens; after flowering it curves downwards into a hoop, bringing the fruit among the leaves (Fig. 63). In *Agapanthus umbellatus* the bud stands vertically upwards, the flower is horizontal, the fruit hangs vertically down. The position of the flower is related to insect visits. Perhaps leverage is lessened by the position of bud and fruit, the heavier fruit hanging down. In Erodium, however, the bud is pendent and the fruit erect.

Goebel (1920) holds that the pre- and post-floral movements are related, based on the same mechanism, and changing in direction with the changing metabolism of bud, flower, and fruit. He does not think that, in most cases, they have much biological significance. There is room for experimental work on the subject. A detailed description of many cases is given by Troll (1922).

Post-floral Changes in Gymnosperms.—The post-floral changes in the gymnosperms may be shortly referred to. In most conifers there is considerable growth of the female cone, both of the axis and of the sporophylls, and this is followed by lignification and complete drying out. Movements may be very conspicuous. The cones of *Pinus sylvestris* are pendent when immature, erect at the pollination stage, and again pendent as they ripen. In a few genera a more or less fleshy " fruit " is formed. In Taxus the seed is provided with a fleshy aril ; in Phyllocladus the ovule becomes completely invested in an arillar outgrowth ; in Podocarpus the sporophylls become fleshy and brightly coloured ; the seeds of Juniperus are invested by the slightly fleshy scales of the female cone.

Induction of Post-floral Changes.—The statement that the post-floral changes in the ovary and other parts of the flower are the consequence of fertilisation is not always correct. We have already seen that an embryo may develop without fertilisation having occurred. Corresponding to this parthenogenesis we have *parthenocarpy*, where a fruit is produced without any seeds. This is the normal condition in the cultivated banana and in seedless oranges, where the fruit is the product of the ovary, and in the pineapple, where the axis of the inflorescence and the bracts are concerned. Where parthenogenetic seeds are produced, we might relate the development of the fruit to the development of the seed, but this cannot apply to parthenocarpic fruits.

PLATE VII

Linaria Cymbalaria.

The flowers are borne towards the light, away from the wall on which the plant grows; to the right two fruiting peduncles have bent away from the light and carried the capsules into a crevice.

The development of the orchid fruit is interesting in this connection. When the flower opens the ovules are quite immature and rudimentary. After pollination the ovary swells, and then the ovules develop and fertilisation occurs, the process often taking some weeks. The swelling of the ovary may be induced by the pollen of species completely sterile with the particular species, *e.g.* the pollen of Cypripedium can cause swelling in the ovary of Orchis. Here the beginning, at least, of fruit formation is the consequence of pollination, and not of fertilisation. On the other hand, we have the familiar fact that an apple often swells very little on a side, the ovules of which, by some chance, have not been fertilised ; and here the growth of the fruit is closely correlated with that of the seed. In marrows and cucumbers basal regions of the fruit, the ovules in which have not been fertilised, also fail to swell.

The development of the fruit is therefore a process which cannot be referred to the same causal factors in all cases. Normally it is related to the growth of the seed, but it may also be connected with pollination. In the case of the parthenocarpic fruits there may be an influence of the wound hormones postulated by Haberlandt.

§ 18. Dispersal

General Remarks.—We may look at seed dispersal from two somewhat different points of view. The spread of a species to new territory takes place almost solely when the seed is cast loose from the parent ; dispersal serves to distribute the species, and the further the seed is carried the more effective is the distribution. But the scattering of seed to a distance of a few feet, or even inches, may be yet more important, for it means that the young seedlings will come up a little apart, and so will not be subject to extreme competition with each other ; moreover, it seems likely that seed scattered is much less conspicuous than seed lying in a heap, and will thus escape to a greater extent the search of foraging birds.

As an example of a seed which secures wide distribution we may take the willow, with the tuft of hairs by means of which it can float on the wind for long distances. As an example of a seed which travels only a very short distance we may take the poppy. The capsule, opening by apical pores, is borne on a stiff, elastic stalk, which jerks back and forwards in the wind or when brushed against by an animal. The seeds do not fall out, they are thrown to a little distance.

The unit of dispersal may be either the seed or the fruit. The latter is the case in dry one-seeded fruits which do not open but become detached from the plant as a whole, like the achene of the buttercups, the nuts of the beech and hazel, and the acorn of the oak. In dehiscent fruits which open and liberate the seed, the seed is the dispersal unit ; and this is also the case in fleshy indehiscent fruits, the walls of which are digested by animals or rot away.

Dispersal by the Plant.—The apt term "censer mechanism" has been applied to the means of dispersal of the poppy. It is capable of greater refinement. In the genus Campanula the capsules, which are again borne on elastic stalks, open by three pores. In some species, *e.g. C. pyrenaica*, the capsules are erect ; in others, *e.g. C. latifolia*, they nod. In the former the pores are apical, in the latter basal, and thus in both cases the capsule opens at the *upper* end, so that the seed cannot fall out but must be thrown out (Fig. 64). In *Lychnis dioica* and many other Caryophyllaceæ the capsule opens by teeth which carry out hygroscopic movements. They curve back and open the capsule in dry air, while in moist or rainy weather they quickly bend up and close the entrance. The seeds are thus saved from the effects of wetting which would tend to mat them together ; they escape only in conditions favourable to their being thrown to some little distance (Fig. 65). In some plants of dry regions the opening movement takes place when the fruit is wet, *e.g.* in Mesembryanthemum.

In pods of the broom and whin the opening, by splitting into two valves, is again connected with a hygroscopic mechanism. Different tissue layers contract to different extents as they dry out, so that considerable tensions are set up between them. At the same time the cells of the separation tissue are weakening. There comes a point when the tensions overcome the resistance of the separation layer, which gives way suddenly. In the whin the valves

FIG. 64.—Campanula capsules: 1, *C. latifolia*, pendent capsules with basal pores; 2, *C. pyrenaica*, erect capsules with apical pores. Nat. size.

bend in, in the broom they flick into a spiral. In both cases the movement is violent and results in the seeds being thrown to a distance of some feet. The minute explosions of the pods of whin and broom, on a hot still day in late summer, is an intimate and characteristic country sound.

In the violets the fruit opens by three valves, and the polished seeds lie in these little boats after opening; gradually the edges of the valves shrink inwards and squeeze the seeds, till finally they are flipped out. In Oxalis the fruit again opens by valves; the seed has an elastic envelope of arillar nature which, often as the result of a slight touch, slips back and squirts the seeds out.

In these cases the tissues concerned with the "catapult"

FIG. 65.—Capsules of *Lychnis Flos-jovis*: 1, dry, with teeth reflexed; 2, wet, teeth closing capsule mouth. Nat. size.

movement are non-living, and the effective shrinkages and tensions are due to drying. In a smaller number of fruits the action of turgor tensions in a living tissue has similar effects. The most familiar example is that of the touch me-not, *Impatiens noli-me-tangere*. The five rather fleshy valves of the capsule possess each a layer, under the epiderm, with a very high turgor pressure and elastic walls. As the fruit ripens, the valves become separated from the internal septa, and are held together only very slightly, so that a touch or breath of wind is sufficient to liberate them. When this happens each valve rolls up inwards like a watch-spring, with great violence, and in doing so whips out the seeds which fly to some distance. In the famous squirting cucumber of Italy, *Ecballium Elaterium*, the internal turgor pressure extends the outer wall of the fruit ; at a certain point the stalk end is forced out, exactly like the cork of a champagne bottle, the fruit wall contracts, and the whole juicy contents including the seeds are squirted out.

Another type of movement, hygroscopic in nature, is exhibited by the awns of Geranium, Erodium, and of some grasses. The awn is derived from the style, which splits off in five strips from a central column. Each awn has attached to its base a bit of the ovarial wall partly enclosing a single seed. The Erodium awn on drying twists into a spiral, that of Geranium curls up ; on wetting they straighten out. These movements may be repeated indefinitely. They have been interpreted in two ways—as enabling the seed to " crawl " slowly over the ground, and as ensuring that it shall be pushed into some crack and so buried.

Dispersal by External Agency.—Three agencies may be effective—animals, the wind, and water currents.

A. Animal transport may be external or internal. In either case it gives the possibility of wide and rapid dispersal. Very evident is the animal relation to those fleshy fruits which are eaten by birds and mammals. The edible part may be the fruit wall or may be an aril, as in the fleshy cup of the yew, or the mace of the nutmeg, which is eaten by pigeons. We may note the frequent occurrence of red in the colouring as related to the fact that mammal and bird both see this in sharp contrast to green. An important point in internal transport is the resistance of the seed to digestion. The kernels of stone fruits must be very resistant both to mastication and digestion, but seeds which seem much less well protected may pass unharmed through the digestive tract, as is the case with those of the tomato or gooseberry. Of course successful resistance depends not only on the seed or fruit, but on the type of animal which eats it. Birds are the chief dispersing agents, at least in temperate countries, and some are much more destructive than others. The bird, we may note, is able to pick out not only fleshy fruits, but many inconspicuous hard seeds and fruits. Many are destroyed by the pecking, and still more in the powerful gizzard, *e.g.* of such birds as the duck. According to Birger (1907) most seeds pass undamaged through the digestive canal of birds like the fieldfare. The efficiency of this type of dispersal is shown by the rapid spread of berry-bearing shrubs in woods and on heaths ; the case of the snowberry, *Symphoricarpus racemosus*, which has spread vigorously in this country since its introduction from North America, is a good example. Little, however, is known of the distances to which plants may be spread in this fashion ; probably scattering over relatively small areas is of most importance.

A special case of considerable interest is the dispersal of the seeds of the mistletoe and of other Loranthaceæ. These parasites can grow only if the seed is deposited on the branch of a suitable tree. The missel-thrush is specially fond of mistletoe berries but does not eat the seed ; it rubs this off against the branch on which it perches, and the seed becomes fixed by the viscid internal flesh of the berry. The same method of dispersal has been described for Singalese Loranthaceæ by Keeble (1895). It is also stated that the mistletoe berries are voided by the missel-thrush on branches and are, in this case too, fixed by the undigested slime, but the other method seems to be the more important.

A very large number, especially of woodland plants, are dispersed by ants. The ant picks out seeds which are provided with special oil bodies or *elaiosomes*, which may be of diverse morphological nature. Most conspicuous is the arillar type seen in the little orange elaiosome of the seeds of whin or broom, and of which a magnificent example is given by the African lucky bean, Afzelia (Fig. 62). Sernander (1906), to whom we owe a detailed monograph of this mode of dispersal, distinguishes between the following types of elaiosome :—

(*a*) In the whin, broom, violet, and others the elaiosome is arillar.

(*b*) The basal part of the fruit wall is differentiated as an elaiosome in Fumaria and *Anemone hepatica*.

(*c*) The basal part of the calyx forms the elaiosome in *Parietaria lusitanica*.

(*d*) Part of the floral axis forms the elaiosome in Ajuga and other Labiatæ, and in some Scrophulariaceæ.

(*e*) A part of the peduncle functions as elaiosome in *Aremonia agrimonioides* and *Thesium alpinum*.

(*f*) In *Carex digitata* and other Carices the elaiosome is bracteal.

(*g*) In *Melica nutans* and other grasses the elaiosome is derived from the inflorescence axis.

(*h*) In Allium and many other monocotyledons the thin outer seed coat is impregnated with oil.

Observations on a large number of seeds and fruits showed that they were vigorously sought for by ants ; seeds from which the elaiosome had been removed were, on the whole, selected less frequently. Transport for a distance of over 70 yds. was observed. Sernander holds that carriage for a small distance from the parent plant is the most important result of dispersal by ants. Plants with ant-dispersed seeds are most numerous in woods, where the bare soil allows the ants freer movement. In the vegetation of a Swedish water meadow, out of 32 species 2, or 6 per cent., were ant-dispersed, while in a neighbouring wood there were 9 out of 35, or 25 per cent. The importance of clear soil is strikingly illustrated by an observation on the dispersal of whin seeds on an English heath by Weiss (1908). The seedlings sprouted only along the margins of tracks on the bare soil of which the ants could move, never among the vegetation.

Ant-dispersal leads us to dispersal by birds or mammals, which carry seeds and fruits sticking on their feathers and fur. The fruits of *Galium aparine*, the goose-grass, are provided with many small hooks which readily become firmly attached to the fur of passing rabbits and sheep. The bracts of the burdock are similarly hooked. The seeds of some species of Juncus have a mucilaginous coat which sticks to the feathers of aquatic birds. Many birds may carry small seeds in mud on their feet or feathers. Birds especially may thus be responsible for transporting seeds over very long distances. Guppy (1917) puts down the dispersal of European species of Juncus and Luzula to the Azores, to the carriage of their sticky seeds on birds' feathers. Nearly 20 per cent. of the species of the flora which has developed on the island of Krakatau since the eruption in 1883, have, according to Ernst (1908), been transported by birds. The distance to the nearest island not affected by the eruption is over 12 miles, and the distance to the Javan coast is some 25 miles.

B. A great many Seeds and Fruits are transported by the Wind.—Many are suited to this mode of dispersal simply by their minute size. Seeds of orchids and of many Ericaceæ are like very fine dust. An orchid seed may weigh only one five-hundredth part of a milligramme. Plants with such seeds were amongst the first colonists of Krakatau, observed three years after the eruption. They must obviously remain in the air for long periods. It is of interest that most epiphytes have seeds of this type. There is no direct method of reaching a favourable station, but minute size and enormous production mean that some of the crop will be able to germinate on suitable trees.

Many fruits and seeds have special "flying" organs. These take the form of plumes of hair, or of wing-like expansions. The latter are well seen in the fruits of the maples and the ash; in the lime the bract functions as a float. Winged seeds in our native flora are those of the pine and the birch. In many tropical trees the seed wings are very large and beautiful, as in Bignonia. Studies by Ridley (1905) on tropical species have shown that the wing does not keep the seed or fruit long afloat; it seldom travels, even in a fair wind, more than two or three times the height of the tree. Rapid spread of trees, which only fruit after many years' growth, is not to be expected (Fig. 66).

The hairy tufts and plumes of such seeds as the cotton, the silk-weeds (Asclepias sps.), the willow herbs, the willow, and such fruits as those of the Compositæ and cotton-grass are much more efficient. Small (1917) has shown that the pappus of the Compositæ enables the fruit to act as a glider in very light breezes. The fruit of the dandelion is kept afloat indefinitely in a wind of only 2 miles per hour, while that of the colt's foot requires a wind of only 0·5 mile per

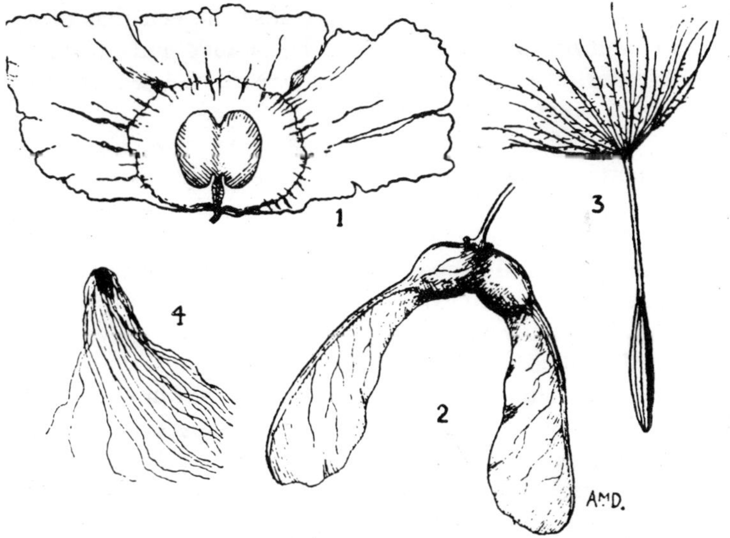

FIG. 66.—Dispersal of fruits and seeds: 1, winged seed (Bignoniaceæ); 2, winged fruit of maple; 3, plumed fruit of goat's-beard; 4, hairy fruit of cotton-grass. Nat. size.

hour. The possibility of dispersal of such fruits is almost limitless and must account for the ubiquity of such genera as Senecio. Possibly twenty-eight species of the new flora of Krakatau (30 per cent. of the whole) have been carried there by wind.

C. Dispersal by Water Currents is well illustrated by the occurrence of alpine plants on river shingles far below their proper level and many miles from their natural stations. Yet dispersal by such means is quite limited in its extent and in the number of species affected. Fresh-water plants have often a very wide distribution, but this is probably in the main due to bird dispersal, and not to the agency of the water.

Dispersal by ocean currents is much more important, though it is also limited almost entirely to the carriage of strand and other shore plants, in particular those of the tropics. It has formed the subject of much investigation, especially by Schimper (1891) and Guppy (1906, 1912, 1917). Perhaps Darwin was the first to realise the fact that many seeds and fruits could float for long periods undamaged in salt water. Seeds and fruits may reach the sea through rivers, floating independently, or wedged in crevices of logs, but it is certain that inland plants, even when they are transported by ocean currents, are practically never cast up in a situation in which they can maintain themselves. The seeds of strand and mangrove plants are frequently carried for hundreds and thousands of miles before being thrown up, and after such voyages they often germinate and may become established. At least forty species have been carried to Krakatau by ocean currents.

An instructive case is afforded by the floras of the two types of mangrove forest. The Eastern Mangrove has a rich flora very uniform along the coasts of East Africa, India, and Malaya. The Western Mangrove has a poor flora, the important species being the same on the west coast of Africa and the east coast of tropical America. The two types have no species in common. Distribution throughout the two regions has been entirely by ocean currents. Guppy states, for the mangroves and their associates of the western region (*Rhizophora Mangle*, *Avicennia nitida*, *Laguncularia racemosa*, *Anona palustris*, *Carapa guaianensis*), that all are capable of floating in sea water for at least two months, and that all could be carried by the main equatorial current from West Africa to Brazil. The case of the mangroves is of special interest, as the fruits are viviparous, and it is the germinating seedling which is carried.

Curiously, the fruit which has for long been the type

example of ocean carriage, the coconut, has been shown recently by O. F. Cook (1902, 1912) to be quite unfitted for this mode of transport. The buoyant fibrous mesocarp and thin cutinised exocarp, which look so well suited to keep the fruit afloat, are really related to germination in arid conditions. The coconut is often cast up in the drift on tropical shores, but the coconut trees seen by Ernst on Krakatau, which had evidently sprung from sea-borne fruits, are perhaps the only authentic instance of the tree establishing itself. It is originally an inland plant which owes its wide distribution along the tropical seaboard solely to the fact that it is a valuable cultivated plant carried everywhere by the native races of the Pacific. A number of fruits which certainly owe their dispersal to ocean currents possess light flotation tissues, *e.g. Barringtonia speciosa*, but many others have no features which are not seen in the fruits of many inland plants.

CHAPTER VI

DEVELOPMENT

§ 1. Vitality of Seeds

The seed, when it is ripe and becomes separated from the parent plant, characteristically contains an embryo in which marked differentiation has taken place. Radicle and cotyledons are present, together with a stem growing-point, which may have already given rise to an axis with several leaf rudiments—the plumule. In the final stages of its formation the seed has lost much water, so that it may contain only about 10 per cent.; sap is no longer present in the vacuoles, and in this condition chemical reactions, even respiration, are reduced to a minimum. In ordinary air, which is not very dry, many seeds continue to respire very slowly, giving off small amounts of carbon dioxide. According to White (1909) the wheat is an example of such a seed, while the oat does not respire appreciably. When seeds are specially desiccated respiration almost always ceases. In such seeds, and in those stored in absence of oxygen, slight anærobic respiration may go on. Becquerel (1907) has shown that some seeds may be desiccated and kept in a vacuum for two years without losing their vitality. It is the essence of the seed's power of rest that reaction is reduced almost to the vanishing point.

There is, through prolonged periods, no appreciable diminution in the food stores, no growth, and no deterioration of the protoplasm. When vitality is finally lost the cause may be, as Crocker (1916) believes, a slow and cumulative coagulation of the proteins of the living matter. White has shown that wheat seeds still contain active enzymes when 20 years old—that is, many years after losing the power of germination.

Seeds in Dry Air.—The seeds of some species of Oxalis germinate as soon as they leave the capsule and die quickly if exposed to dry air. The seeds of the willows retain their power of germination for a few days only. The seeds of the cocoa, the coconut, the para rubber are also short-lived, making their transport difficult. The overwhelming majority of seeds, however, can lie dormant for a few months, at least till the growing season following their formation. In some, *e.g.* the beech, the percentage of germination falls greatly by the second year, but many, perhaps most, can rest for longer. The power of germination of the wheat is lost, according to White (1909), only after from 11 to 16 years, of the barley after 8 to 10, of the oats after 5 to 9, and of the maize and rye after 5 years. The possibility of still longer periods of potential rest is difficult to control. Many exaggerated statements have been made, as that the mummy wheat from Egyptian tombs is still capable of germination, which is certainly not true. Exact investigations on herbarium and stored seed of known age have been made by Ewart (1908) and Becquerel (1907), and a review of the subject has been written by F. F. Blackman (1909). Becquerel tested the germinating capacity of seeds of known age up to 135 years in the herbarium of the Natural History Museum of Paris. He found 7 species, 4 Leguminosæ, and 1 each Malvaceæ, Labiatæ, and Nelumbiaceæ, which retained their power of germination after 50 years, the extreme case being *Cassia bicapsularis* which germinated after 87 years. Of 500 species tested 50 possessed seeds which germinated after 25 years. Of the seeds which had lain in the herbarium from the reign of Louis XVI none germinated, nor did those of the Revolution, nor even those of the 1st Empire. Ewart also found that the longest-lived seeds were Leguminosæ. In general, the long-lived seeds are hard coated with a resistant testa which prevents water absorption while intact and is impervious to gases.

Seeds in Soil and Water.—These are cases of seeds stored in ordinary dry air, but many seeds can remain dormant for long periods in moist soil or in mud. It is well known that on the mud of drained ponds, or on the soil of old pastures and woodlands freshly turned up, a vegetation rapidly appears with many species not occurring under the previous conditions. Some of these may arise from seeds recently transported, but others must spring from seeds which have long lain dormant in the soil. Peter (1893) found that forest soil to the depth of a foot contained seeds capable of germination when the soil was turned up and kept moist. In the soil of primitive forest, seeds of woodland species only, such as the strawberry and raspberry, were found; in the soil of woodland planted on cultivated land 20 to 40 years previously, seeds of such plants as shepherd's purse, charlock, and plantain sprouted; it is difficult to avoid the conclusion that these had lain dormant since planting, or at least since the shade of the trees had sufficed to expel the previous flora. Brenchley (1918) found that when grassland which had replaced arable was ploughed up, seeds of arable weeds (16 species in one case after 10 years in grass) sprouted which had lain dormant in the soil for periods up to 58 years. Such weeds never appeared on grassland turned up for the first time. Darlington (1922) reports on a controlled experiment started by Dr. W. Beal. Seeds in moist sand placed in open bottles were buried 3 feet deep. After 40 years 10 out of 22 species were still viable. These included *Rumex crispus*, *Plantago major*, and *Amarantus retroflexus*. It is of interest that under these conditions Trifolium did not germinate after 5 years, nor Malva after 20 years. G. H. Shull (1914) showed that the seeds of many land plants might remain dormant in glass jars submerged in mud and water for periods of 4 to 7 years. We may recall, too, Darwin's and Guppy's experiments and observations, which showed that seeds of many plants could retain their vitality in salt water for as much as a year.

Retention of vitality in the presence of abundant water, and particularly the failure to germinate, is more difficult to understand than the dormancy of dry seeds. In seeds with impermeable coats it may be due to failure to absorb water while the coat is uninjured. In many cases this explanation fails, and the chief factors must be lack of oxygen and excess of carbon dioxide. The importance of the latter has been emphasised by the work of Kidd (1914, 1917), and Kidd and West (1917, 1920) on *Brassica alba*. They have shown that seeds saturated with water fail to germinate if kept in an atmosphere with a suitable percentage of carbon dioxide, the gas having a narcotic effect. At low temperatures and low oxygen pressures narcosis is secured with less carbon dioxide. Many seeds which showed this effect, *e.g.* beans, cabbage, barley, peas, onions, germinated as soon as they were removed from the inhibiting atmosphere. In the mustard, and presumably it is not unique in this, the inhibition continued indefinitely after the seeds were replaced in normal air, and was only removed by complete drying and re-wetting, or by removal of the testa. Kidd further found that this type of dormancy might be caused by natural conditions in the soil, if abundant organic matter, the decay of which liberates considerable quantities of carbon dioxide, is present. The conditions of low oxygen and high carbon dioxide content in the mud of ponds, or in forest soil, may thus induce dormancy. When the mud or soil is turned up, partially dried, and exposed to free aeration, the inhibiting conditions are removed and germination may take place. We may here recall Kidd's suggestion that the arrest of development of the embryo in the maturing seed is due primarily, not to desiccation, but to the accumulation of a narcotising concentration of carbon dioxide within the testa.

Biological Effects.—The capacity of the dry seed to

retain vitality may be taken as an expression of its powers to resist extreme conditions. Perfectly dry seeds can, in fact, withstand not only the implied drought, but also remarkable extremes of temperature. The temperature of liquid hydrogen (−234° C.) has no harmful effect, and many seeds can withstand a temperature of 100° C. for two days, though the water-saturated seed is quickly damaged by much lower temperatures. In nature it may be necessary to withstand desiccation for long periods, especially in arid climates, where, too, the temperature of the surface soil may rise very high. In temperate countries, however, the seed is probably not thoroughly dried out after it has reached the soil; it may endure low, though not very low, temperatures. The maturation of the seed and its separation from the parent plant entail a cessation of the water supply, and it is in relation to this that the seed has evolved as a dry structure with limited reactivity; from this has arisen its great and prolonged power of resistance as a secondary function which has come to be of great importance.

The forced dormancy of seeds in the soil or under water has a somewhat different significance. While seeds, like those of many leguminous plants, with impermeable seed coats really present cases of inhibition due to desiccation, the others may be regarded as being held dormant in conditions which would be unfavourable to the development of the young plant; the seed immersed in water, or buried deeply in the soil is a case in point. Perhaps it is of even greater importance that the whole of the crop does not germinate at once. The longevity of the seed, combined with dormancy, means that the offspring of a single parent, or year, may go on germinating through a series of seasons; and this may be a real advantage.

§ 2. Dormancy of Seeds

Apart from forced dormancy, as through carbon dioxide narcosis, many seeds, from the nature of the embryo or the structure of the coats, require a more or less prolonged and definite period of "after-ripening" before they can germinate in suitable conditions; the case of the hard-coated seeds is only one of many.

1. **Immature Embryos.**—In some seeds the embryo is not completely formed when separation from the parent occurs. In Ginkgo, when cultivated in Europe, *fertilisation* does not take place till after the seed has fallen. In Japan, its native region, fertilisation has taken place and the embryo is sometimes mature, sometimes not—an unusual instance of plasticity. In Ceratozamia and Gnetum the embryo is small and grows after the fall of the seed. The seeds of *Ranunculus Ficaria*, *Anemone nemorosa*, *Corydalis cava* and a few other dicotyledons contain, when they fall, a small, undifferentiated embryo in which development proceeds slowly through the autumn and winter, and is complete just before germination in the spring. Growth of the embryo also takes place in the ivy and probably in some monocotyledons, *e.g.* *Gagea lutea* and *Paris quadrifolia*; Goebel's account should be consulted. It will be noted that these are woodland plants with a markedly vernal vegetation period, and that this slow after-ripening of the seed leads to germination taking place early in the year following seed formation.

We may here refer to a class of seeds already considered in another connection,—those of the orchids and of such parasites as the broom-rapes. The seeds are minute and contain an undifferentiated embryo capable of further development, and of germination, only in special conditions —in the presence of the appropriate fungus in the orchids, and of the root of a possible host in the parasites.

2. **Other Embryonic Conditions.**—The seeds of *Cratægus mollis*, and probably of other hawthorns, contain a fully developed embryo which is, however, incapable of immediate germination. Davis and Rose (1912) found that, even in the absence of the seed coat, germination did not occur. A slow process of after-ripening takes place, lasting two to three months if the carpel wall is removed, and about one month in the absence of the seed coat.

The same type of dormancy is shown by the apple, the elder, and the lime (Rose, 1919). Eckerson (1913) showed that in the hawthorn the acidity of the embryo increased as after-ripening proceeds, and this must affect metabolism and particularly enzyme action. For *Juniperus virginiana*, Pack (1921) found that during after-ripening acidity increased, as did soluble sugars, phosphatides, and nitrogenous compounds, and the activity of enzymes. An increase in enzymes and acidity was found by Rose in the lime. The low temperature (0° to 5° C.) which is generally favourable to after-ripening may aid the accumulation of cell-building material by keeping down respiration; but that respiration is necessary is suggested by the quicker ripening of the hawthorn seeds with the coats removed. The after-ripening process may evidently be regarded as a period of mobilisation of cell-building material.

3. **Seed Coat Effects.**—Much more numerous are the cases in which the seed coats and fruit walls inhibit germination, which takes place at once when these are removed. The coats may act by preventing absorption of water, or intake of oxygen, or by mechanical restraint. In *Alisma plantago*, Crocker and Davis (1914) found that water absorption takes place readily; the embryo swells and presses against the restraining walls with an imbibition force of about 100 atmospheres, but cannot rupture them and so is unable to germinate. Treatment with acids and bases induces germination as was shown by Fischer (1907) for Sagittaria. It is likely that these act by altering the mechanical nature of the walls. The achenes of Alisma and of many other water plants such as Sagittaria, Potamogeton, Hippuris, Scirpus, and Sparganium, may lie dormant in the mud for many years, and resistance of the fruit wall seems to be the general cause, though oxygen relations may sometimes be involved. The hard endocarps of the bramble druplets delay germination (Rose, 1919). In nature the walls may perhaps be gradually affected by bacterial action or by acids produced in decay.

The hard-coated seeds of the Leguminosæ, Labiatæ,

and other families have already been mentioned. The seeds of the broom, for instance, may lie for months on moist filter paper without change ; a slight scratch on the seed coat is at once followed by absorption of water, the seed swells, and germination follows in a day or so. This behaviour is of practical importance in such commercial hard seeds as the red clover and the spinach, which are sometimes artificially abraded to secure more rapid and complete germination. De Vries (1915) found that the germination of Œnothera seeds was improved by forcing water into them under pressure.

A number of cases are known of seeds and fruits in which the wall interferes with the supply of oxygen. The most fully investigated is that of the fruits of Xanthium, the cocklebur (C. A. Shull, 1911, 1914). Germination takes place if the walls are removed, and may also be secured by placing intact fruits in a high concentration of oxygen. The increased supply of oxygen may be necessary for respiration, or it may remove the narcotic effects of carbon dioxide. Shull and Davis (1923) show that the enzyme catalase increases at germination. In nature germination takes place, curiously, at different times in the two fruits of each bur. The coat of the upper seed is the more resistant and the seed germinates two years after formation, while the lower seed germinates a year earlier.

When dormancy is due to the structure of the seed coat or fruit wall, germination can take place only after some alteration in these. This may be spontaneous, as in Xanthium, probably, as Crocker (1916) thinks, through slow changes in the colloids in the cell wall. External agencies may also be active. There is the possibility of bacterial action. Freezing is important ; in many seeds, as Kinzel (1913, 1915, 1920) has shown, germination in nature takes place only after freezing. The seeds of *Menyanthes trifoliata*, *Teucrium Chamædrys*, *Gentiana lutea*, *G. nivalis*, *Adoxa Moschatellina*, require increasing sharpness of frost in the order given. The last two must be exposed to temperatures of $-20°$ C. If the coat is water-saturated and freezes, it must be stretched ; with hard-coated seeds freezing of the soil may act through the increased pressure of sharp soil particles. Abrasion by soil particles is likely to be important, and may be intensified for those seeds which pass without damage through the gizzards of birds or even of earthworms. Exact investigation of such points is much required.

4. **Light Effects.**—It has been long known that some seeds, such as those of the tobacco, require light for their germination. The mistletoe was thought to require after-ripening which terminated only late in the spring, but Heinricher (1916) has shown that it will germinate at any time if it is artificially illuminated ; it merely requires a favourable temperature and a large quantity of light. In recent years investigation has shown that a considerable number of seeds can germinate only in light or have their germination promoted by illumination. Examples are *Chloris ciliata* (a South American grass), *Epilobium hirsutum*, *Veronica longifolia*, *Lythrum Salicaria*, *Ranunculus sceleratus*, *Rumex crispus*, and many Gesneriaceæ. A smaller number can germinate only in the dark, *e.g.* *Nemophila insignis*, *Phacelia tanacetifolia*, *Veronica Tournefortii*, *Nigella damascena*. The cases now known will certainly be added to, and it may be necessary to revise the general impression that, while darkness is the normal condition during germination, most seeds are indifferent to illumination.

The conditions of germination in such seeds are very complex. We may take first the case of *Chloris ciliata* investigated by Gassner (1910 I and II, 1911). The fruit, enclosed by the tight-fitting glumes, goes through an after-ripening period of about eight months, during which changes in the embryo probably occur. After this it germinates only in the light, unless the glumes are removed, when it germinates equally well in the dark. If after removal of the glumes it is kept moist and dark at a temperature too low for germination (12° C.), then on subsequent transference to a suitable temperature (33° C.) it now requires light, unless the coat at the micropilar end is removed, when it will again germinate in the dark. In the light, after-ripened seeds show good germination in about four days; if after two days in the light, when no germination has occurred, the seeds are transferred to the dark, a considerable percentage germinate subsequently; the effect of the light persists. If after-ripened seeds are kept in suitable conditions for germination (moist at 33° C.), but in the dark, for about a week, subsequent germination in light is very much depressed; a secondary dormancy is induced, very much resembling that of the Brassica seeds already described. The blue end of the spectrum is effective, the red end acts as darkness. The seeds germinate in the dark in rich soil, on Knop's solution, or in presence of nitrogenous compounds, especially of nitrates and nitric acid. They germinate in the dark if exposed to alternating temperatures, *e.g.* 22 hours at 20° C. and 2 hours at 34° C. If the seeds are not fully after-ripened there is a beneficial action of light at 34° C., but not at 16° C.

Lehmann (1911) found that the "light" seeds of *Epilobium roseum*, and some other plants, would germinate in the dark if exposed to a sudden rise of temperature, while Gassner (1915*a*) found that they would germinate with alternating high and low temperatures. Lehmann (1912) found that, within the limits of temperature favourable to germination, certain "light" seeds would germinate in the dark at high temperatures, and certain "dark" seeds in the light at low temperatures. This is illustrated in Table XXXV.

TABLE XXXV

RELATION OF LIGHT AND TEMPERATURE TO GERMINATION

Temperature.	*Nemophila insignis* ("dark" seed). Per cent. germination.		*Epilobium hirsutum* ("light" seed). Per cent. germination.	
	Light.	Dark.	Light.	Dark.
10–12° C.	87	88	—	—
20° C.	—	—	78	10
21° C.	2	75	—	—
24° C.	0	35	60	3
31° C.	0	0	65	55

The presence or absence of light is not necessarily an absolute bar to germination, and the action of light is not independent but is deeply affected by the temperature relation and other conditions.

Ottenwälder (1914) showed that in some cases a surprisingly low intensity of light was effective; germination of *Epilobium roseum* took place at 25° C. in light of 1/400 candle-power. Lehmann (1918*a*) investigated this point more fully and found that continuous illumination was not necessary, and that the amount of light required for the germination of *Lythrum Salicaria* was extremely small, as is shown in Table XXXVI.

TABLE XXXVI

RELATION OF LIGHT TO GERMINATION OF *Lythrum Salicaria*

Temperature.	Illumination.		Per cent. germination.	
	Intensity in Hefner C.P.	Duration.	Light.	Dark.
20° C.	730	15 mins.	12	0
,,	,,	1 min.	5	0
30° C.	,,	1 sec.	30	1–2
,,	,,	1 min.	47	1–2
,,	,,	5 mins.	78	1–2
,,	,,	continuous	99	1–2
,,	2	1 min.	22	2
,,	40	5 secs.	34	2
,,	730	5 secs.	41	2

Ottenwälder (1914), and Lehmann and Ottenwälder (1913), found that acids, and proteolytic enzymes activated the germination of "light" seeds in the dark. Gassner (1915 *b* and *c*) obtained germination in the dark of various "light" seeds with nitrates and nitric acid—which acts as a nitrogen compound, and not as an acid—though with others there was no such result. Lehmann (1919) obtained germination in light of the "dark" seed of *Veronica Tournefortii* in potassium nitrate. Hesse (1923) found all "light" seeds to germinate in the dark with nitrogen compounds, including those with which Gassner had failed. Some few are also activated by acids. Magnus (1920) showed that the germination of the " dark " seed of Phacelia, which takes place in the light with acid, is " false," and is due to a mechanical forcing of dead embryos from the seed coats. Hesse agrees that this is so for most cases of acid activation.

The net result of these investigations seems to be that certain seeds under " ordinary " temperature conditions germinate only in the light, and others only in the dark ; that, in the former case an abnormally high, in the latter a low, temperature makes the seed indifferent to light or dark ; that in the *inappropriate* light condition germination may be secured by treatment with nitrogenous compounds, or with certain enzymes, or in some cases by removing the coats. We may ask what is the mechanism of this relation to light. Three theories have been put forward.

(*a*) Lehmann holds that light acts in the " light " seed by catalysing the conversion of reserve proteids into soluble compounds ; this interpretation is supported by the action of the proteolytic enzymes, and perhaps by the extremely small light exposures which are required to give an improvement in germination. In " dark " seeds the light is supposed to activate fluorescent organic substances which destroy the proteolytic enzymes.

(*b*) Gassner analyses the conditions for Chloris into three divisions : I, The embryo requires a certain after-ripening ; II, the glumes act by preventing the access of oxygen, and light has some effect in counteracting this ; III, light further acts by destroying a substance inhibiting germination, which is produced in the seed coat in conditions of temperature and moisture favourable to germination. Gassner and Hesse apply this explanation, *i.e.* of inhibitors, to light germination in general. It receives some support from experiments of Magnus on the " dark " seed of Phacelia. In low light intensities, germination to the extent of about 30 per cent. of the seeds can be obtained ; this is depressed by placing the seeds in water in which other seeds have been soaked. The conclusion is drawn that an inhibitor is present which acts only in light, or which strengthens the action of light, in this case an inhibiting one.

(*c*) Crocker (1916), in an extensive review of the whole question of dormancy, suggests that light acts by altering the condition of the seed coats in their relation to oxygen and water supply. This view is supported by much of the work of Gassner and Hesse. They have shown definitely that the action of nitrogen compounds, which cause germination in the dark, is on the coats; for concentrations, *e.g.* of nitric acid, can be used which would be fatal if they penetrated to the embryo, and which indeed are fatal if applied to seeds with damaged coats. Treatment with absolute alcohol, subsequently washed off, may also produce dark germination of "light" seeds. Fluctuating temperatures, and sudden rise in temperature, may well act by causing a change in the colloids of the cell walls. The germination in the dark of dark inhibited seeds, with the coats removed, points in the same direction, and a similar result has been obtained by Gardner (1921) for *Rumex crispus*. At the same time it must be admitted that the relations are so complex, and vary so much from species to species, that we have at present no clear evidence decisively in favour of any one of these hypotheses; it is quite likely that the mode of action of light or darkness may be different in different cases, and that more than one effect may be present.

The biological aspect of the light relation resembles that of other types of dormancy. We find again the necessity of special conditions for germination, or, looking at it from the other side, of special conditions in which dormancy occurs. We have again the possibility of secondary dormancy induced by the action of an external factor. Gassner (1910) has related the conditions in Chloris to the environment of the plant and we give his account. Chloris is a typical summer grass of the South American pampas. ("Summer" occurs in the months December to March.) It flowers from December to March and fruits from January to April. Throughout the cold season the seeds are prevented from germinating partly by the low temperature, and partly because of the necessity of after-ripening. In October and November germination takes place in the light; the dark night periods have no inhibiting effect because, although the day temperatures are high, and thus favourable, the night temperatures are low. There is an average difference of 20° C.

Summary.—We have considered this question of dormancy and delayed germination in some detail because it is an admirable example of the variety of ways in which external factors may influence the development of the young plant and of the complexity of the interaction of these factors with the special constitution of the embryo and its protective coverings. In a great variety of ways the germination of the seed is postponed and its vitality retained, often for long periods. Adverse conditions are guarded against. Germination is delayed till a favourable season. The sprouting of a single crop is spread out over a number of years, and the chance of obtaining completely or exceptionally favourable conditions by some of the offspring is greatly increased.

§ 3. Viviparous Seeds

We may here refer to the opposite conditions found in the true viviparous seeds which germinate before separation from the parent. The most striking example is offered by the mangroves, *e.g.* Rhizophora and Bruguiera, in which the seedling has extended its hypocotyl to a length of as much as 18 in. before it falls from the tree. In the mangroves, as in some other cases, *e.g.* Crinum and Melocanna, the integuments are reduced or absent. Kidd (1914) suggests that the immediate germination, or rather the continuous development, is due to the absence of auto-narcosis by the respiratory carbon dioxide, which this condition of the integuments permits. Biologically the vivipary of the mangroves is important in securing the fixation of the seedling in mud subject to tidal submersion.

The seedling drops straight from the tree into the mud, and is, as it were, dibbled in; fixation occurs by the rapid development of a shallow root system.

§ 4. Germination—Conditions

When the seed is finally capable of germination, this can take place only if external conditions are favourable. Most important are water and oxygen supply and temperature, though we may look on light as a necessary condition for the seeds showing the fourth type of dormancy.

The amount of water absorbed by the seed may be very large. In the wheat it ranges from 45 to 60 per cent. of the dry weight, in the maize from 35 to 40 per cent., in the pea from 84 to 106 per cent. The absorption is at first entirely due to the imbibition of the colloids of the embryo, endosperm, and seed coats; later, when the cells have become vacuolate, or rather the collapsed vacuoles have filled, osmotic suction sets in. Though, as we have seen, seeds can absorb water from quite dry soils, it does not follow that germination will occur in such conditions. Even when the supply of moisture is sufficient to saturate the seed and permit the extrusion of the radicle, it may not enable the seedling to establish itself. The decisive effect of a liberal water supply is clearly seen in the great crop of weeds which spring up after summer showers following a dry spell; or still more vividly in the sprouting of desert ephemerals as soon as the rains set in.

A special case of much interest is that of the coconut in which germination takes place very soon after ripening at the expense of the store of water in the seed, evaporation of which is lessened by the thick husk of coir within the pericarp. The leaves appear first, and several may put forth before the adventitious roots are produced in wet weather, and enable the plant to establish itself. In this fruit, which retains its viability only for a short time, it is important that germination should be able to occur in the dry seasons—when the largest crop ripens—and without the necessity of burial. Peculiar modes of germination in other palms are described by Cook (1912).

Temperature affects the rate of water entry into the seed and the growth rate, the latter much more than the former. Temperature may also affect the resistance offered by the seed coats to the extrusion of the radicle. Many determinations have been made of the "minimum," "maximum," and "optimum" temperatures for germination. Thus it has been found that the minimum temperature at which rye will germinate is about 1° C., the maximum 30° C., and the optimum 25° C. For barley the figures are 3° C., 28° C., and 20° C., and for maize 8° C., 44° C., and 32° to 35° C. The minimum points are of some interest in showing at what temperature the seed will lie dormant in the soil without the action of any other factor. They mean that germination can take place rapidly only at considerably higher temperatures. As Blackman (1905) has pointed out, "optimum" points are purely fictitious, varying especially with the time through which the factor in question has operated. This comes out very clearly in the case of germination, where the temperature at which the radicle appears most quickly may be so high as to inhibit or depress further growth; that is, for a short time the high temperature causes rapid growth, which quickly declines owing to the action of the "time factor." With some seed, *e.g.* the barley, and Petunia, a daily alternation of high and low temperatures is required to give good germination (Harrington, 1921); as we have seen, this may be connected with alterations in the seed coats.

§ 5. Germination—Liberation of the Embryo

The emergence of the radicle from the seed takes place against the resistance of the enclosing walls, the seed coats, or, in indehiscent fruits, the fruit walls. This resistance may be quite small, as in thin-coated seeds like the pea, or it may be large, as in the castor-oil bean with its hard testa. Familiar examples of strong walls are such seeds as those of the plum, or the coconut, which remain enclosed in the strong endocarp of the fruit, or of the sunflower and other Compositæ, and the hazel-nut, where the whole fruit wall is hard and persistent. The imbibition forces of the colloids of the swelling seed may amount to hundreds of atmospheres, but do not usually come into play. In indehiscent fruits the ripe seed frequently does not occupy the whole of the available space, so that the swelling embryo has room to expand without exerting pressure on the fruit wall—as in the hazel, or plum, or sunflower. Where the swelling embryo does press on the fruit wall, as in the achenes of *Alisma plantago*, even the great force developed may be powerless to effect rupture, as Crocker and Davis (1914) have shown, and this takes place after changes in the fruit wall have altered its mechanical properties. In the case of many seeds and fruits which swell greatly on soaking in water the envelopes also swell and expand, so that no splitting takes place. This is seen, for example, in the thin-coated pea or bean, in the hard-coated broom, and in the fruits of the cereals.

An exact investigation by G. Müller (1914) deals with the mode of liberation of the embryo in a large number of cases. He finds that rupture of the envelopes by imbibitional swelling of embryo or endosperm is quite uncommon; it takes place, for example, in *Ipomœa purpurea* and in some legumes. After a few hours in water the embryo of Ipomœa swells to thrice its dry volume, and the testa is split by a net-like system of cracks. The rarity of this mode of liberation may seem strange, since the forces of imbibition are the most powerful at the disposal of the plant. Müller sees a biological advantage in this, since damage to the embryo might result if swelling caused rupture of the coats, for absorption of water need not be accompanied by other conditions favourable to the growth of the young plant. We may note that the possibility of seed coat dormancy would be largely lost, and that a dormant embryo would be exposed to bacterial and fungal attack by the bursting of the coat.

In the great majority of plants the forces of growth are

responsible for rupture, which then takes place as the embryo resumes active life. Growth of the endosperm, of the cotyledons, of the hypocotyl or of the radicle may be effective. Growth of the endosperm—as opposed to imbibitional swelling—does not seem to be a common occurrence, but Müller finds that it is responsible for splitting the testa in *Ricinus communis*, *Pinus Pinea*, and other conifers. Much commoner is the growth of the cotyledons; it is effective in such hard fruits as those of the plums, hazel and walnuts. In these the wall splits along predetermined lines of mechanically weak tissue. Among the monocotyledons the Cyperaceæ show this type.

The growth of the hypocotyl or of the radicle is much the commonest cause of rupture. According to the exact way in which the force is applied Müller distinguishes a number of sub-types, regarding which we need only note that in some cases the force acts through the endosperm, in others through the cotyledons, or directly. In the grass *Coix Lachryma* the fruit wall is extremely hard and brittle, but at each end there is a spot occupied by soft fibres through which radicle and shoot make their way. In the coconut, Tradescantia, Potamogeton, Sparganium and others germination takes place by the pushing out of preformed plugs of tissue. Cocos is specially interesting. The endocarp is extremely strong; at its base are three "eyes," one corresponding to each of the three carpels. Only one seed is, however, present, and only the eye corresponding to this seed provides a plug which can be displaced. As in the hazel and plums already mentioned, so in some other fruits, *e.g.* Fumaria and Alisma, predetermined, mechanically weaker lines occur, along which splitting takes place. These may be regarded as homologous with the lines of opening of dehiscent fruits. In certain palms (Lepidocaryeæ), the fruits of which are clad in a scale-like armour, the weak lines form a net-like system. In most seeds there is no special tissue, and rupture is irregular, usually near the point of exit of the radicle.

Müller also measured exactly the force involved in the

rupture of the fruit wall of the hazel-nut and of the seed coat of *Pinus Pinea* and *Ricinus communis* ; in the first the growing cotyledons exert a force of 3·3 atmospheres ; in the others the force developed by the growing endosperm amounts to 3·7 and 3·1 atmospheres respectively. These values are smaller than might have been expected. Müller has shown, however, that they suffice to rupture the walls and coats, though only in the soaked condition ; in the dry state the resistance is much higher. In Corylus the resistance of the fruit wall falls from about eight atmospheres when dry to about three when wet. In *Pinus Pinea* a fall to a third of the value when dry took place on wetting. The same is true of thin-walled seeds ; thus the testa of the bean is six times as resistant when dry as when wet. The force required to expel the plug of *Cocos campestris* is less by 40 per cent. after soaking.

§ 6. Germination—Emergence of the Seedling

With the radicle free from the seed coat germination proceeds, but the rest of the embryo, or some part of it, may remain enclosed longer or permanently. In the germination of the pea the epicotyl frees itself later than the radicle, but the cotyledons remain within the seed coat until both wither or rot away. When the cotyledons remain in the soil we speak of germination as *hypogeal*, when they emerge it is *epigeal*. Both types may occur in a single genus ; thus *Phaseolus multiflorus*, the scarlet runner, is hypogeal, *Phaseolus vulgaris*, the French bean, is epigeal. In the lupin and sunflower the cotyledons carry the coats above ground as a cap, which is got rid of sooner or later as the cotyledons enlarge and spread apart. This happens in many cases, and the young seedlings with their bonnets have a rather quaint appearance. But frequently the carrying up of the seed coat is a matter of chance. If the seed is well buried the resistance of the soil may retain the coat, while if the seed lies on the surface the cotyledons may remain long encased. We may note a remarkable feature of the marrow and other Cucurbitaceæ, where a peg of tissue is formed on the lower side of the hypocotyl (gravitational induction), which levers open the seed coat so that the cotyledons are freed.

Seeds may germinate on the surface of the ground, but we look upon subterranean germination as the normal. How the radicle pushes its way through the soil we have already seen ; the conditions for the shoot are different. It usually bears at its tip a tender bud, and it may be encumbered with the enlarged cotyledons ; in the emergence of the shoot damage is easier, and frictional resistance is greater than with the radicle. The lessening of resistance and the protection of the delicate plumule are achieved in a variety of ways.

Perhaps the simplest case is seen in cereals such as the oat. The first leaf, the *coleoptile*, is a white, closed sheath completely enclosing the younger organs and never forming chlorophyll. Its tip is rather sharp, and, in the dark, it may attain a length of 3 to 5 cm. before it ceases to grow. In fact, it has much more the appearance of a radicle than of a leaf, and it penetrates the soil, growing upwards, nutating slightly, and very sensitive to contact stimulus, much like a root tip. Normally it reaches the surface before the next leaf, completely protected in its passage through the soil, bursts through its tip. In a deeply buried seed the first foliage leaf may have to make its own way through the soil. While this leaf is in the dark it elongates rapidly and remains narrow and rolled into a tube, thus offering a minimum resistance. The importance of this type of etiolation for easy penetration may be shown experimentally. If seed is sown deeply in soil against the glass side of a suitable box, so that the young plant, though still buried, is early illuminated, the first leaf, as soon as it appears, forms chlorophyll, flattens out, and increases in breadth. It cannot make its way upwards ; the friction of the soil keeps it down and it becomes twisted and folded.

An interesting case is that of the onion. After the

fixation of the radicle the sheath-like cotyledon begins to elongate ; it becomes sharply bent just behind the seed, and this bent portion pushes through the soil dragging the seed with it. In some species of the genus (Allium) a special little sharp boring process is developed on the back of the bend. The tip of the cotyledon remains in the seed, withdrawing food from the endosperm, until it withers and the seed falls off. Later the first foliage leaf bursts through the side of the cotyledon.

In the sunflower the elongation of the hypocotyl pulls the cotyledons, protected by the seed coat, through the soil. In the castor bean the elongating hypocotyl pulls the cotyledons from their place between the halves of the endosperm and carries them up. The part which actually pushes through the soil is, in both cases, a sharply bent portion of the hypocotyl ; the mustard and many others behave similarly. In hypogeal seedlings it is of course the elongation of the *epicotyl* which carries the plumule up. The occurrence of a sharp bend is frequent here too, as in the pea and the garden nasturtium. The importance of this curvature is very great. If the plumule were erect the earth would be forced against it and between the leaves, separating them, increasing the resistance, and damaging the most tender parts. Such curvatures are common not only in seedlings but also in young shoots of perennial herbs ; they are not, however, found on all shoots which must penetrate the soil. They also occur on petioles, *e.g.* of the wood anemone. In the dicotyledonous seedling etiolation results in the more rapid elongation of the axis—hypocotyl or epicotyl as the case may be—and in the leaves remaining unexpanded. Here again the normal darkness in the soil is responsible for a condition which aids penetration. The cause of the curvature of the axis or petiole does not seem to be always the same. In general it is geotropic, though the organ straightens out only, or more rapidly, in the light. In some cases, however, an autonomic curvature is present, and in some the mechanical resistance of the soil is effective. The recent papers by Sperlich (1912), Salisbury

(1916) and Leonhardt (1915) and the discussion by Goebel (1920) should be consulted.

Dicotyledons germinating underground in light have the same difficulty in penetrating the soil as the oat. The curvature of the axis tends to straighten out; the leaves and the cotyledons tend to expand and to spread apart so that resistance is increased; growth in length is depressed.

§ 7. Mode of Growth

In a certain sense the seedling penetrating the soil in the dark may be said to grow; it increases in size, and undergoes development, yet it does not increase its substance. In fact, if the dry weight of the seedling, after a few days' growth in the dark, is compared with that of the seed, it is found to be less; organic material has been used up in respiration and no assimilation has occurred to make good the loss; the same is true of those seedlings in which assimilation lags behind chlorophyll formation, even when they are grown in light. It is in general characteristic of the plant that increase in size—extension—is due chiefly to the absorption of water by cells which have completed their divisions, and there is a consequent great increase in volume, and often change in shape, unaccompanied by any local increment in organic matter. The formation of the cells by division and their partial differentiation takes place in special meristematic regions, the growing points of root and shoot, and the cambium; the region of extension lies behind, sometimes well behind, the growing-point. Growth has been measured by increment of length, of area, of volume, and by increase in dry weight. This last measures the net result of the plant's metabolic activities; it sums up the whole of the complex processes of its chemistry, of which assimilation in one direction and respiration in the other are the two most obvious. Though measurements of length, etc., may yield valuable results for particular organs, the real growth of the plant as a whole, the expression of its power to synthesise and construct, is measured only by its increase in dry weight.

The actively assimilating plant in normal conditions grows in a well-defined fashion. After assimilation has become active the power of the young plant to add to itself is proportional to the amount of matter, or, better, to the amount of active protoplasm, already present, and, as this increases momentarily with assimilation and the growth of the assimilating surface, the power of adding new substance also increases constantly. It works, as V. H. Blackman (1919) has pointed out, like money accumulating at compound interest. The interest for a number of periods is reckoned, not on the amount of the original capital, but, for each period, on that capital plus the interest already accrued to the beginning of that period. In the case of the plant there is this difference, that the interest is added to the capital not at the end of each year, or week, or day, but from moment to moment. The grand total, the final dry weight sum, is expressed by the formula

$$W = W_0 \epsilon^{rt}$$

where r is the rate of interest expressed as a fraction, t the time, W and W_0 are the final and initial weights and $\epsilon =$ the base of natural logarithms, 2·718. The rate of interest, which may be taken to represent the efficiency of the particular plant to add to its original capital, the organic material in the seed, is an abstraction with no actual existence; it changes during the period of growth, besides altering with external conditions. The formula is an expression of the way in which growth takes place early in development, and the rate of interest gives a *summarised* idea of the efficiency of the plant's constructive capacity. It may be seen that if dry weight is plotted against time the resulting graph will be a logarithmic curve.

Quite early in development a change takes place. The rate of increase, relative to the dry weight present at any moment, slows down, and the graph connecting growth and time becomes a straight line. The cause of this is partly that, as time goes on, an increasing proportion of the newly formed matter goes to form mechanical tissue which takes no further part in metabolism; the ratio of leaf area to dry weight becomes less. But more important is the fact demonstrated by Kidd, West, and Briggs (1921), and Briggs (1923*b*) that the metabolic activity of the protoplasm, as measured either by respiration or by assimilation, decreases with age. Thus if we put the assimilating capacity of the young leaf of a young sunflower at 100, the capacity of a young leaf of a half-grown plant is only 40, and of a young leaf of a full-grown plant is only 30. The active metabolic matter is therefore less efficient. Towards the end of development the rate of increase slows down still more and the growth-time graph becomes a curve concave to the abscissa axis. This final portion is probably connected with the formation of reproductive organs, in which dissimilation is relatively more important, and with a much diminished formation of new leaves.

The graph representing growth throughout a vegetative period is therefore an S curve. Such a graph for the growth of the sunflower, constructed from the data of Reed and Holland (1919), is given in Fig. 67, with another for the growth of pea roots from the data of Pearsall (1923).

The same mode of growth is found if we examine individual organs instead of the whole plant. Sachs (1887) showed that the daily increment in length in a shoot of Fritillaria increased up to the sixth day and slowly fell away to the twentieth; the same thing was true of a zone of tissue just behind the root tip of the bean, the rate of growth increasing up to the fifth day and falling to zero on the eighth. He called this passing through a "Grand Period of Growth," and it is of course just another expression of the S graph.

Priestley and Pearsall (1922) found that the whole root system formed on cuttings of Tradescantia went through a series of S growth curves. The depression at the end of each corresponds to the time of formation of the roots of next higher order, and is referred to the changes of metabolism, especially increases in respiration, which take place when the new meristems are laid down. The flat portion, representing a period of steady increase, in the middle of each curve may be due to the limits imposed on the supply of material by the conducting system. Gregory (1921) found the same mode of growth followed in the increase in area of the leaves of Cucurbita in favourable conditions.

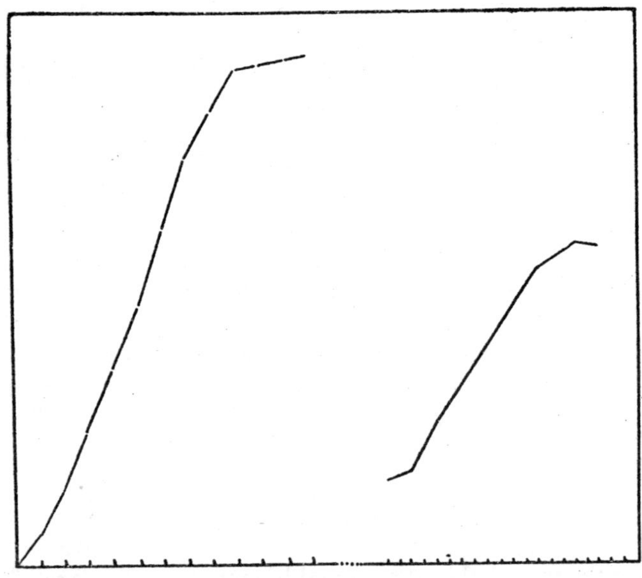

FIG. 67.—Growth curves: left, growth of sunflower in height through twelve weeks from data of Reed and Holland; right, dry weight of root-system of pea through seventeen days from data of Pearsall.

The decline at the end we may here refer to the leaf reaching maturity, or full size. But we must note that this leaves unexplained what factors, presumably controlled largely by the inheritance of the plant, set the limit.

The S growth curve has been compared by Robertson (1924) to that of an autocatalytic reaction (cp. Bayliss, 1920), and the close agreement has been pointed out, too, by Reed and Holland. Robertson holds that the rate of growth is controlled by a catalyst produced by the organism.

§ 8. External Conditions and Growth

The normal mode and rate of growth are subject to modification by the conditions of the environment. The response of the plant to their impact is extremely complex, for, as we have said, growth sums up the whole metabolic activity of the plant, and the external conditions may affect any or all of the links in the chain of processes. We may take as an example the case of light. It acts in the first place through photosynthesis, and if it limits that process it limits also the supply of organic building material. It also acts directly on the process of extension, and in a very complex way. A plant grown in the dark is etiolated, it possesses no chlorophyll, it is drawn, and abnormally elongated, the leaves do not expand. Yet it has been shown by Vogt (1915) that if the coleoptile of an oat, grown in the dark, is illuminated for a few seconds, after a transitory slight decrease in growth rate, there follows a much more marked though also transitory *increase*. Sierp (1917, 1918) found that if the illumination was continued, the increased growth rate was maintained. This apparently contradictory result—that light *increases* growth-rate—is explained by the fact that though the rate of growth is higher in the light than in the dark, the grand period is hastened on and passed through at a lower rate than in the dark : the illuminated seedling actually grows faster than the darkened, but it does not attain the same length. We do not know exactly how illumination affects growth, but it may be through induced changes in the permeability of the cells. The phenomena of etiolation have other causes than the direct effect of light on growth-rate ; the formation of chlorophyll and other chemical changes may be important, and it has been shown by Priestley and Ewing (1923) that changes in the endoderm occur. Sufficient has been said to show the complex action of this single factor.

The influence of temperature on the growth of the pea radicle has already been described ; temperature, too, has an indirect effect through photosynthesis. The supply of salts and of water also affects various metabolic processes.

A. M. Smith (1907), in a study of the growth-rates of plants in Ceylon, has shown that Blackman's conception of limiting factors is applicable to the regulation of growth. Now one factor, now another, is limiting. The growth of Capparis was limited by water supply through the day and by temperature at night; that of Vitis by water supply in July and temperature in January. The same relation has been formulated as the "law of the minimum" by Meyer (1895) for growth factors such as the supply of the various mineral salts. It should be noted that, while the direct effect of such factors as light, temperature, and water supply is immediate, their indirect effect, through assimilation, as well as the effect of the mineral supply, tends to be felt gradually and cumulatively; it is diffused and may be delayed. Balls (1918), in his studies on the cotton in Egypt, found that the action of a particular factor might become visible, *e.g.* on the rate of flower production, weeks after its incidence, and this he calls the "principle of predetermination."

§ 9. Development—The Vegetative Phase

In the course of its development the plant passes through a series of stages the most prominent of which are the vegetative and the reproductive. We have mentioned a case of a plant (*Cardamine chenopodiifolia*) which produces flowers at a very early point, when only two leaves have appeared; but ordinarily a more or less prolonged vegetative phase precedes reproduction. The vegetative phase does not, however, consist in the production of a series of uniform organs. This may be seen by the examination of almost any young plant. If we leave the cotyledons out of account, the first-formed foliage leaves are almost invariably small, and they tend to be smooth in outline; if the adult leaves are lobed or cut the juvenile ones may be almost

PLATE VIII

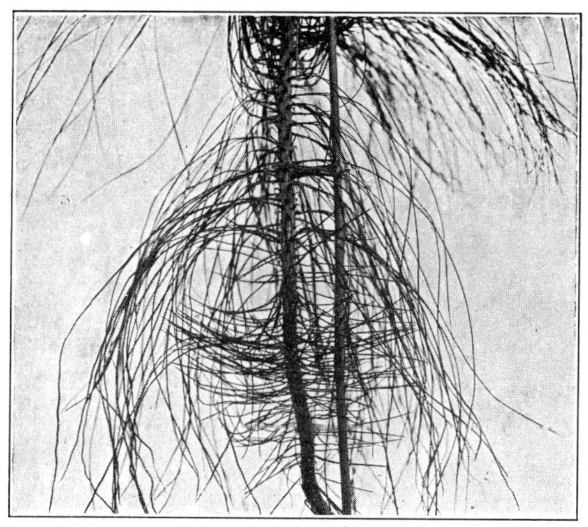

YOUTH AND ADULT LEAF FORMS.

Above *Pinus canariensis*, showing transition from youth leaves ; below *Hedera Helix*, adult shoot (left) and youth shoot (right).

or quite entire; between the two extremes a series of gradations may be traced. On a single shoot, in perennial plants which form buds, the succession of foliage leaves is interrupted by the bud scales, which represent leaves or particular parts of leaves modified by the conditions of their development, and related in structure to their function of protection. Between these bud scales and the foliage leaves transition forms may be traced in many cases, as in the cherry and the horse chestnut. Similarly scale leaves and transition forms are found at the base of the shoots of many perennials. Very frequently the phyllotaxy of the juvenile leaves differs from that on the mature regions of the shoot; in the sunflower the first leaves are opposite and later a two-fifths arrangement is assumed.

One aspect of the change from youth to adult form has already been dealt with in the sun and shade leaves of trees. Examples of more striking differences are the harebell with its round youth leaves and linear adult leaves; the whin, in which the first leaves are trifoliate and the adult leaves spines; many Acacias, which first produce a few pinnate leaves and then phyllodes, sometimes with intermediate forms.

Of great interest is the condition found in various conifers. In the pines the seedling stem bears single, spirally arranged needles; later the long shoots bear only numerous membranous scales, and the needles are borne in pairs, or larger numbers, on special short shoots. The youth needle is longer and softer than the adult, and there are structural differences. In *Pinus sylvestris* the youth needles disappear in the second year; in other species, *e.g. P. canariensis*, they may be produced for several years. In Chamæcyparis and Thuja the seedling has needles, and the mature shoot bears the characteristic cupressoid scales. The youth form in this case, and in the similar case of Thuja, may be *fixed* by using side branches as cuttings, and such plants retain the youth form indefinitely. The fixed youth form of Chamæcyparis is, in fact, cultivated in gardens under a separate generic name—Retinispora. A Chamæcyparis at the transition stage has a highly peculiar appearance (Fig. 68). The youth form very rarely flowers.

The youth form tends to be more mesophytic than the adult, which, in the cases cited, is a pronounced xerophyte. This may be related to the different conditions in which the seedling grows, in the shade of other vegetation, or at a damp season of the year. In many cases, too, the youth form may be looked on as ancestral in type ; in the whin and Acacia, bearing youth leaves of a form characteristic of most members of the family, this is certainly so. Of the Retinisporas Goebel says : " As the youth forms here, as in Pinus, are without doubt to be regarded as primitive, we have thus been able in a sense to bring to life again the ancestral type."

FIG. 68.—Thuja, showing transition from the youth leaf-form. × 1½.

This is not always the case. Goebel regards the adult form of the ivy as ancestral. Growing on a wall, or in the shade of a wood, the leaves, borne in two rows, are deeply lobed and dark green in colour. When the climber grows on to the roof, and receives more uniform and stronger illumination, erect shoots are formed bearing large, pale, slightly lobed, or entire leaves in a one-third or two-fifths spiral ; only on these shoots are flowers produced. Especially in the phyllotaxy the adult shoot resembles the type normal in other members of the Araliaceæ.

Another group of plants with well-marked youth forms are the aquatics. The pond-weeds and water-lilies always produce first a number of linear, or band-shaped, submerged leaves. In *Alisma plantago* and *Sagittaria sagittifolia* the narrow submerged leaves are followed by a few floating leaves, and then by the free oval, or arrow-shaped subaerial leaves. Here again we see the relation of the youth form to the conditions of its environment.

The adult form, as we have seen in the case of sun and shade leaves, may often be modified by external conditions to a greater or less extent, or rather the production of the youth form may be prolonged or reverted to. Glück (1905) found that in water 5 ft. deep Sagittaria produced extremely long band leaves, and only a few floating leaves, while in greater depths, as a rule, submerged leaves only were formed. An adult plant transferred from shallow to deep water starts to form leaves of the band type again. If a harebell, with an erect stem bearing linear leaves, and about to flower, is placed in low light intensity the flower buds wither and side shoots with round leaves are produced. If no flower buds have been formed the main axis may proceed to the formation of round leaves. The whin cultivated in moist air forms broader and softer leaf spines and the formation of shoot spines is suppressed, though no trifoliate leaves are produced. For further details Goebel, Arber, and the monographs of Glück (1905) and Diels (1906) should be consulted.

§ 10. Seasonal Changes, Protection, and Rhythm

Annuals.—Some plants complete their existence in a single season or may run through several generations in a summer; others live for two years, in the first accumulating a store of food, in the second expending it in lavish blossoming; yet others live for many years. The annual habit limits growth to modest dimensions, but enables the plant to pass through seasonal extremes in the best of all protected stages, as a seed; it means, too, the possibility of very rapid distribution. The dangers typically avoided by the annual habit are those of drought. We have noted the presence of short-lived grasses, the only shallow-rooted plants, in the American prairies. In desert conditions annuals are very numerous. On Tumamoc Hill, in the Arizona Desert, Cannon (1911) found 223 annual species, a far greater number than of perennials ; 122 spring up in the winter, and 44 in the summer rainy season. It is as seeds that the plants pass through the dry seasons.

In climates such as ours, where the winter conditions are the least favourable, because of poor light and low temperature, annuals are also numerous, though they form a minority of the flora. Most of our common annuals, however, are weeds of cultivated land, *e.g.* several of the speedwells, the hemp-nettles, the goosefoot, the fumitories, the groundsel, the charlock, the shepherd's purse. These are not really native plants. They have followed the plough, and travelled with man from a centre of distribution which probably lay in Eastern Europe, Asia Minor, and the Mediterranean basin, where semi-arid summer conditions exist. To these they are primarily related, though their periods of vegetation have altered to suit the conditions of their new homes. Frequently their seeds germinate in autumn so that they pass the winter in a state of slow growth. Most can produce several generations in a year. It is of interest to compare the native species of Veronica of the section Chamædrys, typically woodland and marsh plants, and all perennial, with the weed species of the section Omphalospora, growing in waste places, and annuals. Such annuals as *Vicia lathyroides*, *Myosotis collina*, *Erophila verna*, *Teesdalia nudicaulis*, and *Aira praecox* may be true natives. They grow in dry exposed situations, on sand dunes, on the turfy tops of stone walls, and flower in spring, passing the dry summer months as dormant seeds. Even in temperate climates the annual habit may be related to drought rather than to winter frosts.

Arid Conditions.—Plants which live for several years may exist in the nearly uniform conditions of some regions of the tropics, and vegetate continuously, or they may be exposed to a periodic incidence of heat and drought or of cold. We have already seen that several types of tropical and sub-tropical forest and scrub, the savannah forest, the thorn woodland, and the thorn scrub, are characterised by trees and shrubs which cast their leaves in the dry period.

The evergreen trees of arid regions, like the evergreen or live oaks of the Californian Chaparral and of the Mediterranean countries, are even more pronouncedly sclerophyllous than evergreens of temperate countries. Another type of perennial, highly characteristic of semi-arid, open country, is the bulbous geophyte which forms a notable element of the floras of South Africa, Asia Minor, and the Mediterranean lands. During the dry season these rest underground as bulbs or corms, and in the winter or early spring, when moist soil is combined with favourable temperature, extremely rapid growth, made possible by the large preformed buds and the great store of food, results in a sudden wonderful profusion of flower and foliage. There follows a period of active assimilation, the formation of new buds and food stores, and then the leaves wither and die with the onset of the arid season. We have seen that there is reason to believe that our North European bulbous plants are derived from such types, and that their spring growth has enabled them to live in a forest environment by making use of the only season when the conditions as regards light and the other factors necessary for assimilation are favourable. The fact that growth of bulbous plants begins early in winter is, of course, familiar to every gardener.

Frost Resistance.—In temperate climates the unfavourable season is winter, with frost as the chief danger; the cell may be directly damaged by freezing, and indirectly the plant may suffer from temporary excessive transpiration with slow water supply from a frozen soil. The poor light and low temperature in any case reduce assimilation to a minimum, and a covering of snow may stop it entirely for weeks or months.

Many perennial grasses and herbs, especially rosette plants like the daisy and dandelion, retain their leaves throughout the winter. They form what Lidfors (1907) has called the "winter-green flora." Sheltered by their lowly position, existing in the dampest layer of the atmosphere close to the soil, their chief danger lies in direct damage by freezing. Lidfors has shown that in such plants the starch of the leaves is in winter converted into sugar, and by this means the protoplasm is protected from freezing. Maximow (1912) has given experimental demonstration of the fact that in the presence of electrolytes and non-electrolytes the temperature at which death from freezing takes place is lowered, and to a much greater extent than can be accounted for by the actual lowering of the freezing-point. It seems this is rather connected with the eutectic point of the solution, at which all the water and solutes freeze out, and this indicates that death from freezing is due to the complete withdrawal of water from the protoplasm, and the consequent coagulation of the proteins. Solutes in the cell sap protect the protoplasm by lowering the temperature at which this coagulation sets in. The conversion of starch into sugar by the action of cold is familiar in the sweetening of potatoes exposed to frost. It also occurs in the cortical cells of many trees, such as the oak, the elm, and the beech; in other trees, *e.g.* the conifers and the birch, the starch is converted into oil, an emulsion of which is also supposed to have a protective effect against frost. Lewis and Tuttle (1920) have shown that the sugar content of evergreen leaves increases markedly in the winter months. There is no doubt that this chemical protection by sugars and salts in the cell sap is very much more effective than the possession of thick corky bark or of bud-scales; these can do little to prevent the fall of temperature in the underlying tissues during spells of frost.

Bud Protection.—Many perennial herbs die down in winter and pass the cold months well protected underground, under the litter of dead leaves, or under the covering of snow. The parts which most require protection are the buds, and such herbs present a striking contrast to deciduous trees and shrubs with their buds fully exposed to the air, though individually provided with tough bud scales. Bud scales, as we have seen, must be chiefly effective against drought and mechanical damage, the resistance to cold being attained by internal means. The various modes of protection of the buds have been used by Raunkiaer (1910, also

Smith, 1913) as the basis of a classification of the "life forms" or "biological types" of plants. He distinguishes between: *Phanerophytes*, trees and shrubs with exposed dormant buds on branches projecting into the air; *Chamæphytes*, with buds on shoot apices on, or just above, the soil surface, such as *Empetrum nigrum*, or *Stellaria Holostea; Hemicryptophytes*, with dormant buds in the upper crust of the soil, the aerial parts herbaceous and dying away, *e.g. Mercurialis perennis, Chrysanthemum Leucanthemum;* rosette plants may be included here or in the foregoing class; *Geophytes* with subterranean buds, as in the bulbous or tuberous plants; in the *helophytes* (marsh plants) and *hydrophytes* (water plants) buds may be found on perennating rhizomes in the mud, or special detached buds (turions) may be formed, as in the bladderworts; *Therophytes*, which live through the unfavourable season as seeds. There are one or two other classes which need not concern us here, and some of those mentioned are subdivided. Raunkiaer has used the distribution of the species of a flora among his different classes to construct a "biological spectrum," the nature of which is characteristic for different climates. Thus, to take instances which illustrate points already mentioned in this section, it is shown that the therophytes and small phanerophytes predominate in arid regions, *e.g.* Nevada; the chamæphytes, hemicrytophytes, and geophytes in climates with a cold winter, *e.g.* Labrador; the phanerophytes in the moist regions of the tropics, *e.g.* West Indies. Different regions of a small area and different altitudinal zones may show characteristically different spectra.

Leaf-fall and Rhythm.—The periodic climatic change is related to a rhythmical vegetative change very conspicuous in our woods with their autumnal leaf-fall, their winter rest, and their spring awakening. At first sight it seems as if the fall of the leaf were directly due to the storms and early frosts of autumn, and the winter rest to the short, cold day. The relation is in reality not nearly so simple. The fall of the leaf is definitely prepared for by the formation of a special abcission layer of cells weeks before the event, though the actual separation from the tree may be hastened by frost, or may be induced earlier in the season by exceptionally dry and hot weather. Nor is it usually possible to cause the resting buds to open in winter by bringing branches or potted plants into favourable hothouse conditions; growth does not even start till far on in winter, any more than the buds begin to swell after a succession of exceptionally mild November or December days outside.

There are very great differences in the behaviour of various trees, shrubs, and herbs. Thus for woodland herbaceous perennials, which naturally come up in spring, Diels (1918) has shown that some, such as *Asperula odorata* and *Mercurialis perennis*, can be forced to renewed growth at any time of the year by suitable temperature; others, such as *Leucojum vernum* and *Arum maculatum*, can be forced in autumn, but not through the summer; a third type, *Corydalis solida* and *Polygonatum multiflorum*, can be forced only in spring. Such shrubs as the roses and the brambles are capable of continued growth throughout the year, whenever the temperature is favourable, but most of our shrubs and trees form resting buds earlier or later in the summer, and these cannot be forced to open by raising the temperature in winter.

The actual mode of growth through the summer is very different in different trees. In the beech and the oak, after the buds open in May and the leaves and the internodes laid down in them have fully expanded, new buds are formed which, by the end of June, contain the leaf rudiments for the following year. No further growth takes place, with this exception, that a few of the terminal buds open in July and produce the "Lammas shoots," which stand out in the pale green of youth against the darker mature leaves, in August. In the elders, maples, and poplars many shoots, in the hazel the basal shoots, show continued growth till late autumn; in the alder and elm growth goes on through the summer. This summer growth is not necessarily continuous ; buds may be formed, which open immediately without a rest period. Resting buds may be formed early, as in the beech, or at the end of the season. Resting buds may open during the summer, normally as in the " Lammas shoot," or as the result of leaf-fall following very dry, hot weather, or after artificial defoliation, *e.g.* in the ash and beech.

This formation of resting buds, which open normally only in the following spring, and the consequent period of rest, has been ascribed by some authors, *e.g.* Schimper, to an inherent and inherited rhythm. That this does not determine the *annual* periodicity is, however, clear from the behaviour of oaks, beeches, and other trees transplanted to the more uniform conditions in the tropics ; of the tulip tree and oak at Tjibodas in Java, Schimper says that they " reflected winter, spring, and summer on their separate boughs " ; some branches are leafless, some show young leaves or unfolding buds, others have mature leaves. This, Schimper holds, demonstrates an inherent periodicity of growth which is no longer regulated by a definite seasonal climatic cycle. Native tropical trees of the rain forests, where conditions may be very even throughout the year, rarely show continuous growth or a regular succession of unfolding leaves. The conditions have been investigated principally by Wright (1904) for Ceylon, and Volkens (1912) and Simon (1914) for the more uniform climate at Buitenzorg in Java. A number of trees are deciduous, but in general the leafless period is short, and the leaves usually fall more than once in the year. That leaf-fall is not directly dependent on climatic conditions is shown by the fact that in many cases different individuals of the same species lose their leaves at different times. Thus *Ficus fulva* sheds its leaves two or three times in the year, standing bare for one to two weeks. Of seven individuals observed by Volkens, No. 1 and No. 7 were bare on the 20th of January, No. 2 end of January, No. 3 and No. 5 middle of March, No. 4 end of February, No. 6 beginning of June. Most trees are evergreen, but only rarely, *e.g.* in Albizzia and Morinda, was a continuous succession of leaves seen. Generally a race or generation of leaves unfolds itself quickly, and a more or less prolonged period of rest follows. One, two, three, or more such races may be seen on a branch at the same time. The fall of the oldest race frequently occurs just before or just after the unfolding of the youngest; or the old leaves may fall irregularly. Sometimes leaves are to be seen unfolding on one branch, mature on another, while a third is bare. In many trees an occasional "general cleaning" takes place, and the tree stands for a time nearly bare. The general effect of the tropical rain forest is of perennial verdure, but the details of leaf-fall and renewal present a picture of extraordinary complexity, from which it comes out clearly that rhythmical development, apparently unrelated to external conditions, is the rule.

That this rhythm is inherited in the structure of the protoplasm does not follow. Klebs (1914, 1917) has offered an alternative explanation which we may consider in relation to the case he has most fully investigated, that of the beech. Like most of our shrubs and trees the beech cannot be forced by placing it in a hothouse in winter, but Klebs found that, by subjecting it to continuous illumination by electric light, the buds could be made to open at any time, though more easily in September or February (after 10 days) than in November (after 38 days). He further found that in continuous illumination the beech could be prevented from forming buds, and forced to continuous growth during at least four months; by suitable changes it could be made to form buds which at once opened, or to form resting buds.

He considers that *bud formation*, initiated by the suppression of the leaf and the conversion of the stipules into scales, is due to the monopolisation of the salt supply by the cambium and by the vigorously growing and transpiring leaves in early summer, combined with a supply of abundant carbohydrates to the growing-point. The *resting condition* of the bud is determined by inactivation of enzymes through the presence of excess of carbohydrates. This inactivation is most extreme in November, when it may be affected by external conditions, and by such internal changes as the conversion of starch into sugar. Light acts by activating the enzymes ; this occurs naturally when the day lengthens in late spring ; it occurs experimentally with continuous illumination in winter. That continuous growth takes place in continuous illumination is due partly to the demonstrated fact that the quality and intensity of the electric light permitted only of feeble assimilation, and partly to abundant salt supply from rich soil, and to weak transpiration. The Lammas shoot is due to the coincidence of maximum illumination, at the height of summer, with the completion of leaf formation in the buds ; and it is significant that only a few of the best illuminated buds take part.

The hornbeam resembles the beech in its response to continuous illumination. The oak and the ash, on the other hand, cannot be forced in this way, but open their buds in a suitable temperature, with favourable conditions of salt supply. Continuous growth takes place in continuous illumination, but activation is due to high temperature in the dark. Thus the determining factor is not the same in all trees. We have, however, a plausible explanation of periodic development in its most extreme form, the cause being the action of external factors in a peculiarly complex fashion. Klebs lays great stress on the importance of the relation of salt supply to assimilates, and thinks this may be largely responsible for the periodic unfolding of leaves in tropical trees. The supply of salts may accumulate till sufficient is available for the unfolding of the buds and growth, and then a rest period ensue till a further supply is accumulated. It will be seen that this is a particular application of the principle of limiting factors. It is of great interest that changing conditions of illumination can cause the beech to grow in the different modes characteristic of the normal development of other trees ; the precise ratio of supplies of assimilates and salts to the growing points is clearly specific (see also Klebs, 1915).

Klebs's interpretation is supported by analyses of twigs of the apple and peach by Abbott (1923). She found that the carbohydrates increased towards the beginning of the rest period, and that compounds of phosphorus and nitrogen increased towards the end of winter. Howard (1915) found enzymes much more active at the end of the rest period than during it.

Forcing.—Many other methods of overcoming the rest period are known. Müller-Thurgau (1885, 1912) found that potato tubers could be forced to immediate development by subjecting them to cold after harvesting—an imitation of the natural process in winter. Johannsen (1906) forced the lilac and other shrubs by exposing them to ether vapour for one or two days. Molisch (1909, 1921) forced lilac, hazel, willow, and others by plunging them for twelve hours in a bath of warm (30° to 35° C.) water, and the horse chestnut by exposure to radium emanations. Jesenko (1911, 1912) forced various trees by baths or injections of alcohol, acids, and water, and by pressing water into the cut surface of the shoot. Weber (1916) forced the beech by acetylene. Generally it is impossible by these methods to force with success during a middle period of rest, which occurs in September, October, or December; but, as we have seen, Klebs forced the beech at any time (though with difficulty in November), and Lakon (1912) forced various trees at any time by placing the cut shoots in a nutrient salt solution. The ether and the warm bath methods are of economic importance.

These various methods are, of course, unnatural and often drastic. In no case has their action been analysed, but they may very well agree with Klebs's theory that an activation of enzymes is important. While the question of the cause of the resting period and of rhythm in vegetative development is not by any means settled, it may be said that a possible explanation lies in the interaction of external factors with the conditions of growth and development at the vegetative point, even when the external factors are apparently very uniform. We must note that the production of leaves is itself a periodic phenomenon, as is, to go a step further back, cell division also, the basal fact in development, and that the cause of such rhythm must be sought in periodic changes inherent in the complex colloids of the plasma.

§ 11. The Reproductive Phase

The formation of reproductive organs takes place after a period of vegetative growth which in normal conditions is specific. In the ephemeral or annual plant reproduction closes the life of the individual after a few weeks or months; in the biennial in the second year. In perennials the conditions are very varied; in many herbaceous species flowering begins early, after one or several vegetative seasons, and recurs annually, as in the primrose and wood anemone. In bulbous plants the bulb raised from seed may not flower for several years. In shrubs flowering may begin quite early; in trees it is often long delayed. The beech first flowers at an age of 40 to 50 years, the oak at 40 years, the hazel at 10 years, the field maple at 25 years, the elm at 40 years, the fir at 60 years, the pine at 15 years. In most trees the flowers are laid down in the bud nearly a year before they open, and the same is true of bulbous plants. The age of first flowering varies in different climates and conditions; it is always later in trees grown in close canopy. While some trees blossom and fruit every year, in many abundant seed is produced only in periodically recurring "mast" years. The mast years of a beech occur at intervals of 10 to 15 years, while for the pine and the hazel the intervals are 3 to 5 and 2 to 3 years respectively. The period is evidently determined partly by external factors and partly by internal, for different individuals growing together may flourish in different years, and the periods are variable.

In some perennials flowers are produced only once. The famous, but miscalled, century plant, *Agave americana*, grows vegetatively for 8 to 10 years and then produces one immense inflorescence and dies; in the less favourable climate of Europe its life may be prolonged to several decades before flowering takes place. An even more striking instance is offered by *Chusquea abietifolia*, a climbing bamboo described by Seifriz (1920). It flowers in cycles of 32 to 33 years, and as the flowering occurs over the whole of Jamaica simultaneously, and is followed by death, the effect is devastating. Bamboos of South Brazil flower in cycles of 13 years and Indian bamboos show similar conditions.

In the biennial, accumulating food in its first year, in the periodic mast years of the tree, in the long life of the Agave before flowering takes place, we have clear indications that manufacture of a certain minimum of food substance is a necessary preliminary to reproduction. Another indication is given by the familiar fact that in the year *following* an unusually fine summer the flowering shrubs, hawthorn, whin, broom, and the like are exceptionally rich. The favourable conditions for photosynthesis have a delayed effect—an instance of the principle of predetermination. The initiation of the reproductive phase is therefore closely linked with nutritional conditions, and it is, in fact, possible to hasten, or delay, or suppress it altogether by appropriate manipulation of the factors affecting the plant's nutrition.

Many garden plants grown in the shade of trees fail to flower; neither ivy nor honeysuckle flower in the depth of a wood. There is a well-known delay in the ripening of cereals as the result of excess of nitrogenous manure. Klebs (1910, 1918), as the result of extended investigations chiefly on the house-leeks, concludes that the important factor in flower formation is the ratio of assimilates to salts, especially nitrates, at the disposal of the growing-points. Predominance of nitrogenous compounds favours vegetative growth, excess of carbohydrates favours reproduction. The same sort of cause is thus involved as is responsible for the regulation of periodic growth. Hot, dry weather, by favouring photosynthesis and depressing salt absorption, tends to bring on early flowering, while humid conditions with poor light depress assimilation and promote vegetative growth. In his experiments with *Sempervivum Funkii* Klebs found that the plant, grown from a daughter rosette, matures and will flower in the third or fourth year of its life if sufficient light, which is essential to the formation of flower rudiments, is available, as is the case in spring. Such a mature plant placed in a richly manured hotbed in March is prevented from flowering in the following summer, though the same treatment in April is without effect. In April flowering can be prevented by cultivation in the dark or in blue light. By the first method the absorption of nitrates is accelerated, by the second assimilation is depressed, while in both dissimilation predominates over assimilation. Klebs has shown that in the plants prevented from flowering the soluble nitrogen compounds are more and the soluble carbohydrates less abundant than in flowering plants of the same age. His experiments make it probable that the nutritive conditions determine whether the plant shall develop reproductively or vegetatively. We know little of the intimate reactions concerned ; we do not know what nitrogen compounds are involved, nor do we know in any case what the critical ratio is. Other authors, *e.g.* Sachs, have postulated the formation of special flower-forming compounds ; there is no experimental evidence for this theory, which does not also support the clearer hypothesis of Klebs.

§ 12. Senescence, Death, and Individuality

The death of the annual, of the biennial, and of some perennials is closely associated with the achievement of reproduction ; the plant produces seed and then dies. The connection of the one event with the other may be emphasised by experiment. The mignonette is an annual, but, by carefully removing the flower buds, it may be made to persist through several years, growing into a little shrub (Molisch, 1921). The turnip is a biennial, but, by keeping it over winter in a hothouse, it may be forced to vegetate for two or three years, evidently because the high rate of respiration and growth maintained depletes the food store to a certain extent.

In perennials there is rarely this relation, and their life

may be very long. The span of life, which in some cases seems so very definite, may be easily modified. Where the span is longer, as in trees, it also seems to be less definite. Its investigation is difficult, for the tree so often outlives many generations of men ; a Sequoia has been felled during the present century which was a seedling about 1317 B.C., over 3000 years ago (Douglas, 1919). In the case of herbaceous, bulbous, and aquatic perennials we meet with another difficulty in determining the length of life.

Vegetation Multiplication.—Most perennials multiply vegetatively. In a creeping plant like the ground ivy the ordinary stems root at the nodes, and may ultimately become separated from the parent. Underground rhizomes, as in the wood anemone, or rootstocks, as in the primrose, branch and, by the dying off of the old parts, give rise to new plants. The strawberry sends out special runners, axillary shoots with very long internodes producing one or more rosettes of leaves, which root and become separated from the parent and from each other. Runners of a simpler type are produced by the bugle and willow herbs. Axillary subterranean shoots in the potato or artichoke form swollen, food-storing tubers, which separate on the death of the parent. In bulbs and corms axillary buds grow into daughter bulbs and corms which replace or multiply the old. Many shrubs and trees form suckers, as the rose and willow ; trailing branches take root as in the Rhododendron, and bramble. Buds may also arise on the roots, as in the sheep's sorrel and the elm, the roots of which, after a tree has been felled, may send up shoots many yards distant from the parent trunk. Adventitious buds may be formed on leaves and take root, as in *Cardamine pratensis*, Begonia, Tolmeia, or the famous *Bryophyllum calycinum*. Finally we may mention the so-called " viviparous " production of bulbils, or shoots, in place of seeds by varieties of *Poa alpina*, *Festuca ovina*, *Polygonum alpinum*, by *Cardamine bulbifera*, and by many other plants, which Ernst (1918) regards as an expression of hybrid vigour. The possibilities of vegetative propagation have been taken advantage of, and greatly extended, by the gardener, who, besides all the methods of cutting, layering, and so on, has in his hands the possibilities of grafting and budding.

Individuality.—In all cases of vegetative multiplication it is a matter of difficulty to decide where one *individual* ends and another begins. We can root a cutting from a willow, but is the plant so obtained really any more an independent individual than the other shoots which we left on the tree ? The rhizome of an anemone branches ; while the parent rhizome lives the branches are undoubtedly part of it ; do they become new individuals just because it dies away ? We can define the individual as the product of a reproductive cell ; but we have restricted the term " reproduction " to these cases in which a fresh start is made with the highly specialised spore or zygote. If we hold to this, and it seems the only logical course, we exclude vegetative multiplication from the conception of reproduction, and we deny the rank of individual to the separate plants produced by it. Thc products of the original vegetative point, whether they remain attached to a parent stem, or separate from it in a thousand parts, are members of one individual.

Senescence.—This conception has the further advantage that it focuses our attention once more on the significance of sexual reproduction. If the sexual process has some necessary rejuvenating action, we should expect to find the span of life of one of the " greater individuals " restricted, as is the span of a tree. The evidence is conflicting, but on the whole tells against any limitation whatever. On the one side we have the belief, never sufficiently investigated, that fruit trees degenerate after a certain period of propagation by grafting ; and that many cultivated plants, usually propagated vegetatively, such as the potato and the sugar-cane, are very subject to disease. Pallis (1916) believes that in the " plav," the great floating reed-swamp of the Danube, the masses of rhizome show a distinct succession, ending, after a period of vegetative multiplication, in the production of stunted shoots, senility, and death. Benedict (1915) has given evidence that the amount of assimilating tissue in vine leaves, measured by the area of the mesh between veins, gets less as the plant ages, and that this is an indication of senility. Ensign (1921), however, disputes Benedict's observations. We have seen that in the case of an annual plant, the sunflower, there is an undoubted decrease in the metabolic activity of the protoplasm as the plant grows older. It remains to be seen whether any definite evidence can be obtained that the same thing takes place in the successive years of the life of a perennial. A decrease in respiring power has been found in older leaves of the olive by Nicolas (1918), and in leaves of older trees of the olive by Ruby (1917), but the data are scanty.

On the other side we have the fact that many cultivated plants have been reproduced from prehistoric times by cuttings only, and, in the banana, for example, which sets no seed, show perfect vigour. The sweet-flag, *Acorus Calamus*, was introduced into Europe from Asia over 500 years ago, and since its introduction it has multiplied vegetatively and spread vigorously; no seed has been set, for the insects necessary for pollination are absent. Similarly the water thyme, *Elodea canadensis*, was introduced once or twice into the waters of this country in the 'forties of last century; only the female plant grows here, and the enormous vigour of its spread has been due entirely to new branches breaking away from old stems. In any locality, after a few years' growth, its vigour seems to decline, but this looks more like the effect of external conditions than of internal change. Many native plants are reproduced almost or entirely by propagation—the lesser celandine, the periwinkle, and some meadow grasses and fescues; and we may recall the case of the parthenogenetic hawkweeds. The evidence is ambiguous, but on the balance it seems to point to the possibility of a very long and, perhaps, in some cases, indefinite activity of the plasma, which originated in a single reproductive cell and has since split up in innumerable growing-points, without the intervention of a rejuvenating process.

Causes of Death.—The cause of death would thus

always lie outside the properties of the protoplasm. In annuals and biennials it would seem to be nutritive. It can be readily seen that, as a perennial of the type of a forest tree waxes old, the difficulties of supply to the buds, reaching further and further from the root system, must increase. In its heart-wood, dead and functionless except as a support, the tree bears within it a source of danger; bacterial and fungal infection through broken branches can spread more readily in dead tissue, destroying the supporting material, and from there may invade the living regions. The tree becomes more susceptible to disease, and less able to withstand the force of the wind. If, after a life of centuries, it must ultimately fall, we may still ascribe its death, not to the ageing of the protoplasm, but to effects, in their essence, secondary. Discussions of this difficult subject will be found in Arber, Molisch (1921), Child (1915), Dofflein (1919), and Pearl (1922).

We may conclude by drawing attention to the beautiful illustration of the rôles and relative importance of vegetative multiplication and sexual reproduction offered by the practice of the gardener and breeder. The stock of a plant is rapidly and easily increased by propagating it from cuttings; and by this means the gardener makes sure of a uniform progeny. He knows that if he raises a fine snapdragon from seed, he may get a bright mixture of offspring, due to crossing, but that if he uses cuttings he will preserve his fine variety true. The breeder looking for new varieties uses seed. "Bud" sports do crop up now and again, *e.g.* those which originated the weeping varieties of trees, but they are infrequent in comparison with the mutations, which are prepared for in the intimate processes of nuclear division prior to reproduction, and which appear in the offspring from seed. Here, too, is given the further and most valuable power of redistribution and combination in hybridising. Evolution in nature and the art of breeding may be said to depend on sexual reproduction. Vegetatively the plant multiplies, but remains the same; through sex it changes.

BIBLIOGRAPHY

A few general works, which are frequently quoted and are referred to in the text by the author's name without a date, will be found in the short list under A. Papers and books which are referred to by the author's name and a date will be found under B.

Volumes are indicated by heavy type. Thus **76**, 167, means volume 76, page 167.

A.

ARBER, A. *Water Plants.* Cambridge, 1920.

CZAPEK, F. *Biochemie der Pflanzen.* 3 Vols., 2nd ed. Jena, 1913–21.

GOEBEL, K. *Organographie der Pflanzen.* 3 Vols., 2nd ed. Jena, 1913–23. (Engl. Trans. of 1st ed., 2 Vols. Oxford, 1900–05.)

JOST, L. *Vorlesungen über Pflanzenphysiologie*, 3rd ed. Jena, 1913. (4th ed. as JOST-BENECKE, *Pflanzenphysiologie.* 2 Vols. Jena, 1923–24. Engl. Trans. of 1st ed., 1 Vol. with Supplement. Oxford, 1907.)

KERNER VON MARILAUN, A. *The Natural History of Plants.* London, 1894–95.

KIRCHNER, O. VON, LOEW, E. and SCHROETER, C. *Lebensgeschichte der Blütenpflanzen Mitteleuropas.* Jena, 1908.

KNUTH, P. *Handbook of Flower Pollination.* Oxford, 1906.

HABERLANDT, G. *Physiological Plant Anatomy.* London, 1914.

MÜLLER, H. *The Fertilisation of Flowers.* London, 1883.

NEGER, F. W. *Biologie der Pflanzen auf experimenteller Grundlage.* Stuttgart, 1913.

PFEFFER, W. *Physiology of Plants.* 3 Vols. Oxford, 1900–06.

RUSSELL, E. J. *Soil Conditions and Plant Growth*, 4th ed. London, 1921.

SCHIMPER, A. F. W. *Plant-geography upon a Physiological Basis.* Oxford, 1903.

WARMING, E. *Œcology of Plants.* Oxford, 1909.

B.

ABBOTT, O. (1923). Chemical Changes at Beginning and Ending of Rest Period in Apple and Peach. Bot. Gaz., **76**, 167.

ADAMS, J. (1916). On the Germination of the Pollen Grains of Apple and other Fruit Trees. Bot. Gaz., **61**, 131.

ANDREÆ, E. (1903). Inwiefern werden Insekten durch Farbe und Duft der Blumen angezogen. Beihft. Bot. Centrlbl., **15**, 427.

ANGELSTEIN, U. (1911). Ueber die Kohlensaüreassimilation submerser Wasserpflanzen in Bikarbonat- und Karbonatlösungen. Beitr. z. Biol. d. Pflanz., **10**, 87.

ANTHONY, S. and HARLAN, H. V. (1920). Germination of Barley Pollen. Journ. Agr. Res., **18**, 525.

APPLEMAN, C. O. (1916). Biochemical and Physiological Study of the Rest Period in the Tubers of *Solanum tuberosum.* Bot. Gaz., **61**, 265.

ARBER, A. (1918). The Phyllode Theory of the Monocotyledonous Leaf, with Special Reference to Anatomical Evidence. Ann. Bot., **32**, 465.

ASKENASY, E. (1875). Ueber die Temperatur welche Pflanzen im Sonnenlicht annehmen. Bot. Zeitng., **33**, 441.

ASO, K. (1910). Können Bromeliaceen durch die Schuppen der Blätter Salze aufnehmen? Flora, **100**, 447.

ATKINS, W. R. G. (1922). Some Factors affecting the Hydrogen Ion Concentration of the Soil and its Relation to Plant Distribution. Sc. Proc. Roy. Soc. Dublin, **16**, 369.

BAILEY, I. W. (1922). Notes on Neotropical Ant-plants. I, *Cecropia angulata* sp. nov. Bot. Gaz., **74**, 369.

—— (1923). Notes on Neotropical Ant-plants. II, *Tachigalia paniculata* Aubl. Bot. Gaz., **75**, 27.

BAKKE, A. L. (1914). Studies on the Transpiring Power of Plants as Indicated by the Method of Standardised Hygrometric Paper. Journ. Ecol., **2**, 145.

BAKKE (1915). The Index of Foliar Transpiring Power as an Indicator of Permanent Wilting in Plants. Bot. Gaz., **60**, 314.

—— (1918). Determination of Wilting. Bot. Gaz., **66**, 81.

BALLS, W. L. (1912). The Cotton Plant in Egypt. London, 1912.

—— (1918). Analyses of Agricultural Yield. Part III, The Influence of Natural Environmental Factors upon the Yield of Egyptian Cotton. Phil. Trans. Roy. Soc., **208 B**, 157.

BARNES, C. R. (1910) in *A Textbook of Botany*, by J. M. Coulter, C. R. Barnes and H. C. Cowles. 2 Vols. New York, 1910, 1911.

DE BARY, A. (1884). Comparative Anatomy of the Phanerogams and Ferns. Oxford, 1884.

BATESON, W. (1913). Problems of Genetics. Newhaven, 1913.

——, and GREGORY, R. P. (1905). On the Inheritance of Heterostylism in Primula. Proc. Roy. Soc., **76 B**, 581.

——, and PELLEW, C. (1920). The Genetics of "Rogues" among Culinary Peas (*Pisum sativum*). Proc. Roy. Soc., **91 B**, 186.

BATTEN, L. (1918). Observations on the Ecology of *Epilobium hirsutum*. Journ. Ecol., **6**, 162.

BAYLISS, W. M. (1920). Principles of General Physiology. London, 1920.

BECQUEREL, P. (1907). Recherches sur la vie latente des grains. Ann. Sc. Nat. Bot., IX, **5**, 193.

BEIJERINCK, M. W. (1888). Die Bakterien der Papilionaceen-Knöllchen. Bot. Ztng., **46**, 725.

BEISSNER, L. (1888). Ueber Jugendformen von Pflanzen speciell von Coniferen. Ber. deutsch. bot. Ges., **6**, lxxxiii.

BELT, T. (1874). The Naturalist in Nicaragua. London, 1874.

BENEDICT, H. M. (1915). Senile Changes in Leaves of *Vitis vulpina* L, and Certain Other Plants. Cornell Univ. Agr. Expt. Stat., 281.

BERGEN, J. Y. (1903). The Transpiration of *Spartium junceum* and other Xerophytic Shrubs. Bot. Gaz., **36**, 464.

—— (1904). Transpiration of Sun-leaves and Shade-leaves of *Olea europea* and other Broad-leaved Evergreens. Bot. Gaz., **38**, 285.

BERGMAN, H. F. (1920). The Relation of Aëration to the Growth and Activity of Roots, and its Influence on the Ecesis in Plants in Swamps. Ann. Bot., **34**, 13.

BERNARD, N. (1909). L'évolution dans la symbiose. Ann. Sc. Nat. Bot., IX, **9**, 1.

—— (1911). Les mycorhizes des Solanum. Ann. Sc. Nat. Bot., IX, **14**, 235.

BIRGER, S. (1907). Ueber endozoische Samenverbreitung durch Vögel. Svensk. Bot. Tidskrft., **1,** 1.

BITTER, G. (1909). Zur Frage der Geschlechtsbestimmung von *Mercurialis annua* durch Isolation weiblicher Pflanzen. Ber. deutsch. bot. Ges., **27,** 120.

BLACKMAN, F. F. (1895). On the Path of Gaseous Exchange between Aerial Leaves and the Atmosphere. Phil. Trans. Roy. Soc., **186 B,** 503.

—— (1905). Optima and Limiting Factors. Ann. Bot., **19,** 281.

—— (1909). The Longevity and Vitality of Seeds. New Phyt., **8,** 31.

—— and MATTHÆI, G. C. L. (1905). A Quantitative Study of Carbon-dioxide Assimilation and Leaf-temperature in Natural Illumination. Proc. Roy. Soc., **76 B,** 402.

—— and SMITH, A. M. (1911). On Assimilation in Submerged Water-plants and its Relation to the Concentration of Carbon Dioxide and other Factors. Proc. Roy. Soc., **83 B,** 389.

BLACKMAN, V. H. (1914). The Wilting Coefficient of the Soil. Journ. Ecol., **2,** 43.

—— (1919). The Compound Interest Law and Plant Growth. Ann. Bot., **33,** 353.

BONNIER, G. (1895). Recherches expérimentales sur l'adaptation des plantes au climat alpin. Ann. Sc. Nat. Bot., VII, **20,** 217.

BOTTOMLEY, W. B. (1907). The Structure of the Root Tubercles in Leguminous and other Plants. Rep. Brit. Assoc., 693.

—— (1912*a*). The Structure and Physiological Significance of the Root-nodules of *Myrica Gale*. Proc. Roy. Soc., **84 B,** 215.

—— (1912*b*). The Root-nodules of *Myrica Gale*. Ann. Bot., **26,** 111.

—— (1915). The Root-nodules of *Ceanothus americanus*. Ann. Bot., **29,** 605.

BOUYOUCOS, G. (1921*a*). A New Classification of Soil Moisture. Soil. Sc., **11,** 33.

—— (1921*b*). The Amount of Unfree Water in Soils at Different Moisture Contents. Soil Sc., **11,** 255.

—— and MCCOOL, M. M. (1915). The Freezing Point Method as a New Means of Measuring the Concentration of the Soil Solution directly in the Soil. Mich. Agr. Coll. Expt. Sta. Techn. Bull., No. 24, 5.

BOWER, F. O. (1908). The Origin of a Land Flora. London, 1908.

BOYSEN-JENSEN, P. (1918). Studies on the Production of Matter in Light- and Shadow-plants. Bot. Tidskrft., **36,** 219.

BRANNON, J. M. (1923). Influence of Certain Sugars on Higher Plants. Bot. Gaz., **75,** 370.

BRENCHLEY, W. E. (1916). The Effect of Concentration of the Nutrient Solution on the Growth of Barley and Wheat in Water Cultures. Ann. Bot., **30**, 77.

—— (1918). Buried Weed Seeds. Journ. Agr. Sc., **9**, 1.

—— and JACKSON, V. G. (1921). Root Development in Barley and Wheat under Different Conditions of Growth. Ann. Bot., **35**, 533.

BRIGGS, G. E. (1920). The Development of Photosynthetic Activity during Germination. Proc. Roy. Soc., **91 B**, 249.

—— (1923*a*). The Development of Photosynthetic Activity during Germination of Different Types of Seeds. Proc. Roy. Soc., **94 B**, 12.

—— (1923*b*). The Characteristics of Subnormal Photosynthetic Activity resulting from Deficiency of Nutrient Salts. Proc. Roy. Soc., **94 B**, 20.

——, KIDD, F. and WEST, C. (1920). A Quantitative Analysis of Plant Growth. Ann. Appl. Biol., **7**, 103.

BRIGGS, L. J. and SHANTZ, H. L. (1912*a*). The Wilting Coefficient and its Indirect Determination. Bot. Gaz., **53**, 20.

—— —— (1912*b*). The Relative Wilting Coefficients for Different Plants. Bot. Gaz., **53**, 229.

BRISTOL, B. M. (1920). On the Alga-flora of some Desiccated English Soils: an Important Factor in Soil Biology. Ann. Bot., **34**, 35.

BROCHER, F. (1911). Le problème de l'utriculaire. Ann. de Biol. lacustre, 5.

BROWN, H. T. and ESCOMBE, F. (1900). Static Diffusion of Gases and Liquids in Relation to the Assimilation of Carbon and Translocation in Plants. Phil. Trans. Roy. Soc., **190 B**, 223.

—— —— (1905*a*). Researches on the Physiological Processes of Green Leaves, with Special Reference to the Interchange of Energy between the Leaf and its Surroundings. Proc. Roy. Soc., **76 B**, 29.

—— (1905*b*). On a New Method for the Determination of Atmospheric Carbon Dioxide based on the Rate of its Absorption by a Free Surface of a Solution of Caustic Alkali. Proc. Roy. Soc., **76 B**, 112.

—— and WILSON, W. A. (1905). On the Thermal Emissivity of a Green Leaf in Still and Moving Air. Proc. Roy. Soc., **76 B**, 122.

BROWN, P. E. (1913). A Study of Bacteria at Different Depths in some Typical Iowa Soils. Centrbl. f. Bakt., II, **37**, 497.

BROWN, W. H. (1913). The Relation of the Substratum to the Growth of Elodea. Philippine Journ. Sc. (Bot.), **8**, 1.

—— (1916). The Mechanism of Movement and the Duration

of the Effect of Stimulation in the Leaves of Dionæa. Amer. Journ. Bot., **3,** 68.

BROWN (1918). The Theory of Limiting Factors. Philippine Journ. Sc. (Bot.), **13,** 345.

—— and SHARP, L. (1910). The Closing Response in Dionæa. Bot. Gaz., **49,** 290.

BUCKLE, P. (1923). On the Ecology of Soil Insects. Journ. Ecol., **11,** 93.

BURCK, W. (1900). Preservation of the Stigma against the Germination of Foreign Pollen. Proc. Akad. Weten. Amsterdam, **3,** 264.

BURDON-SANDERSON, J. (1882). On the Electromotive Properties of the Leaf of Dionæa in the Excited and Unexcited States. Phil. Trans. Roy. Soc., **173,** 1.

—— (1889). On the Electromotive Properties of the Leaf of Dionæa in the Excited and Unexcited States (second paper). Phil. Trans. Roy. Soc., **179 B,** 417.

BURGEFF, H. (1909). Die Wurzelpilze der Orchideen. Jena, 1909.

BURGERSTEIN, A. (1876). Untersuchungen über die Beziehungen der Nährstoffe zur Transpiration der Pflanzen. Sitzber. d. kais. Akad. d. Wiss. Wien (Math.-Nat. Kl.), **73,** 191.

—— (1904). Die Transpiration der Pflanzen. Jena, 1904.

BÜSGEN, M. (1883). Die Bedeutung des Insektenfanges für *Drosera rotundifolia*. Bot. Ztng., **41,** 569, 585.

—— (1905). Studien über die Wurzelsysteme einiger dicotyler Holzpflanzen. Flora, **95,** 58.

CALDWELL, J. S. (1913). The Relation of Environmental Conditions to the Phenomenon of Permanent Wilting in Plants. Physiol. Res., **1,** 1.

CANNON, W. A. (1910). Under MACDOUGAL and CANNON.

—— (1908). The Topography of the Chlorophyll Apparatus in Desert Plants. Carneg. Inst. Wash., Publ. No. 98.

—— (1911). The Root Habits of Desert Plants. Carneg. Inst. Wash., Publ. No. 131.

—— (1913). Botanical Features of the Algerian Sahara. Carneg. Inst. Wash., Publ. No. 178.

—— (1915). On the Relation of Root Growth and Development to the Temperature and Aeration of the Soil. Amer. Journ. Bot., **2,** 211.

—— (1920). Effect of a Diminished Oxygen-supply in the Soil on the Rate of the Growth of Roots. Carneg. Inst. Wash. Yearbook, **19,** 59.

—— and FREE, E. E. (1920). Anaerobic Experiments with Helium. Carneg. Inst. Wash. Yearbook, **19,** 61.

Cavers, F. (1914). Gola's Osmotic Theory of Edaphism. Journ. Ecol., **2**, 209.

Chemin, E. (1920). Observations anatomiques et biologiques sur le genre " Lathræa." Ann. Sc. Nat. Bot., X, **2**, 125.

Child, C. M. (1915). Senescence and Rejuvenescence. Chicago, 1915.

Church, A. H. (1904). On the Relation of Phyllotaxis to Mechanical Laws. London, 1904.

—— (1919). Thalassiophyta and the Subaerial Transmigration. Oxford, 1919.

Clark, W. M. (1920). The Determination of Hydrogen Ions. Baltimore, 1920.

Clements, F. E. (1905). Research Methods in Ecology. Lincoln, Neb., 1905.

—— (1921). Aeration and Air Content. The Rôle of Oxygen in Root Activity. Carneg. Inst. Wash., Publ. No. 315.

Compton, R. H. (1909). The Morphology and Anatomy of *Utricularia brachiata* Oliver. New. Phyt., **8**, 117.

—— (1913). Phenomena and Problems of Self-sterility. New. Phyt., **12**, 197.

Cook, O. F. (1902). The Origin and Distribution of the Coco Palm. Contrib. U.S. Nation. Herb., **7**, 257.

—— (1912). The History of the Coconut Palm in America. Contrib. U.S. Nation. Herb., **14**, 271.

Correns, C. (1904). Experimentelle Untersuchungen über die Gynodiœcie. Ber. deutsch. bot. Ges., **22**, 506.

—— (1905). Weitere Untersuchungen über die Gynodiœcie. Ber. deutsch. bot. Ges., **23**, 452.

—— (1907). Die Bestimmung und Vererbung des Geschlechtes nach neuren Versuchen mit höheren Pflanzen. Berlin, 1907.

—— (1908). Die Rolle der männlichen Keimzellen bei der Geschlechsbestimmung der gynodiœcischen Pflanzen. Ber. deutsch. bot. Ges., **26 A**, 686.

—— (1913). Selbststerilität und Individualstoffe. Biol. Centrlbl., **33**, 389.

—— (1917). Ein Fall experimenteller Verschiebung des Geschlechtsverhältnisses. Sitzber. Akad. Wiss. Berlin (Math.-Phys. Kl.), 685.

—— (1922). Geschlechtsbestimmung und Zahlenverhältnis der Geschlechter beim Sauerampfer (*Rumex Acetosa*). Biol. Centrlbl., **42**, 465.

Costantin, J. and Magrou, J. (1922). Applications industrielles d'une grande découverte française. Ann. Sc. Nat. Bot., X, **4**, 1.

Coville, F. V. (1920). The Influence of Cold in Stimulating the Growth of Plants. Proc. Nat. Acad. Sc., **6**, 434.

CRAIB, W. G. (1918). Regional Spread of Moisture in the Wood of Trees. Notes Roy. Bot. Gard. Edin., No. 51, 1.

CREMER, H. (1923). Untersuchungen über die periodischen Bewegungen der Laubblätter. Ztschrft. f. Bot., **15**, 593.

CHRISTOPH, H. (1921). Untersuchungen über die mykotrophen Verhältnisse der " Ericales " und die Keimung von Pirolaceen. Beihft. Bot. Centrlbl., **38**, 115.

CROCKER, W. (1916). Mechanics of Dormancy in Plants. Amer. Journ. Bot., **3**, 99.

—— and DAVIS, W. E. (1914). Delayed Germination in Seed of *Alisma plantago*. Bot. Gaz., **58**, 285.

CRUMP, W. P. (1913*a*). The Coefficient of Humidity : a New Method of Expressing the Soil Moisture. New Phyt., **12**, 125.

—— (1913*b*). Notes on Water Content and the Wilting Point. Journ. Ecol., **1**, 96.

CUNNINGHAM, D. D. (1889). On the Phenomena of Fertilisation in *Ficus Roxburghii*. Calcutta, 1889.

CUTLER, D. W., CRUMP, L. and SANDON, H. (1922). A Quantitative Investigation of the Bacterial and Protozoan Population of the Soil, with an Account of the Protozoan Fauna. Phil. Trans. Roy. Soc., **211 B**, 317.

CZAJA, A. T. (1922). Die Fangvorrichtung der Utriculariablase. Ztschrft. f. Bot., **14**, 705.

DACHNOWSKI, A. (1908). The Toxic Property of Bog Water and Bog Soil. Bot. Gaz., **46**, 130.

DARLINGTON, H. T. (1922). Dr. W. J. Beal's Seed-viability Experiment. Amer. Journ. Bot., **9**, 266.

DARWIN, C. (1862). The Various Contrivances by which British and Foreign Orchids are Fertilised by Insects. London, 1862.

—— (1868). The Variation of Animals and Plants under Domestication. London, 1868.

—— (1875*a*). Insectivorous Plants. London, 1875.

—— (1875*b*). The Movements and Habits of Climbing Plants. London, 1875.

—— (1876). The Effects of Cross and Self Fertilisation in the Vegetable Kingdom. London, 1876.

—— (1877). The Different Forms of Flowers on Plants of the Same Species. London, 1877.

—— (1880). The Power of Movement in Plants. London, 1880.

—— (1881). The Formation of Vegetable Mould through the Action of Earthworms, with Observations on their Habits. London, 1881.

DARWIN, F. (1880). Experiments on the Nutrition of *Drosera rotundifolia*. Journ. Linn. Soc. Bot., **17**, 17.
—— (1898). Observations on Stomata. Phil. Trans. Roy. Soc., **190 B**, 531.
—— (1914). The Effect of Light on the Transpiration of Leaves. Proc. Roy. Soc., **87 B**, 281.
—— (1916). On the Relation between Transpiration and Stomatal Aperture. Phil. Trans. Roy. Soc., **207 B**, 413.
—— and PERTZ, D. M. (1912). A New Method of Estimating the Aperture of Stomata. Proc. Roy. Soc., **84 B**, 136.
DAVEY, A. J. and GIBSON, C. M. (1917). Notes on the Distribution of Sexes in *Myrica Gale*. New Phyt., **16**, 147.
DAVIS, W. C. and ROSE, R. C. (1912). The Effects of External Conditions upon the After-ripening of the Seeds of *Cratægus mollis*. Bot. Gaz., **54**, 49.
DELF, E. M. (1911). Transpiration and Behaviour of Stomata in Halophytes. Ann. Bot., **25**, 485.
—— (1912). Transpiration in Succulent Plants. Ann. Bot., **26**, 409.
—— (1915). The Meaning of Xerophily. Journ. Ecol., **3**, 110.
DETTO, C. (1905). Über die Bedeutung der Insektenähnlichkeit der Ophrysblüte nebst Bemerkungen über die Mohrenblüte bei *Daucus Carota*. Flora, **94**, 287.
DIELS, L. (1906). Jugendform und Blütenreife im Pflanzenreich. Berlin, 1906.
—— (1918*a*). Über Wurzelkork bei Pflanzen stark erwärmter Boden. Flora, **111, 112**, 490.
—— (1918*b*). Das Verhältniss von Rhythmik und Verbreitung bei den Perennen des europäischen Sommerwaldes. Ber. deutsch. bot. Ges., **36**, 337.
DINGLER, H. (1915). Die Flugfähigkeit schwerster geflügelter Dipterocarpus-Früchte. Ber. deutsch. bot. Ges., **33**, 348.
DIXON, H. H. (1914). Transpiration and the Ascent of Sap. London, 1914.
DOFLEIN, F. (1919). Das Problem des Todes und der Unsterblichkeit bei den Pflanzen und Tieren. Jena, 1919.
DOUGLASS, A. E. (1919). Climatic Cycles and Tree Growth. Carneg. Inst. Wash., Publ. No. 289.
DRABBLE, E. and DRABBLE, H. (1907). The Relation between Osmotic Strength of the Cell Sap of Plants and their Physical Environment. Biochem. Journ., **2**, 117.

EAST, E. M. and JONES, D. F. (1919). Inbreeding and Outbreeding, their Genetic and Sociological Significance. Philadelphia, 1919.

ECKERSON, S. (1913). A Physiological and Chemical Study of After-ripening. Bot. Gaz., **55**, 286.

EMERSON, F. W. (1921). Subterranean Organs of Bog Plants. Bot. Gaz., **72**, 359.

ENGELMANN, T. W. (1883). Farbe und Assimilation. Bot. Zeitng., **41**, 17.

ENGLER, A. (1918). Tropismen und exzentrisches Dickenwachstum der Baüme. Zurich, 1918.

ENSIGN, M. R. (1921). Area of Vein-islets in Leaves of Certain Plants as an Age Determinant. Amer. Journ. Bot., **8**, 433.

ERBAN, M. (1916). Über die Verteilung der Spaltöffnungen in Beziehung zur Schlafstellung der Blätter. Ber. deutsch. bot. Ges., **34**, 880.

ERNST, A. (1918). Bastardierung als Ursache der Apogamie im Pflanzenreich. Jena, 1918.

ESCHERICH, K. (1911). Zwei Beiträge zum Kapitel " Ameisen und Pflanzen." Biol. Centrlbl., **31**, 44.

EWART, A. J. (1908). On the Longevity of Seeds. Proc. Roy. Soc., Victoria, **21**, 1.

—— and MASSON-JONES, A. J. (1906). The Formation of Red Wood in Conifers. Ann. Bot., **20**, 201.

FABER, F. C. VON (1912). Das erbliche Zusammenleben von Bakterien und tropischen Pflanzen. Jahrb. wiss. Bot., **51**, 285.

—— (1913). Über Transpiration und osmotischen Druck bei den Mangroven. Ber. deutsch. bot. Ges., **31**, 277.

—— (1914). Die Bakteriensymbiose der Rubiaceen. Jahrb. wiss. Bot., **54**, 243.

FARMER, J. B. (1919). On the Quantitative Differences in the Water-conductivity of the Wood in Trees and Shrubs. Part I, Evergreens. Part II, Deciduous Plants. Proc. Roy. Soc., **90 B**, 218, 232.

FISCHER, A. (1907). Wasserstoff- und Hydroxylionen als Keimungsreize. Ber. deutsch. bot. Ges., **25**, 108.

FITTING, H. (1903). Untersuchungen über den Haptotropismus der Ranken. Jahrb. wiss. Bot., **38**, 545.

—— (1903). Weitere Untersuchungen zur Physiologie der Ranken, nebst einigen neuen Versuchen über den Reizleitung bei Mimosa. Jahrb. wiss. Bot., **39**, 424.

—— (1909*a*). Die Beeinflussung der Orchideenblüten durch die Bestaübung und durch andere Umstände. Ztschrft. f. Bot., **1**, 1.

—— (1909*b*). Entwickelungsphysiologische Probleme der Fruchtbildung. Biol. Centrlbl., **29**, 193.

—— (1911). Die Wasserversorgung und die osmotischen Druckverhältnisse der Wüstenpflanzen. Ztschrft. f. Bot., **3**, 209.

FITTING (1915). Untersuchungen über die Aufnahme von Salzen in die lebende Zelle. Jahrb. wiss. Bot., **56**, 1.

FRANK, B. (1885). Über die auf Wurzelsymbiose beruhende Ernährung gewisser Baüme durch unterirdische Pilze. Ber. deutsch. bot. Ges., **3**, 128.

—— (1893). Die Assimilation des freien Stickstoffs durch die Pflanzenwelt. Bot. Ztng., **51**, 139.

FREIDENFELDT, T. (1902). Studien über die Wurzeln kraütiger Pflanzen. Flora, **91**, 115.

FRISCH, K. VON (1914). Der Farbensinn und Formensinn der Biene. Zool. Jahrb. (Abt. f. allg. Zool. u. Phys.), **35**. Also as separate book, Jena, 1914.

—— (1919*a*). Zur Streitfrage nach dem Farbensinn der Bienen. Biol. Centrlbl., **39**, 122.

—— (1919*b*). Über den Geruchsinn der Biene. Zool. Jahrb. (Abt. f. allg. Zool. u. Phys.), **38**. Also as separate book, Jena, 1919.

GALLAUD, I. (1905). Études sur les mycorhizes endotrophes. Rev. Gén. Bot., **17**, 1.

GARDNER, W. A. (1921). Effects of Light on Germination of Light-sensitive Seeds. Bot. Gaz., **71**, 249.

GASSNER, G. (1910). Über Keimungsbedingungen einiger Sudamerikanischer Gramineensamen. I and II Ber. deutsch. bot. Ges., **28**, 350, 504.

—— (1911). Vorlaüfige Mitteilung neuerer Ergebnisse meiner Keimungsuntersuchungen mit *Chloris ciliata*. Ber. deutsch. bot. Ges., **29**, 708.

—— (1915*a*). Altes und neues zur Frage des Zusammenwirkens von Licht u. Temperatur bei der Keimung lichtempfindlicher Samen. Ber. deutsch. bot. Ges., **33**, 203.

—— (1915*b*). Einige neue Fälle von keimungsauslösender Wirkung der Stickstoffverbindungen. Ber. deutsch. bot. Ges., **33**, 217.

—— (1915*c*). Über die Keimungsauslösende Wirkung der Stickstoffsalze auf lichtempfindliche Samen. Jahrb. wiss. Bot., **55**, 259.

GATES, F. C. (1914). Winter as a Factor in the Xerophily of Certain Evergreen Ericads. Bot. Gaz., **57**, 445.

GERICKE, F. W. (1921). Root Development of Wheat Seedlings. Bot. Gaz., **72**, 404.

GERTZ, O. (1915). Über die Schutzmittel einiger Pflanzen gegen schmarotzende Cuscuta. Jahrb. wiss. Bot., **56**, 123.

GILTAY, E. (1898). Über die vegetabilische Stoffbildung in den Tropen und in Mitteleuropa. Ann. Bot. Gard. Buitenzorg, **15**, 43.

GILTAY (1904). Über die Bedeutung der Krone bei den Blüten und über das Unterscheidungsvermögen der Insekten I. Jahrb. wiss. Bot., **40**, 368.

—— (1906). Über die Bedeutung der Krone bei den Blüten und über das Farbenunterscheidungsvermögen der Insekten. Jahrb. wiss. Bot., **43**, 468.

GLÜCK, H. (1905–24). Biologische und morphologische Untersuchungen über Wasser- und Sumpfgewächse. I, Die Lebensgeschichte der europäischen Alismaceen, 1905. II, Untersuchungen über die mitteleuropäischen Utriculariaarten, über die Turionenbildung bei Wasserpflanzen, sowie über Ceratophyllum, 1906. III, Die Uferflora, 1911. IV, Submerse und Schwimmblattflora, 1924. Jena, 1905–24.

GOEBEL, K. (1889–93). Pflanzenbiologische Schilderungen. 2 Vols. Marburg, 1889, 1893.

—— (1904). Die kleistogamen Blüten und die Anpassungs theorien. Biol. Centrlbl., **24**, 673, 737, 769.

—— (1910). Über sexuellen Dimorphismus bei Pflanzen. Biol. Centrlbl., **30**, 657.

—— (1920). Die Entfaltungsbewegungen der Pflanzen und deren teleologische Deutung. Jena, 1920.

—— (1922). Erdwurzeln mit Velamen. Flora, **115**, 1.

GOLDING, J. (1905). The Importance of the Removal of the Products of Growth in the Assimilation of Nitrogen by the Organism of the Root Nodules of Leguminous Plants. Journ. Agr. Sc., **1**, 59.

GOLDSCHMIDT, R. (1923). The Mechanism and Physiology of Sex Determination. London, 1923.

GRADMANN, H. (1921). Die Bewegungen der Windepflanzen. Ztschrft. f. Bot., **13**, 337.

—— (1923). Die Windschutzeinrichtungen an den Spaltöffnungen der Pflanzen. Jahrb. wiss. Bot., **62**, 449.

GRAFE, V. (1914). Ernährungsphysiologisches Praktikum der höheren Pflanzen. Berlin, 1914.

GREGORY, F. G. (1921). The Increase in Area of the Leaves and Leaf Surface of *Cucumis sativus*. Ann. Bot., **35**, 93.

GUPPY, H. B. (1903–06). Observations of a Naturalist in the Pacific. 2 Vols. London, 1903–06. (Vol. ii, Plant Dispersal.)

—— (1912). Studies in Seeds and Fruits. London, 1912.

—— (1917). Plants, Seeds, and Currents in the West Indies and Azores. London, 1917.

HAAS, A. R. (1916). The Excretion of Acids by Roots. Proc. Nat. Acad. Sc., **2**, 561.

HABERLANDT, G. (1893). Eine botanische Tropenreise. Leipzig, 1893.

HABERLANDT (1905). Die Lichtsinnesorgane der Laubblätter. Leipzig, 1905.

—— (1906). Sinnesorgane im Pflanzenreich. Leipzig, 1906.

—— (1922). Über Teilungshormone und ihre Beziehungen zur Wundheilung, Befruchtung, Parthenogenesis und Adventivembryonie. Biol. Centrlbl., **42**, 145.

HALES, S. (1727). Statical Essays: Containing Vegetable Staticks; or an Account of some Statical Experiments on the Sap of Vegetables. London, 1727.

HALL, A. D., BRENCHLEY, W. E. and UNDERWOOD, L. M. (1914). The Soil Solution and the Mineral Constituents of the Soil. Phil. Trans. Roy. Soc., **204 B**, 179.

HANNIG, E. (1907). Über Pilzfreies *Lolium temulentum*. Bot. Ztng., **65**, 25.

—— (1912). Untersuchungen über die Verteilung des osmotischen Drucks in der Pflanze in Hinsicht auf die Wasserleitung. Ber. deutsch. bot. Ges., **30**, 194.

HANSGIRG, A. (1904). Pflanzenbiologische Untersuchungen. Vienna, 1904.

HARDER, R. (1921). Kritische Versuche zu Blackman's Theorie der "begrenzenden" Faktoren. Jahrb. wiss. Bot., **60**, 531.

—— (1923). Bemerkungen über die Variationsbreite des Kompensationspunktes beim Gaswechsel der Pflanzen. Ber. deutsch. bot. Ges., **41**, 194.

HARRINGTON, G. T. (1921). Optimum Temperatures for Flower Seed Germination. Bot. Gaz., **72**, 337.

HARRIS, J. A. (1918). On the Osmotic Pressure of the Tissue Fluids of Phanerogamic Epiphytes. Amer. Journ. Bot., **5**, 490.

—— and LAWRENCE, J. V. (1916). On the Osmotic Pressure of the Tissue Fluids of Jamaican Loranthaceæ Parasitic on Various Hosts. Amer. Journ. Bot., **3**, 438.

—— —— (1917). Cryoscopic Determinations on Tissue Fluids of Plants of Jamaican Coastal Deserts. Bot. Gaz., **64**, 285.

HASSELBRING, H. (1914*a*). The Relation between the Transpiration Stream and the Absorption of Salts. Bot. Gaz., **57**, 72.

—— (1914*b*). The Effect of Shading on the Transpiration and Assimilation of the Tobacco Plant in Cuba. Bot. Gaz., **57**, 257.

HAYES, H. K. and GARBER, R. J. (1921). Breeding Crop Plants. New York, 1921.

HEINRICHER, E. (1910). Die Aufzucht und Kultur der parasitschen Samenpflanzen. Jena, 1910.

—— (1916). Die Krummungsbewegungen des Hypokotyls von Viscum album . . . Beziehungen zwischen Lichtgenuss und Keimung. Jahrb. wiss. Bot., **57**, 321.

HEINRICHER (1916). Über die geotropischen Reaktionen unserer Mistel (*Viscum album* L.). Ber. deutsch. bot. Ges. **34**, 818.

HELLRIEGEL, H. u. WILFARTH, H. (1888). Untersuchungen über die Stickstoffnahrung der Gramineen und Leguminosen Beilage zu der Ztschrft. d. Vereins f.d. Rübenzucker-Industrie Nov. 1888.

—— —— (1889). Erfolgt die Assimilation des freien Stickstoffs durch die Leguminosen unter Mitwirkung niederer Organismen. Ber. deutsch. bot. Ges., **7**, 138.

HENNING, H. (1916). Der Geruch. Leipzig, 1916.

HENRICI, M. (1921). Zweigipfelige Assimilationskurven. Mit spezieller Berücksichtigung der Photosynthese von alpinen phanerogamen Schattenpflanzen und Flechten. Verh. Naturforsch. Ges. Basel, **32**, 107.

HEPBURN, J. S. (1919). A Study of the Protease of the Pitcher Liquid of Nepenthes. Contrib. Bot. Lab. Univ. Pennsyl., **4**, 442.

—— and JONES, F. M. (1919). Occurrence of Anti-Proteases in the Larvæ of Sarcophaga Associates of *Sarracenia flava*. Contrib. Bot. Lab. Univ. Pennsyl., **4**, 460.

——, —— and ST. JOHN, E. Q. (1920). The Absorption of Nutrients and Allied Phenomena in the Pitchers of Sarraceniaceæ. Journ. Franklin Inst., **189**, 147.

HESS, C. VON (1913). Experimentelle Untersuchungen über die angeblichen Farbensinn der Bienen. Zool. Jahrb. (Abt. f. allg. Zool. und Phys.), **34**, 81.

—— (1917). Der Farbensinn der Vögel und die Lehre von den Schmuckfarben. Arch. f. d. ges. Physiol., **166**, 381.

—— (1918). Beiträge zur Frage nach einem Farbensinne bei Bienen. Arch. f. d. ges. Physiol., **170**, 337.

HESSE, O. (1923). Über die Keimungsauslösende Wirkung chemischer Stoffe auf lichtempfindliche Samen. Ber. deutsch. bot. Ges., **41**, 316.

HESSELMAN, H. (1904). Zur Kenntnis des Pflanzenlebens schwedischer Laubwiesen. Beihft. Bot. Centrlbl., **17**, 311.

HILL, T. G. (1908). Observations on the Osmotic Properties of the Root Hairs of Certain Salt-marsh Plants. New Phyt., **7**, 133.

HOAGLAND, D. R. (1917). The Effect of Hydrogen and Hydroxyl Ion Concentration on the Growth of Barley Seedlings. Soil Sc., **3**, 547.

HOWARD, W. L. (1915). An Experimental Study of the Rest Period in Plants. Physiological Changes accompanying Breaking of the Rest Period. Mo. Agr. Expt. Stat., Res. Bull., 21.

HUBER, B. (1923). Transpiration in verschiedener Stammhohe. I, *Sequoia gigantea*. Ztschrft. f. Bot., **15**, 465.

IHERING, H. VON (1907). Die Cecropien und ihre Schutzameisen. Bot. Jahrb., **39**, 666.

ILJIN, V. S. (1915). Die Regulierung der Spaltöffnungen in Zusammenhang mit der Veränderung des osmotischen Druckes. Beihft. Bot. Centrlbl., **32**, 15.

—— (1916). Relation of Transpiration to Assimilation in Steppe Plants. Journ. Ecol., **4**, 65.

—— (1922). Über den Einfluss des Welkens der Pflanzen auf die Regulierung der Spaltöffnungen. Jahrb. wiss. Bot., **61**, 670.

—— NAZAROVA, P. and OSTROVSKAJA, M. (1916). Osmotic Pressure in Roots and in Leaves in Relation to Habitat Moisture. Journ. Ecol., **4**, 160.

IRVING, A. A. (1910). The Beginning of Photosynthesis and the Development of Chlorophyll. Ann. Bot., **24**, 805.

JANSE, J. M. (1897). Les endophytes radicaux de quelques plantes javanaises. Ann. Bot. Gard. Buitenzorg, **14**, 53.

JEFFREY, E. C. (1917). Anatomy of Woody Plants. Chicago, 1917.

JESENKO, F. R. (1911). Einige neue Verfahren die Ruheperiode der Holzgewächse abzukürzen. Ber. deutsch. bot. Ges., **29**, 273.

—— (1912). Einige neue Verfahren die Ruheperiode der Holzgewächse abzukürzen. (2te Mitteilung), Ber. deutsch. bot. Ges., **30**, 81.

JOHANNSEN, W. (1906). Das Ätherverfahren beim Frühtreiben mit besonderer Berücksichtigung der Fliedertreiberei. Jena, 1906.

JOHNSTON, T. (1888). *Arceuthobium oxycedri*. Ann. Bot., **2**, 157.

JONES, W. N. (1918). On the Nature of Fertilisation and Sex. New Phyt., **17**, 167.

JONES, L. H. and SHIVE, J. W. (1922). Influence of Wheat Seedlings upon Hydrogen Ion Concentration of Nutrient Solutions. Bot. Gaz., **73**, 391.

JOST, L. (1905). Zur Physiologie des Pollens. Ber. deutsch. bot. Ges., **23**, 504.

—— (1907). Über die Selbststerilität einiger Blüten. Bot. Ztng., **65**, 77.

KAMERLING, Z. (1914). Verdunstungsversuche mit tropischen Loranthaceen. Ber. deutsch. bot. Ges., **32**, 17.

KARSTEN, G. (1918). Über Kompasspflanzen. Flora, **111, 112**, 48.

KEEBLE, F. W. (1895). Observations on the Loranthaceæ of Ceylon. Trans. Linn. Soc. Bot., II, **5**, 91.

KEEN, B. A. (1914). The Evaporation of Water from the Soil. Journ. Agr. Sc., **6**, 456.

—— (1922). The System Soil—Soil-moisture in *Physico-chemical Problems relating to the Soil.* (Reprinted from the Transactions of the Faraday Society, London.) London, 1922.

KELLERMAN, K. F. (1910). Nitrogen-gathering Plants. Yearbk. Dept. Agr. U.S.A., 213.

KIDD, F. (1914). The Controlling Influence of Carbon Dioxide on the Maturation, Dormancy, and Germination of Seeds. Parts I and II. Proc. Roy. Soc., **87 B**, 408, 609.

—— (1917). The Controlling Influence of Carbon Dioxide. Part III, The Retarding Effect of Carbon Dioxide on Respiration. Proc. Roy. Soc., **89 B**, 136.

—— and WEST, C. (1917). The Controlling Influence of Carbon Dioxide. Part IV, On the Production of Secondary Dormancy in the Seeds of *Brassica alba* following Treatment with Carbon Dioxide, and the Relation of this Phenomenon to the Question of Stimuli in Growth Processes. Ann. Bot., **31**, 456.

—— —— (1920). The Rôle of the Seed-coat in Relation to the Germination of Immature Seed. Ann. Bot., **34**, 439.

—— —— and BRIGGS, G. E. (1921). A Quantitative Analysis of the Growth of *Helianthus annuus*, Part I, The Respiration of the Plant and of its Parts throughout the Life Cycle. Proc. Roy. Soc., **92 B**, 368.

KINZEL, W. (1913, '15, '20). Frost und Licht als beeinflussende Kräfte bei der Samenkeimung., 2 Nachträge. Stuttgart, 1913, '15, '20.

—— (1917). Teleologie der Wirkungen von Frost, Dunkelheit, und Licht auf die Keimung der Samen. Ber. deutsch. bot. Ges., **35**, 581.

KIRCHNER, O. VON (1922). Über Selbstbestaübung bei den Orchideen. Flora, **115**, 103.

KLEBS, G. (1910). Alterations in the Development and Forms of Plants as a Result of Environment. Proc. Roy. Soc., **82 B**, 547.

—— (1914). Über das Treiben der einheimischen Baüme speziell der Buche. Abhand. d. Heidelberger Akad. d. Wiss. (Mathnat. Kl.), **3**.

—— (1915). Über Wachstum und Ruhe tropischer Baumarten. Jahrb. wiss. Bot., **56**, 734.

KLEBS (1917). Über das Verhältnis von Wachstum und Ruhe bei den Pflanzen. Biol. Centrlbl., **37**, 373.

—— (1918). Über die Blütenbildung von Sempervivum. Flora, **111**, **112**, 128.

KLIMMER, W. and KRÜGER, R. (1914). Sind die bei den verschiedenen Leguminosen gefundenen Knöllchenbakterien artverschieden? Bakt. Centrlbl., II, **40**, 256.

KNIEP, H. and MINDER, F. (1909). Über den Einfluss verschiedenfarbigen Lichtes auf die Kohlensaüreassimilation. Ztschrft. f. Bot., **1**, 619.

KNIGHT, R. C. (1917*a*). The Inter-relations of Stomatal Aperture, Leaf Water-content and Transpiration Rate. Ann. Bot., **31**, 221.

—— (1917*b*). "Relative Transpiration" as a Measure of the Intrinsic Transpiring Power of the Plant. Ann. Bot., **31**, 351.

—— (1922). Further Observations on the Transpiration, Stomata, Leaf Water-content and Wilting of Plants. Ann. Bot., **36**, 361.

KNOLL, F. (1914*a*). Über die Ursache des Ausgleitens der Insektenbeine an wachsbedeckten Pflanzenteilen. Jahrb. wiss. Bot., **54**, 448.

—— (1914*b*). Zur Ökologie und Reizphysiologie des Andrœcium von *Cistus salvifolius* L. Jahrb. wiss. Bot., **54**, 498.

—— (1921). *Bombylius fuliginosus* und die Farbe der Blumen. Abhand. d. zool.-bot. Ges. in Wien, **12**, Hft. 1.

—— (1922*a*). Lichtsinn und Blumenbesuch des Falters von *Macroglossum stellatarum*. Abhand. d. zool.-bot. Ges. in Wien, **12**, Hft. 2.

—— (1922*b*). Der Tierversuch im Dienste der Blütenökologie. Ber. deutsch. bot. Ges., **40**, (30).

KNUDSON, L. (1922). Nonsymbiotic Germination of Orchid Seeds. Bot. Gaz., **73**, 1.

KOPELOFF, N. and COLEMAN, D. A. (1917). A Review of Investigations in Soil Protozoa and Soil Sterilization. Soil. Sc., **3**, 197.

KOSTYSCHEW, S. (1922). Über die Ernährung der grünen Halbschmarotzer. Ber. deutsch. bot. Ges., **40**, 273.

—— (1923). Die Photosynthese der Insektivoren. Ber. deutsch. bot. Ges., **41**, 277.

KRAUS, G. (1911). Boden und Klima auf kleinstem Raum. Jena, 1911.

KRAEPELIN, H. (1920). Die Sprengel'sche Saftmal-Theorie. Biol. Centrlbl., **40**, 120.

KÜMMLER, A. (1922). Über die Funktion der Spaltöffnungen weissbunter Blätter. Jahrb. wiss. Bot., **61**, 610.

KUSANO, S. (1911). *Gastrodia elata* and *Armillaria mellea.* Journ. of Agr., Tokio, **4**, 1.

LAIDLAW, C. G. P. and KNIGHT, R. C. (1916). A Description of a Recording Porometer and a Note on Stomatal Behaviour during Wilting. Ann. Bot., **30**, 47.

LAKON, G. (1912). Die Beeinflussung der Winterruhe der Holzgewächse durch die Nährsalze. Ztschrft. f. Bot., **4**, 561.

LEITCH, I. (1916). Some Experiments on the Influence of Temperature on the Rate of Growth in *Pisum sativum.* Ann. Bot., **30**, 25.

LEITGEB, H. (1886). Beiträge zur Physiologie der Spaltöffnungs-apparate. Mittheil. a. d. bot. Inst. zu Graz., 123.

LEHMANN, E. (1911). Temperatur und Temperaturwechsel in ihrer Wirkung auf die Keimung lichtempfindlicher Samen. Ber. deutsch. bot. Ges., **29**, 577.

—— (1912). Über die Beeinflussung der Keimung lichtempfindlicher Samen durch die Temperatur. Ztschr. f. Bot., **4**, 465.

—— (1918*a*). Über die minimale Belichtungszeit welche die Keimung der Samen von *Lythrum Salicaria* auslöst. Ber. deutsch. bot. Ges., **36**, 157.

—— (1918*b*). Über die Selbststerilität von *Veronica syriaca.* Ztschrft. f. Ind. Abst. u. Vererbl., **21**, 1.

—— (1919). Über die Keimfördernde Wirkung von Nitrat auf licht gehemmte Samen von *Veronica Tournefortii.* Ztschrft. f. Bot., **11**, 161.

—— (1922). Über Selbststerilität von *Veronica syriaca*, II, Ztschrft. f. Ind. Abst. u. Vererbl., **27**, 161.

—— and OTTENWÄLDER, A. (1913). Über katalytische Wirkung des Lichtes bei der Keimung lichtempfindlicher Samen. Ztschrft. f. Bot., **5**, 337.

LEONHARDT, W. (1915). Über das Verhalten von Sprossen bei Widerstand leistender Erdbedeckung. Jahrb. wiss. Bot., **55**, 91.

LEWIS, F. and TUTTLE, G. M. (1920). Osmotic Properties of some Plant Cells at Low Temperatures. Ann. Bot., **34**, 405.

LIDFORS, B. (1896). Zur Biologie des Pollens. Jahrb. wiss. Bot., **29**, 1.

—— (1899*a*). Über den Chemotropismus der Pollenschlaüche. Ber. deutsch. bot. Ges., **17**, 236.

—— (1899*b*). Weitere Beiträge zur Biologie des Pollens. Jahrb. wiss. Bot., **33**, 232.

—— (1907). Die Wintergrüne Flora. Eine biologische

Untersuchung. Lund. Univ. Arsskrft. Afd. 2, N. F. 2, No. 13.

LIDFORS (1909). Untersuchungen über die Reizbewegungen der Pollenschlaüche. Ztschrft. f. Bot., **1,** 443.

LINSBAUER, K. (1917). Beiträge zur Kenntnis der Spaltöffnungsbewegungen. Flora, **109,** 100.

LIVINGSTON, B. E. (1905). Relation of Transpiration to Growth in Wheat. Bot. Gaz., **40,** 178.

—— (1906). The Relation of Desert Plants to Soil Moisture and Evaporation. Carneg. Inst. Wash., Publ. No. 50.

—— and BROWN, W. H. (1912). Relation of the Daily March of Transpiration to Variations in the Water Content of Foliage Leaves. Bot. Gaz., **53,** 309.

LLOYD, F. E. (1908). The Physiology of Stomata. Carneg. Inst. Wash., Publ. No. 82.

LOEW, O. (1892). Über die physiologische Functionen der Calcium- und Magnesiumsalze im Pflanzenorganismus. Flora, **75,** 368.

LOFTFIELD, J. G. V. (1921). The Behaviour of Stomata. Carneg. Inst. Wash., Publ. No. 314.

LONG, H. C. (1917). Plants Poisonous to Live Stock. Cambridge, 1917.

LOTSY, J. P. (1916). Evolution by Means of Hybridisation. The Hague, 1916.

LOVELL, J. H. (1920). The Flower and the Bee. London, 1920.

LUBIMENKO, W. (1908). La concentration du pigment vert et l'assimilation chlorophyllienne. Rev. Gén. Bot., **20,** 162.

LUNDEGARDH, H. (1921). Ecological Studies in the Assimilation of Certain Forest-plants and Shore-plants. Svensk. bot. Tidskrft., **15,** 46.

—— (1922). Zur Physiologie und Ökologie der Kohlensaüre-assimilation. Biol. Centrlbl., **42,** 337.

LUNDSTROM, A. N. (1884). Pflanzenbiologische Studien. I, Die Anpassungen der Pflanzen an Regen und Tau. Upsala, 1884.

MACDOUGAL, D. T. (1911). An Attempted Analysis of Parasitism. Bot. Gaz., **52,** 249.

—— (1912). The Water Balance of Desert Plants. Ann. Bot., **26,** 71.

—— and CANNON, W. A. (1910). The Conditions of Parasitism in Plants. Carneg. Inst. Wash., Publ. No. 129.

——, RICHARDS, H. M. and SPOEHR, H. A. (1919). Basis of Succulence in Plants. Bot. Gaz., **67,** 405.

—— and SPALDING, E. S. (1910). The Water Balance of Succulent Plants. Carneg. Inst. Wash., Publ. No. 141.

McDougal, W. B. (1914). On the Mycorhiza of Forest Trees. Amer. Journ. Bot., **1**, 51.

—— (1916). The Growth of Forest Tree Roots. Amer. Journ. Bot., **3**, 384.

McLean, F. T. (1920). Field Studies of the Carbon Dioxide Absorption of Coco-nut Leaves. Ann. Bot., **34**, 367.

McLean, R. C. (1919). Studies in the Ecology of Tropical Rain-forest ; with Special Reference to the Forests of South Brazil. Journ. Ecol., **7**, 5, 121.

McLuckie, J. (1923). Studies in Parasitism. A Contribution to the Physiology of the Loranthaceæ of New South Wales. Bot. Gaz., **75**, 333.

Magnus, W. (1900). Studien an der endotrophen mykorrhiza von *Neottia Nidus avis* L. Jahrb. wiss. Bot., **35**, 205.

—— (1913). Der physiologische Atavismus unserer Eichen und Buche. Biol. Centrlbl., **33**, 309.

—— (1920). Über Hemmungsstoffe und falsche Keimung. Ber. deutsch. bot. Ges., **38**, (19).

Magrou, J. (1921). Symbiose et tubérisation. Ann. Sc. Nat. Bot., X, **3**, 181.

Martin, J. N. (1913). The Physiology of the Pollen of *Trifolium pratense*. Bot. Gaz., **56**, 112.

Matthaei, G. C. L. (1904). On the Effect of Temperature on Carbon-dioxide Assimilation. Phil. Trans. Roy. Soc., **197 B**, 47.

Mayer, A. (1901). Lehrbuch der Agrikulturchemie. Heidelberg, 1901.

Mayr, F. (1915). Hydropoten an Wasser- und Sumpfpflanzen. Beihft. Bot. Centrlbl., **32**, 278.

Maximow, N. A. (1912). Chemische Schutzmittel der Pflanzen gegen Erfrieren. Ber. deutsch. bot. Ges., **30**, 52, 293, 504.

Melin, E. (1921). On the Mycorhizas of *Pinus sylvestris*, L. and *Picea abies*, Karst. Journ. Ecol., **9**, 254.

Mendiola, N. B. (1922). Effects of Different Rates of Transpiration on the Dry Weight and Ash Content of the Tobacco Plant. Philipp. Journ. Sc., **20**, 639.

Merl, E. M. (1922). Biologische Studien über die Utriculariablase. Flora, **115**, 59.

Mevius, W. (1921). Beiträge zur Physiologie " kalkfeindlicher " Gewächse. Jahrb. wiss. Bot., **60**, 147.

Mez, C. (1904). Die Wasser-Oekonomie der extrem atmospharischen Tillandsien. Jahrb. wiss. Bot., **40**, 157.

Miehe, H. (1911). Die Bakterien-Knoten an den Blatträndern der *Ardisia crispa*, A.DC. Abhandl. Kgl. Sächs. Ges. Wiss. (Math.-Phys. Kl.), **32**, 399.

MIEHE (1914). Weitere Untersuchungen über die Bakteriensymbiose bei *Ardisia crispa*. Jahrb. wiss. Bot., **53**, 1.

—— (1918). Anatomische Untersuchung der Pilzsymbiose bei *Casuarina equisetifolia* nebst einigen Bemerkungen über das Mykorhizaproblem. Flora, **111, 112**, 431.

—— (1919). Weitere Untersuchungen über die Bakteriensymbiose bei *Ardisia crispa*. II, Die Pflanze ohne Bakterien. Jahr. wiss. Bot., **58**, 29.

MIRANDE, M. (1900). Recherches physiologiques et anatomiques sur les Cuscutacées. Bull. Sc. de la Fr. et de la Belg., **35**.

—— (1905). Recherches sur le développment et l'anatomie des Cassythacées. Ann. Sc. Nat. Bot., IX, **2**, 181.

MIYOSHI, M. (1894). Über Reizbewegungen der Pollenschlaüche. Flora, **78**, 76.

MÖBIUS, M. (1912). Beiträge zur Blütenbiologie und zur Kenntnis der Blütenfarbstoffe. Ber. deutsch. bot. Ges., **30**, 365.

—— (1918). Über Orientierungsbewegungen von Knospen, Blüten, und Früchten. Flora, **111, 112**, 396.

MOELLER, H. (1890). Beitrag zur Kenntnis der *Frankia subtilis*, Brunchorst. Ber. deutsch. bot. Ges., **8**, 215.

MOLISCH, H. (1893). Zur Physiologie des Pollens, mit besonderer Rucksicht auf die chemotropischen Bewegungen der Pollenschlaüche. Sitzber. kais. Akad. Wiss. Wien. (Math.-Nat. Kl.), **102**, 423.

—— (1909). Das Warmbad als Mittel zum Treiben der Pflanzen. Jena, 1909.

—— (1921). Pflanzenphysiologie als Theorie der Gärtnerei. Jena, 1921.

MOLLIARD, M. (1908). Cultures saprophytiques de *Cuscuta monogyna*. Comptes rend. Acad. Sc., **147**, 685.

—— (1913). Le *Lepidium sativum* rendu semi-parasite expérimentalement. Comptes rend. Acad. Sc., **156**, 1694.

MONTFORT, C. (1921). Die aktive Wurzelsaugung aus Hochmoorwasser im Laboratorium und am Standort und die Frage seiner Giftwirkung. Jahrb. wiss. Bot., **60**, 184.

MORGAN, J. F. (1917). The Soil Solution obtained by the Oil Pressure Method. Soil Sc., **3**, 531.

MORGAN, T. H. (1919). The Physical Basis of Heredity. Philadelphia, 1919.

MUENSCHER, W. C. (1922). The Effect of Transpiration on the Absorption of Salts by Plants. Amer. Journ. Bot., **9**, 311.

MÜLLER, A. (1904). Die Assimilationsgrösse bei Zucker- und Stärkeblättern. Jahrb. wiss. Bot., **40**, 443.

MÜLLER, G. (1914). Beiträge zur Keimungsphysiologie. Untersuchungen über die Sprengung der Samen- und Fruchthüllen bei der Keimung. Jahrb. wiss. Bot., **54**, 529.

MÜLLER, H. (1873). Probosces capable of Sucking the Nectar of *Angræcum sesquipedale*. Nature, **8**, 223.

MÜLLER-THURGAU, H. (1885). Beitrag zur Erklärung der Ruheperiode der Pflanzen. Landwirth. Jahrb., **14**, 851.

—— and SCHNEIDER-ORELLI, O. (1912). Beiträge zur Kenntnis der Lebensvorgänge in ruhenden Pflanzenteilen. Flora, **104**, 387.

MURBECK, S. (1901). Parthenogenetische Embryobildung in der Gattung Alchemilla. Lunds. Univ. Arskrft., **36**, No. 7.

NATHANSOHN, A. (1910). Der Stoffwechsel der Pflanzen. Leipzig, 1910.

NEGER, F. W. (1918). Die Wegsamkeit der Laubblätter für Gase. Flora, **111**, **112**, 152.

—— and FUCHS, J. (1915). Untersuchungen über den Nadelfall der Koniferen. Jahrb. wiss. Bot., **55**, 608.

NEWCOMBE, F. C. (1922). Significance of the Behaviour of Sensitive Stigmas. Amer. Journ. Bot., **9**, 99.

NICOLAS, G. (1918). Contribution a l'étude des variations de la respiration des végétaux avec l'age. Rev. Gén. Bot., **30**, 209.

NOBBE, F. (1862). Über die feinere Verästelung der Pflanzenwurzel. Landwirth. Versuchs-Stat., **4**, 212.

—— (1875). Beobachtungen und Versuche über die Wurzelbildung der Nadelhölzer. Landwirth. Versuchs-Stat., **18**, 279.

NOBBE, F. und HILTNER, L. (1896). Über die Anpassungsfähigkeit der Knöllchenbakterien ungleichen Ursprungs an verschiedenen Leguminosen Gattungen. Landwirth. Versuchs-Stat., **47**, 257.

NORDHAUSEN, M. (1903). Über Sonnen- und Schattenblätter. Ber. deutsch. bot. Ges., **21**, 30.

—— (1912). Über Sonnen- und Schattenblätter. Ber. deutsch. bot. Ges., **30**, 483.

NOYES, H. A., TROST, J. F. and YODER, L. (1918). Root Variations induced by Carbon Dioxide Gas Additions to the Soil. Bot. Gaz., **66**, 364.

OEHLKERS, F. (1921). Zur Reizphysiologischen Analyse der postfloralen Krümmungen des Blütenstiels von *Tropæolum majus*. Ber. deutsch. bot. Ges., **39**, 20.

OLSEN, C. (1923). Studies on the Hydrogen Ion Concentration of the Soil and its Significance to the Vegetation, especially to the Natural Distribution of Plants. Comptes rend. d. Trav. d. Labor. Carlsberg., **15,** 1.

ORR, M. Y. (1923). The Leaf Glands of *Dioscorea macroura*, Harms. Notes Roy. Bot. Gard. Edin., **14,** 57.

OSTERHOUT, W. J. V. (1906). On the Importance of Physiologically Balanced Solutions for Plants. I, Marine Plants. Bot. Gaz., **42,** 127.

—— (1907). On the Importance of Physiologically Balanced Solutions for Plants. II, Fresh-water and Terrestrial Plants. Bot. Gaz., **44,** 259.

—— (1922). Injury, Recovery and Death in Relation to Conductivity and Permeability. Philadelphia, 1922.

OTIS, C. H. (1914). The Transpiration of Emersed Water Plants. Bot. Gaz., **58,** 457.

OTTENWÄLDER, A. (1914). Lichtintensität und Substrat bei der Lichtkeimung. Ztschrft. f. Bot., **6,** 785.

OYE, P. VON (1921). Zur Biologie der Kanne von *Nepenthes melamphora*. Biol. Centrlbl., **41,** 529.

PACK, D. A. (1921). After-ripening and Germination of Juniperus Seeds. Bot. Gaz., **71,** 32.

PALLIS, M. (1916). The Structure and History of Plav: the Floating Fen of the Delta of the Danube. Journ. Lin. Soc. Bot., **43,** 233.

PAMMEL, L. H. (1911). A Manual of Poisonous Plants. Cedar Rapids, 1911.

PANTANELLI, E. (1915). Über Ionenaufnahme. Jahrb. wiss. Bot., **56,** 689.

PARKER, G. H. (1922). Smell, Taste and Allied Senses in the Vertebrates. Philadelphia, 1922.

PEARL, R. (1922). The Biology of Death. Philadelphia, 1922.

PEARSALL, W. H. (1923). Studies in Growth. IV. Correlations in Development. Ann. Bot., **37,** 261.

PEIRCE, G. J. (1894). A Contribution to the Physiology of the Genus Cuscuta. Ann. Bot., **8,** 53.

—— (1904). Artificial Parasitism: a Preliminary Notice. Bot. Gaz., **38,** 214.

—— (1905). The Dissemination and Germination of *Arceuthobium occidentale*, Eng. Ann. Bot., **19,** 99.

PEKLO, J. (1910). Die pflanzenlichen Aktinomykosen. Bakt. Centrlbl., II, **27,** 451.

PETER, A. (1893). Culturversuche mit "ruhenden" Samen. Nachricht. Kgl. Ges. Wiss. Göttingen, No. 17, 673.

PEYRONEL, M. B. (1921). Nouveaux cas de rapports mycorhiziques entre phanérogames et basidiomycètes. Bull. trim. Soc. Myc. de France, **37,** 4.

PITRA, A. (1861). Über die Anheftungsweise einiger phanerogamen Parasiten an ihre Nährpflanzen. Bot. Ztng., **19,** 53.

PLATEAU, F. (1895, '96). Comment les fleurs attirent les insectes. Recherches experimentales. Bull. Acad. Roy. Belg., **30,** 466 ; **32,** 505.

PRAZMOWSKI, A. (1890, '91). Die Wurzelknöllchen der Erbse. Landw. Versuchs-Stat., **37,** 161 ; **38,** 5.

PRIESTLEY, J. H. and EWING, J. (1923). Physiological Studies in Plant Anatomy. VI, Etiolation. New Phyt., **22,** 30.

—— and PEARSALL, W. H. (1922). An Interpretation of Some Growth Curves. Ann. Bot., **36,** 239.

RACIBORSKI, M. (1898). Biologische Mitteilungen aus Java. Flora, **85,** 325.

RAMSBOTTOM, J. (1922). Orchid Mycorrhiza. Trans. Brit. Myc. Soc., **8,** 28.

RAUNKIAER, C. (1910). Statistik der Lebensformen als Grundlage für die biologische Pflanzengeographie. Beihft. Bot. Centrlbl., **27,** 171.

RAYNER, M. C. (1913). The Ecology of *Calluna vulgaris*. New Phyt., **12,** 59.

—— (1915). Obligate Symbiosis in *Calluna vulgaris*. Ann. Bot., **29,** 97.

—— (1916*a*). Recent Developments in the Study of Endotrophic Mycorrhiza. New Phyt., **15,** 161.

—— (1916*b*). The Pollen of *Echeveria retusa*, Lindl. as Laboratory Material. New Phyt., **15,** 136.

—— (1921). The Ecology of *Calluna vulgaris*. II, The Calcifuge Habit. Journ. Ecol., **9,** 60.

—— (1922*a*). Nitrogen Fixation in Ericaceæ. Bot. Gaz., **73,** 226.

—— (1922*b*). Mycorrhiza in the Ericaceæ. Trans. Brit. Myc. Soc., **8,** 61.

REDFERN, G. M. (1922). On the Absorption of Ions by the Roots of Living Plants. I, The Absorption of the Ions of Calcium Chloride by Pea and Maize. Ann. Bot., **36,** 167.

REED, H. G. and HOLLAND, R. H. (1919). Growth Rate of an Annual Plant, Helianthus. Proc. Nat. Acad., **5,** 135.

RENNER, O. (1909). Zur Morphologie und Ökologie der pflanzlichen Behaarung. Flora, **99,** 127.

—— (1910). Beiträge zur Physik der Transpiration. Flora, **100,** 451.

RIDLEY, H. N. (1905). On the Dispersal of Seeds by Wind. Ann. Bot., **19,** 351.

—— (1910). Symbiosis of Ants and Plants. Ann. Bot., **24,** 457.

RIEDE, W. (1922). Die Abhängigkeit des Geschlechtes von den Aussenbedingungen. Flora, **115**, 259.

RIGG, G. B. (1916*a*). Decay and Soil Toxins. Bot. Gaz., **61**, 295.

—— (1916*b*). The Toxicity of Bog Water. Amer. Journ. Bot., **3**, 436.

RIMBACH, A. (1921). Über die Verkürzung des Hypocotyls. Ber. deutsch. bot. Ges., **39**, 285.

ROBERTSON, T. B. (1923). The Chemical Basis of Growth and Senescence. Philadelphia, 1923.

ROBBINS, W. J. (1922). Cultivation of Excised Root-tips and Stem-tips under Sterile Conditions. Bot. Gaz., **73**, 376.

ROSE, D. H. (1915). A Study of Delayed Germination in Economic Seeds. Bot. Gaz., **59**, 425.

ROSE, R. C. (1919). After-ripening and Germination of Seeds of Tilia, Sambucus and Rubus. Bot. Gaz., **67**, 281.

ROSENBERG, O. (1906). Über die Embryobildung in der Gattung Hieracium. Ber. deutsch. bot. Ges., **24**, 157.

RUBY, J. (1917). Recherches sur l'olivier. Ann. Sc. Nat. Bot., IX, **20**, 1.

RUHLAND, W. (1915). Untersuchungen über die Hautdrüsen der Plumbaginaceen. Ein Beitrag zur Biologie der Halophyten. Jahrb. wiss. Bot., **55**, 409.

RUSSELL, E. J. and APPLEYARD, A. (1915). The Atmosphere of the Soil. Journ. Agr. Sc., **7**, 1.

RUTTNER, F. (1921). Das elektrolytische Leitvermögen verdünnter Lösungen unter dem Einflusse submerser Gewächse. Sitzber. Akad. Wiss. Wien (Math.-Nat. Kl.), **130**, 71.

SACHS, J. VON (1887). Physiology of Plants. Oxford, 1887.

SALISBURY, E. J. (1916). The Emergence of the Aerial Organs in Woodland Plants. Journ. Ecol., **4**, 121.

—— (1920). The Significance of the Calcicolous Habit. Journ. Ecol., **8**, 202.

—— (1921). Stratification and Hydrogen-ion Concentration with Special Reference to Woodlands. Journ. Ecol., **9**, 220.

SARGENT, O. (1918). Fragments of the Flower Biology of Westralian Plants. Ann. Bot., **32**, 215.

SAYRE, J. D. (1923). Physiology of Stomata of *Rumex patientia*. Science, **57**, 205.

SCHAFFNER, J. H. (1919). The Nature of the Diœcious Condition in *Morus alba* and *Salix amygdaloides*. Ohio Journ. Sc., **19**, 409.

—— (1922). Control of the Sexual State in *Arisæma triphyllum* and *Arisæma Dracontium*. Amer. Journ. Bot., **9**, 72.

SCHENCK, H. (1892, '93). Beiträge zur Biologie und Anatomie

der Lianen, im besonderen der in Brasilien einheimischen Arten. 2 vols. Jena, 1892, '93.

SCHIMPER, A. F. W. (1891). Die indo-malayische Strandflora. Jena, 1891.

SCHIPS, M. (1914). Zur Öffnunsmechanik der Antheren. Beihft. Bot. Centrlbl., **31**, 119.

SCHMID, G. (1912). Beiträge zur Biologie der Insektivoren Pflanzen. Flora, **104**, 335.

SCHMITT, E. M. (1922). Beziehungen zwischen der Befruchtung und den postfloralen Blüten-bzw. Fruchtstielbewegungen bei *Digitalis purpurea*, *Digitalis ambigua*, *Althæa rosea*, und *Linaria Cymbalaria*. Ztschrft. f. Bot., **14**, 625.

SCHRAMM, R. (1912). Über die anatomischen Jugendformen der Blätter einheimischer Holzpflanzen. Flora, **104**, 225.

SCHREINER, O. and SKINNER, J. J. (1910). Some Effects of a Harmful Organic Soil Constituent. Bot. Gaz., **50**, 161.

—— —— (1912). The Toxic Action of Organic Compounds as Modified by Fertilizer Salts. Bot. Gaz., **54**, 31.

SCHULOW, I. (1913). Versuche mit sterilen Kulturen höherer Pflanzen. Ber. deutsch. bot. Ges., **31**, 97.

SCHWARZ, F. (1883). Die Wurzelhaare der Pflanzen, ein Beitrag zur Biologie und Physiologie dieser Organe. Unters. a. d. Bot. Inst. z. Tübingen., **1**, 140.

SCHWEIDLER, E. and SPERLICH, A. (1922). Die Bewegung der Primärblätter bei etiolierten Keimpflanzen von *Phaseolus multiflorus*. Ztschrft. f. Bot., **14**, 477.

SCHWENDENER, S. (1874). Das mechanische Prinzip im anatomischen Bau der Monokotylen. Leipzig, 1874.

—— (1881). Über Bau und Mechanik der Spaltöffnungen. Monatsber. d. Berlin. Akad. d. Wiss., 833.

SEIFRIZ, W. (1920). The Length of the Life Cycle of a Climbing Bamboo. A Striking Case of Sexual Periodicity in *Chusquea abietifolia*, Grisb. Amer. Journ. Bot., **7**, 83.

SERNANDER, R. (1906). Entwurf einer Monographie der europäischen Myrmekochoren. Kungl. Svensk. Vetenskapsakad. Handl., **41**, No. 7.

SHERFF, E. E. (1912). The Vegetation of Skokie Marsh, with Special Reference to Subterranean Organs and their Interrelationships. Bot. Gaz., **53**, 415.

SHIBATA, K. (1902). Cytologische Studien über die endotrophen Mykorrhizen. Jahrb. wiss. Bot., **37**, 643.

SHIVE, J. W. (1915). A Study of Physiological Balance in Nutrient Media. Physiol. Res., **1**, 327.

—— and LIVINGSTON, B. E. (1914). The Relation of Atmospheric Evaporating Power to Soil Moisture Content at Permanent Wilting in Plants. Plant World, **17**, 81.

SHREVE, E. B. (1914). The Daily March of Transpiration in a Desert Perennial. Carneg. Inst. Wash., Publ. No. 201.

—— (1916). An Analysis of the Causes of Variations in the Transpiring Power of Cacti. Physiol. Res., **2,** 73.

—— (1920). Causes of the Seasonal Changes in the Transpiration of *Encelia farinosa*. Yearbk. Carneg. Inst. Wash., **19,** 73.

SHREVE, F. (1914*a*). A Montane Rain Forest. Carneg. Inst. Wash., Publ. No. 199.

—— (1914*b*). The Direct Effects of Rainfall on Hygrophilous Vegetation. Journ. Ecol., **2,** 82.

SHULL, C. A. (1911). The Oxygen Minimum and the Germination of Xanthium Seeds. Bot. Gaz., **52,** 453.

—— (1914). The Rôle of Oxygen in Germination. Bot. Gaz., **57,** 64.

—— (1916). Measurement of the Surface Forces in Soils. Bot. Gaz., **62,** 1.

—— and DAVIS, W. B. (1923). Delayed Germination and Catalase Activity in Xanthium. Bot. Gaz., **75,** 268.

SHULL, G. H. (1910). Inheritance of Sex in Lychnis. Bot. Gaz., **49,** 110.

—— (1911). Reversible Sex-mutants in *Lychnis dioica*. Bot. Gaz., **52,** 329.

—— (1914). The Longevity of Submerged Seeds. Plant World, **17,** 329.

SIERP, H. (1917). Über den Einfluss des Lichts auf das Wachstum der Pflanzen. Ber. deutsch. bot. Ges., **35,** (8).

—— (1918). Ein Beitrag zur Kenntnis des Einflusses des Lichts auf das Wachstum der Koleoptile von *Avena sativa*. Ztschrft. f. Bot., **10,** 641.

SIMON, S. V. (1914). Studien über die Periodicität der Lebensprozesse der in dauernd feuchten Tropengebieten heimischen Baüme. Jahrb. wiss. Bot., **54,** 71.

SKENE, M. (1915). The Acidity of Sphagnum and its Relation to Chalk and Mineral Salts. Ann. Bot., **29,** 65.

SMALL, J. (1917, '18, '19). The Origin and Development of the Compositæ. New Phyt., **16, 17, 18.**

SMITH, A. M. (1907). On the Application of Limiting Factors to Measurements and Observation of Growth in Ceylon. Ann. Bot. Gard. Peradeniya, **3,** 303.

—— (1919). The Temperature Coefficient of Photosynthesis: a Reply to Criticism. Ann. Bot., **33,** 517.

SMITH, W. G. (1913). Raunkiaer's "Life-forms" and Statistical Methods. Journ. Ecol., **1,** 16.

SNELL, K. (1908). Untersuchungen über die Nahrungsaufnahme der Wasserpflanzen. Flora, **98,** 213.

SOLMS-LAUBACH, H. ZU (1882). Die Herkunft, Domestication und Verbreitung d. gewöhnlichen Feigenbaüme. Abhandl. d. kön. Ges. d. Wiss. zu Göttingen, **28**.

—— (1885). Die Geschlechterdifferenzierung bei den Feigenbäumen. Bot. Ztng., **43**, 513, 529, 545, 561.

SNOW, L. M. (1914). Contribution to the Knowledge of the Diaphragms of Water Plants. I, *Scirpus validus*. Bot. Gaz., **58**, 495.

—— (1920). Diaphragms of Water Plants. II, Effect of certain Factors upon Development of Air Chambers and Diaphragms. Bot. Gaz., **69**, 297.

SPERLICH, A. (1912). Über die Krummungsursache bei Keimstengeln und beim Monokotylenblatte. Jahrb. wiss. Bot. **50**, 502.

SPISAR, K. (1910). Beiträge zur Physiologie der *Cuscuta Gronovii* Willd. Ber. deutsch. bot. Ges., **28**, 329.

SPRATT, E. R. (1912*a*). The Morphology of the Root Tubercles of Alnus and Eleagnus, and the Polymorphism of the Organism causing their Formation. Ann. Bot., **26**, 119.

—— (1912*b*). The Formation and Physiological Significance of Root Nodules in the Podocarpineæ. Ann. Bot., **26**, 801.

—— (1915). The Root Nodules of the Cycadaceæ. Ann. Bot., **29**, 618.

—— (1919). A Comparative Account of the Root Nodules of the Leguminosæ. Ann. Bot., **33**, 189.

SPRENGEL, C. K. (1793). Das entdeckte Geheimniss der Natur im Bau und in der Befruchtung der Blumen. Berlin, 1793.

STAEDTLER, G. (1923). Über Reduktionserscheinungen im Bau der Antherenwand von Angiospermen-Blüten. Flora, **116**, 85.

STAHL, E. (1888). Pflanzen und Schnecken. Jena, 1888.

—— (1893). Regenfall und Blattgestalt. Ann. Bot. Gard. Buitenzorg, **11**, 98.

—— (1894). Einige Versuche über Transpiration und Assimilation. Bot. Ztng., **52**, 117.

—— (1897). Über die Pflanzenschlaf und verwandte Erscheinungen. Bot. Ztng., **55**, 71.

—— (1900). Der Sinn der Mycorhizenbildung. Jahrb. wiss. Bot., **34**, 539.

—— (1909). Zur Biologie des Chlorophylls. Jena, 1909.

STALFELT, M. G. (1922). Zur Kenntnis der Kohlenhydratproduktion von Sonnen- und Schattenblättern. Meddel. fr. Statens. Skogsförsöksanstalt, H 18, Nr. 5.

STARK, P. (1915). Untersuchungen über Kontaktreizbarkeit. Ber. deutsch. bot. Ges., **33**, 389.

—— (1917). Experimentelle Untersuchungen über das Wesen

und die Verbreitung der Kontaktreizbarkeit. Jahrb. wiss. Bot., **57**, 189.

STEINBRINCK, C. (1906). Über Schrumpfungs- und Kohäsionsmechanismen von Pflanzen. Biol. Centrlbl., **26**, 657, 721.

STILES, W. (1915). On the Relation between the Concentration of the Nutrient Solution and the Rate of Growth of Plants in Water Cultures. Ann. Bot., **29**, 90.

—— (1916). On the Interpretation of the Results of Water-culture Experiments. Ann. Bot., **30**, 427.

STILES, W. and JORGENSEN, I. (1914). The Nature and Methods of Extraction of the Soil Solution. Journ. Ecol., **2**, 245.

—— —— (1917). Observations on the Influence of Aeration of the Nutrient Solution in Water Culture Experiments, with Some Remarks on the Water Culture Method. New Phyt., **16**, 181.

—— and KIDD, F. (1919*a*). The Influence of the External Concentration on the Position of the Equilibrium attained in the Intake of Salts by Plant Cells. Proc. Roy. Soc., **90 B**, 448.

—— —— (1919*b*). The Comparative Rate of Absorption of Various Salts by Plant Tissue. Proc. Roy. Soc., **90 B**, 487.

STOCKER, O. (1923). Die Transpiration und Wasserökologie nordwestdeutscher Heide- und Moorpflanzen am Standort. Ztschrft. f. Bot., **15**, 1.

STOPPEL, R. (1910). Über den Einfluss des Lichts auf das Öffnen und Schliessen einiger Blüten. Ztschrft. f. Bot., **2**, 309.

—— (1912). Über die Bewegungen der Blätter von Phaseolus bei Konstanz der Aussenbedingungen. Ber. deutsch. bot. Ges., **30**, (29).

—— (1916). Die Abhängigkeit der Schlafbewegungen von *Phaseolus multiflorus* von verschiedenen Aussenbedingungen. Ztschrft. f. Bot., **8**, 609.

—— (1920). Die Pflanze in ihren Beziehung zur atmosphärischen Elektrizitat. Ztschrft. f. Bot., **12**, 529.

—— and KNIEP, H. (1911). Weitere Untersuchungen über das Öffnen und Schliessen der Blüten. Ztschrft. f. Bot., **3**, 369.

STOUT, A. B. (1919). Intersexes in *Plantago lanceolata*. Bot. Gaz., **68**, 109.

—— (1923). Alternation of Sexes and Intermittent Production of Fruit in the Spider Flower (*Cleome spinosa*). Amer. Journ. Bot., **10**, 57.

STRASBURGER, E. (1909). Das weitere Schicksal meiner isolierten weiblichen *Mercurialis annua* Pflanzen. Ztschrft. f. Bot., **1**, 507.

—— (1910). Über Geschlechtsbestimmende Ursachen. Jahrb. wiss. Bot., **48**, 427.

TANSLEY, A. G. (1917). On Competition between *Galium saxatile* L. (*G. hercynicum* Weig.) and *Galium sylvestre* Poll. (*G. asperum* Schreb.) on Different Types of Soil. Journ. Ecol., **5,** 175.

TERNETZ, C. (1907). Über die Assimilation des atmosphärischen Stickstoff durch Pilze. Jahrb. wiss. Bot., **44,** 353.

THATCHER, K. M. (1921). The Effect of Peat on the Transpiration and Growth of Certain Plants. Journ. Ecol., **9,** 39.

THODAY, D. (1910). Some Experiments on Assimilation in the Open Air. Proc. Roy. Soc., **82 B,** 421.

—— (1921). On the Behaviour during Drought of Leaves of Two Cape Species of Passerina, with Some Notes on their Anatomy. Ann. Bot., **35,** 585.

—— and SYKES, M. G. (1909). Preliminary Observations on the Transpiration Current in Submerged Water Plants. Ann. Bot., **23,** 635.

THOMPSON, D'A. W. (1917). Growth and Form. London, 1917.

TISCHLER, G. (1917). Pollenbiologische Studien. Ztschrft. f. Bot., **9,** 417.

TOTTINGHAM, W. E. and RANKIN, E. J. (1922). Nutrient Solutions for Wheat. Amer. Journ. Bot., **9,** 270.

TRELEASE, S. F. and LIVINGSTON, B. E. (1916). The Daily March of Transpiring Power as indicated by the Porometer and by Standardised Hygrometric Paper. Journ. Ecol., **4,** 1.

TREUB, M. (1883). Sur les urnes du *Dischidia Rafflesiana* Wall. Ann. Bot. Gard. Buitenzorg, **3,** 13.

TROLL, W. (1922). Über Staubblatt und Griffelbewegungen und ihre teleologische Deutung. Flora, **115,** 191.

TRUE, R. H. (1922). The Significance of Calcium for Higher Plants. Science, **54,** 1.

TRUOG, E. (1918). Soil Acidity : I, Its Relation to the Growth of Plants. Soil. Sc., **5,** 169.

—— and MEACHAM, M. R. (1919). Soil Acidity : II, Its Relation to the Acidity of the Plant Juice. Soil Sc., **7,** 469.

TSCHIRCH, A. (1905). Über die Heterorhizie bei Dikotylen. Flora, **94,** 68.

—— (1911). Die Feigenbäume Italiens und ihre Beziehungen zu einander. Ber. deutsch. bot. Ges., **29,** 83.

—— and RAVASINI, G. (1911). Le type sauvage du figuier et ses relations avex le caprifiguier et le figuier femelle domestique. Comptes Rend. Acad. Sc., **152,** 885.

TUBEUF, K. VON (1923). Monographie der Mistel. Munich, 1923.

ULE, E. (1906). Ameisenpflanzen. Bot. Jahrb., **37**, 335.

UNGER, F. (1862). Neue Untersuchungen über die Transpiration der Pflanzen. Sitzber. d. kais. Akad. d. Wiss. Wien (Math.-Nat. Kl.), **44**, 181.

URSPRUNG, A. (1917). Über die Stärkebildung im Spektrum. Ber. deutsch. bot. Ges., **35**, 44.

—— and BLUM, G. (1916). Zur Kenntnis der Saugkraft. Ber. deutsch. bot. Ges., **34**, 539.

—— —— (1918*a*). Zur Kenntnis der Saugkraft. II, Ber. deutsch. bot. Ges., **36**, 577.

—— —— (1918*b*). Besprechung unserer bisherigen Saugkraftmessungen. Ber. deutsch. bot. Ges., **36**, 599.

—— —— (1921*a*). Zur Kenntnis der Saugkraft. IV, Der Endodermissprung. Ber. deutsch. bot. Ges., **39**, 70.

—— —— (1921*b*). Zur Kenntnis der Saugkraft. V, Eine Methode zur Bestimmung des Widerstandes den der Boden der Wasserabsorption durch die Wurzel entgegensetzt. Ber. deutsch. bot. Ges., **39**, 139.

VINES, S. H. (1905). The Proteases of Plants. Ann. Bot., **19**, 171.

VOGT, E. (1915). Über den Einfluss des Lichts auf das Wachstum der Koleoptile von *Avena sativa*. Ztschrft. f. Bot., **7**, 193.

VOLKENS, G. (1887). Flora der ägyptisch-arabischen Wüste. Berlin, 1887.

—— (1912). Laubfall und Lauberneurung in den Tropen. Berlin, 1912.

DE VRIES, H. (1915). Über künstliche Beschleunigung der Wasseraufnahme in Samen durch Druck. Biol. Centrlbl., **35**, 161.

WALLACE, A. R. (1870). Under Wallace, 1891.

—— (1878). Under Wallace, 1891.

—— (1891). Natural Selection and Tropical Nature. London, 1891.

—— (1892). Island Life. London, 1892.

—— (1907). Fertilisation of Flowers by Insects. Nature, **75**, 320.

WARBURG, O. (1919,'20). Über die Geschwindigkeit der photochemischen Kohlensaürezersetzung in lebenden Zellen. Biochem. Ztschrft., **100**, 230, **103**, 188.

WARMING, E. (1881). Familie Podostemaceæ. Copenhagen, 1881.

WATERMAN, W. G. (1919). Development of Root Systems under Dune Conditions. Bot. Gaz., **68**, 22.

WEAVER, J. E. (1919). Ecological Relations of Roots. Carneg. Inst. Wash., Publ. No. 286.

—— (1920). Root Development in the Grassland Formation. Carneg. Inst. Wash., Publ. No. 292.

——, FRANK, C. and CRIST, J. W. (1922). Development and Activities of Roots of Crop Plants. Carneg. Inst. Wash., Publ. No. 316.

WEBER, C. A. (1879). Über spezifische Assimilationsenergie. Arbeit. d. bot. Inst. Würzburg., **2**, 343.

WEBER, F. (1916). Über das Treiben der Buche. Ber. deutsch. bot. Ges., **34**, 7.

WEIR, J. R. (1918). Experimental Investigations on the Genus Razoumowskya. Bot. Gaz., **66**, 1.

WEISS, F. E. (1908). The Dispersal of Fruits and Seeds by Ants. New Phyt., **7**, 23.

WERTH, E. (1915). Kurzer Überblick über die gesamtfrage der Ornithophilie. Bot. Jahrb., **53** (Beibl. 116), 314.

WHITAKER, E. S. (1923). Root Hairs and Secondary Thickening in the Compositæ. Bot. Gaz., **76**, 30.

WHITE, J. (1909). The Ferments and Latent Life of Resting Seeds. Proc. Roy. Soc., **81 B**, 417.

—— (1910). The Proteolytic Enzyme of Drosera. Proc. Roy. Soc., **83 B**, 134.

WIESNER, J. VON (1907). Der Lichtgenuss der Pflanzen. Leipzig, 1907.

WIGGANS, R. G. (1921). Variations in the Osmotic Concentration of the Guard Cells during the Opening and Closing of the Stomata. Amer. Journ. Bot., **8**, 30.

WILLE, N. (1884). Kritische Studien über die Anpassungen der Pflanzen an Regen und Tau. Beitr. z. Biol. d. Pflanz., **4**, 285.

WILLIS, J. C. (1902). Studies in the Morphology and Ecology of the Podostemaceæ of Ceylon and India. Ann. Bot. Gard. Peradeniya, **1**.

WILLSTÄTTER, R. and STOLL, A. (1913). Untersuchungen über Chlorophyll. Berlin, 1913.

—— —— (1918). Untersuchungen über die Assimilation der Kohlensaüre. Berlin, 1918.

WILMOTT, A. J. (1921). Assimilation by Submerged Plants in Dilute Solutions of Bicarbonate and of Acids: an Improved Bubble Counting Technique. Proc. Roy. Soc., **92 B**, 304.

WINKLER, H. (1920). Verbreitung und Ursache der Parthenogenesis im Pflanzen- und Tierreiche. Jena, 1920.

WRIGHT, H. Foliar Periodicity of Endemic and Indigenous Trees in Ceylon. Ann. Bot. Gard. Peradeniya, **2**.

WYLIE, R. B. (1904). The Morphology of *Elodea canadensis*. Bot. Gaz., **37**, 1.
—— (1917). The Pollination of *Vallisneria spiralis*. Bot. Gaz., **63**, 135.

YAMPOLSKI, C. (1919). Inheritance of Sex in *Mercurialis annua*. Amer. Journ. Bot., **6**, 410.
—— (1920). The Occurrence and Inheritance of Sex Intergradation in Plants. Amer. Journ. Bot., **7**, 21.
YAPP, R. H. (1908). Wicken Fen. New Phyt., **7**, 61.
—— (1909). On Stratification in the Vegetation of a Marsh and its Relation to Evaporation. Ann. Bot., **23**, 275.
—— (1912). *Spiræa Ulmaria* L. and its bearing on the Problem of Xeromorphy in Marsh Plants. Ann. Bot., **26**, 815.

ZIMMERMANN, A. (1902). Über Bakterienknoten in den Blättern einiger Rubiaceen. Jahrb. wiss. Bot., **37**, 1.

INDEXES

I. INDEX OF PLANT NAMES
II. SUBJECT INDEX

INDEX OF PLANT NAMES

SUBJECT INDEX